# Technology, Violence, and War

# History of Warfare



VOLUME 125

The titles published in this series are listed at *brill.com/hw*

# Technology, Violence, and War

*Essays in Honor of Dr. John F. Guilmartin, Jr.*

*Edited by*

Robert S. Ehlers, Jr.
Sarah K. Douglas
Daniel P.M. Curzon

BRILL

LEIDEN | BOSTON

Cover illustration: Fol. 102r, Kriegstechnik (Bilderhandschrift), Zentralbibliothek, Zürich, Ms. Rh hist. 33b. Papier, 168 ff,. 30 × 21 cm, Oberrhein um 1420–1440. With kind permission of the Zentralbibliothek Zürich. Photo Dr Guilmartin: courtesy of Lore Guilmartin.

Library of Congress Cataloging-in-Publication Data

Names: Guilmartin, John F., Jr. (John Francis), 1940-2016, honouree. | Ehlers, Robert S., editor. | Douglas, Sarah K., editor. | Curzon, Daniel P. M., editor.
Title: Technology, violence, and war : essays in honor of Dr. John F. Guilmartin, Jr. / Edited by Robert S. Ehlers, Jr., Sarah K. Douglas, Daniel P.M. Curzon.
Description: Leiden ; Boston : Brill, [2019] | Series: History of warfare, ISSN 1385-7827 ; Volume 125 | Includes bibliographical references and index.
Identifiers: LCCN 2018055829 (print) | LCCN 2018056711 (ebook) | ISBN 9789004393301 (ebook) | ISBN 9789004383418 (hardback : alk. paper)
Subjects: LCSH: Military research. | Military art and science--Technological innovations.
Classification: LCC U390 (ebook) | LCC U390 .T 2019 (print) | DDC 355/.07--dc23
LC record available at https://lccn.loc.gov/2018055829

Typeface for the Latin, Greek, and Cyrillic scripts: “Brill”. See and download: brill.com/brill-typeface.

ISSN 1385-7827
ISBN 978-90-04-38341-8 (hardback)
ISBN 978-90-04-39330-1 (e-book)



This book is printed on acid-free paper and produced in a sustainable manner.

*To Dr. John F. ("Joe") Guilmartin, Jr. from his students.*
*He modeled strength in adversity, humility in achievement,*
*and wisdom in action. And to Lore Guilmartin,*
*the gracious and formidable woman behind Joe's successes and ours.*

# Contents

# Acknowledgements

The editors of this volume would like to thank everyone who contributed to its creation. Without the generous support of those mentioned here, we could not have held the commemorative conference that took place at the Mershon Center for International Security Studies on 23rd and 24th September of 2016, nor could we have compiled the presented papers into this volume, which seeks to honor the extraordinary legacy of Dr. John F. Guilmartin, Jr.

First, we thank the Lynde and Harry Bradley Foundation of Milwaukee. The Foundation members were tireless in their efforts to support Joe's various contributions to the study of warfare, and were generous with funding his research and also in assisting graduate students at The Ohio State University who were completing their own military history studies. Not only did they support the conference held in Joe's honor, but many essays in this volume were written by students who benefitted from their assistance.

We would also like to thank Scott Laidig, a great friend of Dr. Guilmartin's and a champion of The Ohio State University's Department of History. Scott's direct support to the conference has been heartwarming and most welcome. So has his unstinting advocacy of student development and scholarships. Scott's service as a Marine officer during the Vietnam War helped to make him a man of great character, an outstanding advocate for students and scholars of military history, and a student and author of military history in his own right.

Next, we would like to thank the Department of History and the Mershon Center for International Security Studies at The Ohio State University. Not only did the Department of History provide essential financial support for the conference, but military history professors Geoffrey Parker, Peter Mansoor, Jennifer Siegel, Mark Grimsley, and Bruno Cabanes were immediately willing to assist with the event in any way needed, which they did to great effect. The Mershon Center was also willing to host the conference event, which is no surprise given their decades of support to both Joe's own scholarship and that of his students.

We would like to thank other contributors to the creation of both the conference and volume, including Dr. Kelly DeVries, Dr. Alan Millett, Dr. Williamson Murray, and all those scholars who will contribute to a second volume of essays to commemorate Joe's life and academic work. Drs. Millett and Murray were immediately eager to help celebrate their former colleague, and Dr. DeVries was essential, not only to the planning of the conference, but in its success and the compiling of this volume.

Lastly but most importantly, we would like to thank Joe's wife, Lore, and the rest of his wonderfully supportive family. Joe would be the first to say that Lore's love and support were instrumental to his success as a scholar and mentor. Her unwavering encouragement to those who compiled this volume was equally beneficial. She was an intrinsic part of this project from beginning to end, and Joe's family stood both behind and beside her through all her endeavors. We thank them for that, and submit this volume as a token of our thanks for all they, and those mentioned above, have done to make it a success.

# Illustrations, Tables, and Maps

## Illustrations



## Tables

## Maps

## Notes on Contributors

*Alan Beyerchen*
Ph.D., University of California, Santa Barbara, taught modern German history at The Ohio State University until his retirement (2012). Known for his work on science and technology around World War II, on the theories of Carl von Clausewitz, and on German plays.

*Robert H. Clemm*
Ph.D., The Ohio State University (2012), is an associate professor at Grove City College. Teaches African, imperial, and military history. Significant contributor to the *Encyclopedia of African and Colonial Conflicts* and held fellowships from the *Mellon* and the *Bradley* Foundations.

*Edward Coss*
Ph.D., The Ohio State University (2005), is a Professor of Military History in the Army Command and General Staff College's Department of Military History, and a Fellow of the Royal Historical Society. Known for his book *All for the King's Shilling* and several other works on European military history.

*Sebastian Cox, OBE*
M.A., King's College London (1979), has been the Head of the Air Historical Branch (RAF) since 1996. Member of the Board of Trustees of the Society for Military History, where he befriended Dr. Guilmartin, and of the Royal Air Force Museum's Research Board.

*Daniel P.M. Curzon*
Ph.D. candidate, The Ohio State University, is working on a dissertation entitled *Pacific Triumvirate,* concerning the relations among Great Britain, Japan, and the United States of America.

*Sarah K. Douglas*
Ph.D., The Ohio State University (2015), is a Lecturer at OSU, teaching a range of military history courses from the middle ages to Vietnam. Her first book, *Partus Pestilentiae,* is currently being reviewed for publication.

*Robert S. Ehlers, Jr.*
Ph.D., The Ohio State University (2005), is the Senior Mentor for the Information Environment Advanced Analysis Course administered by the Department

of Defense. Known for his works on the Allied air forces during World War II, including *Targeting the Third Reich* and *The Mediterranean Air War.*

*Andrew de la Garza*
Ph.D., The Ohio State University (2010), is currently at the University of Louisiana, Lafayette. Teaching covers Asian, world, and military history. His monograph, *The Mughal Empire at War,* expands the complexity of the military revolutions debate to South Asia.

*John F. Guilmartin, Jr.*
Ph.D., Princeton University (1975), was a Professor of Military History at OSU. He is best known for his work on naval warfare in the Mediterranean during 15th and 16th centuries, including *Gunpowder and Galleys* and *Galleons and Galleys.* Also contributed greatly to the fields of airpower history and the history of technology and warfare.

*Matthew M. Hurley*
Ph.D., The Ohio State University (2009), is a Senior Non-Resident Fellow with the Mitchell Institute for Aerospace Studies in Arlington, VA. His recent publication, "*Beyond the Iron Triad: The Future of Airborne C2ISR,* was published by the Mitchell Institute in 2017.

*Peter R. Mansoor*
Ph.D., The Ohio State University (1995), is the General Raymond E. Mason, Jr., Chair of Military History at OSU. He has produced several monographs on the effectiveness of soldiers and leaders from World War II to the current war in Iraq.

*Edward B. McCaul, Jr.*
Ph.D., The Ohio State University (2005), is a retired Army lieutenant colonel. He has served in a variety of academic and public positions and has published works on military technology and the American Civil War.

*Michael Pavelec*
Ph.D., The Ohio State University (2004), is a Professor of Airpower History at the Air Force Air Command and Staff College. He has published several works on air power and air technology covering the 20th Century.

*William Roberts*
Ph.D., The Ohio State University (1999), specializes in U.S. Civil War naval history. A retired U.S. Navy commander, he has published three books and

numerous articles and book reviews regarding the American Civil War's maritime aspects.

*Robyn Rodriguez*
Ph.D., The Ohio State University (2011), is an historian working for the Defense POW/MIA Accounting Agency. Her research and publications have centered around Sino-German military interaction during the Interwar Period.

*Clifford J. Rogers*
Ph.D., The Ohio State University (1994), is a Professor of History at the United States Military Academy and the Editor of *The Journal of Medieval Military History*. Best known for his monographs *War Cruel and Sharp* and *The Military Revolution Debate.*

*William Waddell*
Ph.D., The Ohio State University (2014), is an Assistant Professor of Strategy at the Air Force Air War College. His first book, *In the Year of the Tiger: the War for Cochinchina, 1945–1951*, is slated for publication in 2018.

*Corbin Williamson*
Ph.D., The Ohio State University (2015), is an Assistant Professor of Strategy at the Air Force Air War College. Published a pair of works on Anglo-American naval cooperation at the beginning of World War II and anticipates a related monograph.

# Introduction

This volume honors the life and work of Dr. John F. Guilmartin, Jr., an Air Force Lieutenant Colonel, a Professor of History, and our esteemed mentor. Joe was a model servant of the Republic, serving through deed and word. For 23 years he protected the citizenry in the United States Air Force, and, for 40 years afterwards, dedicated himself to shepherding young minds in academe. Joe's scholarly efforts ranged widely from mentoring thousands of students to advancing knowledge in a broad geographic area extending from Colorado Springs, CO, to Marrakesh, Morocco.

We, his students and colleagues, knew him during his 30 years of service to The Ohio State University as a Professor of Military History. It was in that role that Joe inspired us, taught us, and supported us. He modeled for us how to think deeply and meaningfully. He approached every topic, including the linkage between technology and warfare, holistically, interweaving the material fibers with socio-cultural ones on an expansive and highly intelligent methodological loom. This approach encouraged us to seek meaning within context—as Joe fondly remarked "Just because *we* don't understand why a particularly society did something, does not mean that way was unwise." It was just a method we had yet to understand as that society had. Joe also educated us about the practical side of our profession, and assisted us by connecting us with other universities, with publishers, and with other mentors. From those experiences, we matured as scholars in service to the common weal.

When Joe asked us to plan and superintend the conference in which this book had its genesis, he requested that its theme be technology. He believed deeply in technology's central importance to warfare, to human history, and to the discipline of military history. His definition of technology embraces this strongly held belief. He viewed it as "the application of knowledge to achieve a physical effect by means of an artifact, object or thing (distinct from art, which involves the application of knowledge to achieve aesthetic effect). (2): the artifact itself; the class of artifacts to which it belongs. (3): the knowledge needed to design, manufacture, operate, sustain and logistically support the artifact or thing and its user(s)."[1] This definition informs and is, in fact, at the core of the following chapters.

When Joe unexpectedly died shortly before the conference that was to mark his retirement from The Ohio State University, we vowed to publish a work

1 John F. Guilmartin, Jr., "Technology and Strategy: What Are the Limits?," Strategic Studies Institute, U.S. Army War College, Carlisle Barracks, PA, 1994, 1.

 | DOI:10.1163/9789004393301_002

containing the collected writings, works, and wisdom from the conference. To our departed mentor, we humbly present this fifteen-chapter salute.

Gandhi once noted, "The best way to find yourself is to lose yourself in the service of others." Joe's service began at an early age and his journey revealed his worth. He graduated from the Air Force Academy in 1962 and from pilot training in 1964, during which he earned ratings in both fixed-wing aircraft and helicopters. Joe chose to fly the latter. He earned an outstanding reputation as an officer and a pilot during two tours of duty in Southeast Asia. Joe flew the HH-3 "Jolly Green Giant" helicopter in a search-and-rescue role from 1965–1966 and the HH-53 "Super Jolly Green Giant" from 1974–1975. During his first tour, Joe saved many airmen from captivity or death, earning two silver stars for valor. Returning eight years later, Joe led the helicopter unit that rescued 1,457 Americans, nearly 6,000 Vietnamese, 205 Cambodians, and over 100 third-country citizens during the fall of Phnom Penh and Saigon to communist forces. Many of the Vietnamese and Cambodians he rescued became American citizens; in some cases, their children, and now grandchildren, serve in the Armed Forces, defending the Republic and rescuing other people in peril, as Joe did. The aircrews Joe saved returned home to their families, often developing correspondence and lifelong friendships with him. Through Joe's deeds in service, the men he rescued knew his worth, and they cherished new opportunities in living lives that may well have been cut short without Joe's help.

As Joe grew into a warrior airman of some renown, he also matured as a lover of history. Joe's youth was filled with tales of retired warriors' deeds. His father had been a pilot during World War II, and Joe was surrounded by those veterans and their stories. Additionally, he listened to the recollections of Great War veterans, and of veterans who survived the "forgotten war" in Korea. Those stories fueled Joe's two life-long passions: flying and history, in particular teaching military history. In his case, Clio and Mars proved to be close partners indeed.

Those of us who organized his professional papers and books for donation to various archives were awed by the range of historical subjects Joe studied and taught during his life. His best-known books are about galley warfare in the Mediterranean Sea during the 16th century—unexpected from an Airman! His collection of teaching and research documents embraced the world ranging from monographs on samurai to technical articles on engine construction and testing. There were also some touching reminders of Joe's early love for flying and history, including a copy of the children's book *Airplane Andy,* lovingly preserved, as well as a young person's history of Texas, Joe's home-state. Clearly, airmanship, scholarship, and service through those mediums were deeply ingrained in Joe's character by the time he became a young man.

Joe's first foray into teaching occurred at the Air Force Academy—his Alma Mater—from 1973 to 1975. During this phase of his military service, Joe also earned an Air Force-sponsored Ph.D. at Princeton University. Under the mentorship of Professor Ted Rabb, Joe honed his scholarship and produced the dissertation that eventually became his first book—a classic of early modern naval history—*Gunpowder and Galleys*. Joe carried his deep passion for flying and all things military forward to the Air University at Maxwell Air Force Base, Alabama, where he edited the *Air Power Review*, now titled the *Air and Space Power Journal*. His teaching inspired students, while his scholarship garnered frequent in-service accolades.

After Joe retired from the Air Force in 1985 at the rank of Lieutenant Colonel with 23 years of service, he quickly transitioned from his first love, flying in military service, to his second, furthering growth in students and scholarship. His first non-military position was as the official historian of the National Aeronautics and Space Administration (NASA) at Rice University. After a few years, Joe moved to a new position at The Ohio State University, which became his professional home for the next 30 years.

To say that Joe built a large family at The Ohio State University would be akin to calling water wet. During his time at the university, Joe advised 36 doctoral dissertations and over 200 master's theses—more than any other history professor since the founding of the Graduate School in 1912. He taught a range of graduate and undergraduate courses immensely popular with majors and non-majors alike, sharing with them his brilliant insights as a combat veteran and a student of military history. Joe's course on the Vietnam War is still the most popular and consistently over-enrolled history course at Ohio State, and several of his others are not far behind. Vietnam was Joe's war, so it makes perfect sense that he taught it with great discernment; although, he was equally engaging when discussing any armed conflict from any time in human history due to his far-flung studies. Rather than presenting history as "one damned thing after another," Joe brought it alive for his students. He asked deep questions, sought good answers from all sides of the political and social spectra, and, in the process, taught young adults how to conduct mature, civil, and useful dialogues with one another.

The students with whom he worked most closely and intensively were his graduate students. All of us who studied for the Ph.D. with Joe received an almost daily barrage of questions, insights, humorous anecdotes, and challenges. His questions taught us by exposing our analytical blind-spots, while his humor lifted our spirits. His anecdotes highlighted key concepts, while his challenges inspired us to excel at our work and to pierce the veil of time. We visited Joe's home in Worthington, Ohio, often—sometimes weekly—to discuss our dissertation research and writing, to run ideas by him, and to hear in return a rich mixture of advice, insights, questions, and counterpoints that sharpened our work as well as our thinking. Cookouts and parties at Joe's house, commonplace during the summer and nearly every holiday, brought everyone together for fun and a bit of relaxation, but also for a mutual honing of the collective mind. Joe knew very well that such a dynamic was taking place, and he fueled it with a mixture of exceptional humor, deep wisdom, and useful advice. While we cannot by definition be Joe, the graduate students who studied with this fine man and brilliant mentor came away more than just better educated. Joe guided us toward maturing into better and wiser people, who are harder workers, more intellectually curious, and deeply committed to the twin principles of service to others and duty to our students and our fellow citizens.

In addition to being an outstanding teacher and mentor, Joe was a prolific author with a huge range of professional interests. After publishing *Gunpowder and Galleys* in 1974 (a revised edition appeared in 2003), he wrote a number of airpower-related articles while serving as the Editor of the *Air Power Journal* from 1982 to 1985. In 1988, he and Michael O'Leary co-authored *Helicopters: The Illustrated History of the Vietnam War, Vol. 11*. Three years later, in 1991, he authored *America in Vietnam: The Fifteen-Year War,* followed in 1995 by *A Very Short War: The Mayaguez Incident and the Battle of Koh Tang*. In 2002, Joe published *Galleons and Galleys: Gunpowder and the Changing Face of Warfare at Sea, 1300–1650*. In addition to writing these major works, Joe co-authored NASA's multi-volume history of the development of the Space-Shuttle program. Further, he co-authored the massive, seven-volume *Gulf War Airpower Survey* with a distinguished team of military and civilian researchers. Numerous scholarly articles and reviews rounded out his impressive publication effort. Sadly, Joe died before he was able to complete a major work on the Vietnam War, but a small group of his former students is planning to bring this project to completion.

Besides his written scholarship, Joe was in demand worldwide as a lecturer. He often visited Spain, Portugal, Turkey, Italy, and other countries to speak about the development of galley warfare and its importance in the historical development and direction of the Mediterranean Basin and the wider world. Joe served as a visiting scholar for a year at the National Air and Space Museum in Washington, D.C., and for another year at the U.S. Military Academy at West Point, New York. These broad-ranging activities typify his many service-focused activities.

As one might expect, Joe performed a great deal of service during his long career. He was a member of the Society for Military History (SMH) board for several years. Joe's efforts ensured that the SMH maintained the highest level of effectiveness and its reputation as the world's premier military history professional association. In addition, Joe was a longstanding member of the Jolly Green Association, where he and his fellow rescue pilots, the pararescuemen who flew with them, and many of the people who they rescued gathered each year to remember their tours of duty. Joe was particularly touched to have been made an honorary pararescueman—something granted to very few people—as a result of his extraordinary exploits and valor in upholding the Air Rescue Service motto: "These Things We Do That Others May Live."

In recognition of Joe and his service in and out of uniform, his close friends and colleagues established a scholarship at The Ohio State University in his honor. A large number of current and former students, fellow scholars, and comrades from Joe's military days have contributed so much to the scholarship that it is now fully vested. It funds travel and all other expenses for undergraduate military history students from The Ohio State University to travel to battlefields and archives worldwide to study past armed conflicts. The fifth set of students to benefit from the scholarship is preparing to travel to Europe later this year. This scholarship commemorates Joe's dedication to his students.

While the scholarship continues Joe's support of his students, this volume honors Joe's professional reputation and service by highlighting former students' wide range of topical expertise and adept scholarship. It is abundantly clear in the various chapters that Joe instilled in his students his own belief in the importance of studying military history broadly defined, and of assessing technology's role in warfare. Every one of the contributors' chapters is a reflection of Joe's absolute focus on mentorship.

This book begins, fittingly, with Joe's own thoughts on technology, warfare, and strategy in "Technology and Strategy: What Are the Limits?" He presented this piece in 1994 at the Strategic Studies Institute, collocated with the U.S. Army War College at Carlisle Barracks, Pennsylvania. Along with Sir Michael Howard, Joe presented a keynote speech—a mark of his high scholarly reputation. Joe wrote and delivered this piece during a time of great ferment and uncertainty, in the heady days just after the end of the Cold War and the stunning military success of Operation DESERT STORM. Joe's message was straightforward and cautionary: while technology has always been a central component of armed conflict, it remains—as do all other military elements—subordinate to the strategic aims of a given war and the strategic acumen of those who declare and wage it. He warned his listeners to place technology and the larger questions surrounding warfare within a proper historical, cultural, social, and economic context lest a fascination with advanced weapons and a reliance on overwhelming force blind us to the complex and nonlinear realities of armed conflict. This admonition includes allusions to the advantages that weaker and technologically inferior adversaries may nonetheless have on their side when fighting a major power. Joe's essay is still a fascinating and immensely useful read today—a perennial warning about the absolute necessity of placing technology and war within their larger strategic context.

The contributors' 15 chapters follow Joe's paper, highlighting the necessity of understanding technology and warfare contextually and with sufficient depth. They are divided into four broad epochs and cover a global range of topics and wars. The first part's chapters discuss the integration and understanding of technology in a new framework between 1300 and 1800. The next segment's chapters evaluate the use of technology in the 1800s, incorporating the impact of organizational and governmental inputs to technological development. The following piece's chapters continue the trend of investigating the interpretations of technology during the world wars in an institutional and conflict-based context. The final portion's chapters expand the bounds of technology in warfare, delving into the abstract relation between information and technology, while examining the human response. Collectively, these chapters represent the breadth of Joe's definition of technology, while arguing his essential point that mental and physical tools must be understood from the perspective of those who used them, and that technology does not spring fully formed like Athena from Zeus' brow, but rather evolves as we use it.

The first section, covering the period from 1300 to 1800, begins with Dr. Clifford J. Rogers' "Gunpowder Artillery in Europe, 1326–1500: Innovation and Impact." He examines the development of gunpowder artillery between 1326 and 1500, explaining with great force why gunpowder artillery replaced mechanical forms of artillery, and how that change contributed to the centralization of state power. From there, it moves to Dr. Sarah K. Douglas' "To The Seas: The Genesis of Ship Rigging in the Medieval and Early Modern Worlds," which examines how the "age of sail" got its start in Europe and elsewhere. The next chapter, Dr. Andrew de la Garza's "The Lost World: Change and Continuity in Mughal Military Technology," moves us away from Europe to the Indian subcontinent and gives us a very different set of insights into technological development and war. His chapter leaves the reader with a clear understanding that people adapt technologies to geographical, social, cultural, and other overarching contextual factors.

The second segment explores key developments during the 1800s, beginning with two pieces on technological developments during the American Civil War. Dr. Edward McCaul's "If You Can Be Seen, You Can Be Killed: The Technological Increase in the Killing Zone during the American Civil War" and Dr. William Roberts' "Without Experience or Precedent: Transformational Technology and the Light Draft Monitors" give us land and sea versions, respectively, of technology at play during that massive conflict. The killing range of rifled muskets and the Union's problematic efforts to build an ironclad fleet give the reader an understanding of the evolutionary process of adopting and adapting to newer forms of technology. Dr. Robert H. Clemm then shifts us to the continent of Africa with his "The Ugandan Railway and the Fabrication of Kenya," which discusses how the technology known as "the railroad" facilitated the British colonization of Kenya as well as its emergence as a state, placing technology within the larger context and topic of imperialism.

The third period in this work covers the world wars. Dr. Robyn Rodriguez leads off with "German Technology and the Origins of World War II in East Asia," connecting German technological and advisory talent to the war in China. Sebastian Cox, Director of the Royal Air Force Air Historical Branch, and Joe's longtime friend and colleague from the SMH board, follows with "Freeman's Folly—The Debate over the Development of the 'Unarmed Bomber' and the Genesis of the de Havilland Mosquito, 1935–1940." His work gives us deep insights into how and why the iconic and extraordinarily successful Mosquito developed as it did. Dr. Corbin Williamson then transitions us from the skies to the sea with "The Impact of Institutional Context: Anglo-American Naval Fire Control," addressing how radar and related technologies came together with Navy organizational structures and processes to drive the development of fire-control capabilities. Dr. Robert Ehlers concludes this section with "Strategy, Technology, and Timing: Aircraft, the Mediterranean Air War, and the Turning of the Tide in the European Theater." This chapter addresses why "airpower technologies" are so much more than just aircraft, and how larger contextual factors such as grand strategy, geography, climate, and organizational innovations drove the employment of airpower technologies in this crucial but overlooked theater of operations.

The book's final selection surveys the period from 1945 to the present. Dr. Matthew Hurley's "'The War Is Lost': Technological Surprise and the Collapse of Portugal's Colonial Airpower Strategy, Guinea 1963–1974," initiates our march toward the present. It discusses how the SA-7 surface-to-air missile changed the conduct and course of the war. From there, Dr. Edward Coss discusses "The Vicissitudes of Violence: Fear, Physiology, and Behavior under Fire," which focuses on combat experiences during the Iraq and Afghanistan Wars. This chapter helps us to understand how human beings in combat survive, adapt to, and afterwards deal with the violence they experience on the battlefield, and by extension the full range of technologies involved in that violence. Dr. William Waddell then gives us "Des Fusils aux idées: Technological skepticism and masculinity in the French Army, 1954–1962." His work gives us deep insights into the tensions that existed in the French military based on competing views of technology's "masculine" and "feminizing" characteristics. Dr. Michael Pavelec follows with "Cyber: War?" This chapter brings us to one of the latest and most contentious debates about both technology and war and also technology in war—namely, how the cyber domain, which is not a physical but rather a logical domain, relates to and influences the conduct of armed conflict. Dr. Peter Mansoor, one of Joe's fellow professors at Ohio State, concludes this portion of the work with "The Precision-Information Revolution in Military Affairs and the Limits of Technology." This chapter builds upon the admonition Joe gave us in his paper that we not place too much faith in technology as a war winner.

Dr. Alan Beyerchen concludes the work with an epilogue titled "Reflections on A Warrior and Scholar." Dr. Beyerchen, who also spent an entire career teaching and mentoring history students at Ohio State and spent six years in the U.S. Army as a military intelligence officer, worked with Joe closely to sharpen military history students' insights and skills. His words about Joe convey a deep understanding about his colleague's depth of dedication to his students, his brilliance as a scholar of history and technology, and his tireless service to others.

Working with Joe was one of the great honors and pleasures of our professional lives. Being his students and friends was a source of great satisfaction. Individuals like Joe Guilmartin make the world a better place, because they strive to bring wisdom, learning, and civility into every one of their endeavors. Whether saving fellow airmen on the battlefield, teaching a room full of students, writing a book, or working in a wide variety of service-focused roles, Joe was a role model for us. We will strive to continue his legacy of superb teaching and scholarship, unexcelled leadership and mentorship, and gallant humility and good humor. In the spirit of this celebratory writing—this *Festschrift* as the Germans would call it—we offer Joe a warm and well-known toast in their language, which Joe himself used many times to toast us: "Prosit!"

CHAPTER 1

# Technology and Strategy: What Are the Limits?

*John F. Guilmartin, Jr.*

technology, n., (1): the application of knowledge to achieve a physical effect by means of an artifact, object or thing (distinct from art, which involves the application of knowledge to achieve aesthetic effect). (2): the artifact itself; the class of artifacts to which it belongs. (3): the knowledge needed to design, manufacture, operate, sustain and logistically support the artifact or thing and its user(s). Author's working definition.

∴

Sometime around the beginning of the fourth millennium BC an anonymous metalworker in the Fertile Crescent, no doubt after consultation with his village hero, the defender of fields and agricultural surplus, smelted a quantity of imported ore and cast a small sphere of copper pierced with a hole tapered to accept a haft. Breaking the object from its mold, he dressed and polished it and inserted the haft, fashioned from wood carefully chosen for strength, resilience, and perhaps magical properties. He then fixed the sphere securely in place with wedges, thongs, or perhaps glue, a chore of vital importance, for the bond between haft and sphere would have to reliably resist unprecedented accelerational and vibrational forces in the harsh arena of combat, and structural failure would expose the user to lethal risk. Invoking the blessings of the deities who presided over the smelting and hafting processes and had permitted him to penetrate their mysteries, he presented the finished object to the hero. In so doing, he completed the research and development (R&D) cycle of the acquisition program of the first weapons system–or for that matter any system–to use metal for other than ceremonial or ornamental purposes: the copper-headed mace.[1]

1 The dating is speculative; see my introduction to "War, Technology of," Encyclopedia Britiannica (1992) and Yigael Yadin, *The Art of War in Biblical Lands in the Light of Archaeological Study*, 2 vol. (London: McGraw-Hill, 1963), Vol. 1, 40. Though now dated, *The Art of War in Biblical Lands* remains the best single source on very early military technology and is valuable by virtue of having been written by a competent military professional. The Chalcolithic era, by definition the period in which man had learned to smelt copper and precious metals but

 | DOI:10.1163/9789004393301_003

To this point, the production of food, clothing, and shelter had been the leading edge of technology. The development by selective breeding in late Paleolithic and Neolithic times of plant and animal sources of food and fiber was perhaps the most impressive achievement, though the development of seafaring, fired pottery and polished stone implements for cutting and grinding was not far behind.[2] The earlier development of the bow, no doubt for hunting, though arrows could kill humans as well as game, was a major benchmark as well. Whether the application of animal traction to agriculture came before or after the development of the stone-headed mace is not known. What is clear is that the appearance of sedentary agriculture and storable food surpluses made the specialist combatant both possible and necessary.[3]

We know little about the first military specialists, but it is clear that their ideas about weaponry, its production and uses, diverged sharply from those of hunter-gatherers who might supplement their larder by raiding neighboring tribes. From that change in attitudes emerged the specialized technology of war, and with the appearance of the copper-headed mace military applications assumed a leading role in the development of technology which they have rarely relinquished since. We can surmise the rest of the story from the archaeological and historical record. The hero, skilled in the use of the stone-headed mace, adjusted his swing and footwork to the new and more powerful weapon, no doubt with the aid of a sparring partner. The results were sufficiently encouraging to prompt adoption of the new weapon. Operational test and evaluation (OT&E) was a smashing success, both figuratively and literally. Invoking the protection of the gods of the village and battle, the hero demonstrated the effectiveness of the new weapon, laying out for divine and human inspection in Chalcolithic body-count the barbarian warriors whose crushed skulls were proof of its power–and of that of the gods which sponsored its development. As fabricator and wielder hoped, the bold use of a new technology had yielded important advantages in combat. But copper was scarce and smiths able to work it scarcer still. Copper-headed maces were wielded only

still relied on stone implements for grinding and cutting, is generally held to have begun in southern Anatolia and the Fertile Crescent around the middle of the fifth millennium BC. Though archaeological evidence for mace heads of stone is abundant, that for copper is understandably scant since copper mace heads would have been melted down for bronze after the mace became obsolescent.

2 As indicated in the definition above, I consider technology to encompass human manipulation of biological processes. In this light, the development of natural grasses into food grains, perhaps best understood with regard to the Amerindian development of maize, was a technological achievement of the first order.

3 William Hardy McNeill, *The Pursuit of Power: Technology, Armed Force and Society since A.D. 1000* (Chicago: University of Chicago Press, 1982), vii.

by the select of the gods and the possession and use of such weapons became closely associated with divine approbation and earthly power. Significantly, the earliest known inscription in which an individual human being is identified by name, the Palette of King Narmer dating from around 3,100 BC, depicts the first Pharaoh of Upper and Lower Egypt ceremonially smashing the forehead of a prostrate enemy with a mace.[4]

The copper-headed mace in the hands of a powerful champion remained the world's premier weapons system for a very long time, but then as now both technology and the political ends to which it was applied were subject to change. In Egypt and regions of the Fertile Crescent, blessed with broad river valleys and fertile soil, villages gave way to cities. Competition for resources and favor in the sight of the gods increased and the defenders of cities fought one another as well as barbarians. The barbarians responded to increasing prosperity among the civilized polities with increased aggression and to advances in civilized weaponry by augmenting the power of their traditional weapons, notably the bow. Proliferation became a reality as barbarians adopted metal points for their spears and arrows. The smith provided the hero with protective armor of leather-backed Electrum to shield his body from stones and arrows and to protect his head from the crushing blow of a mace. Early in the third millennium, Sumerian smiths responded to the appearance of effective head protection by developing a whole new technology, first manifested in the piercing axe of arsenic bronze. The appearance of the piercing axe accelerated the race between offensive and defensive weaponry which continues to this day and ushered in a whole new age, the Bronze Age.

The above summary, of course, is simplified. Examined in detail, the interplay in late prehistoric and early historic times among developments in the design and production of weapons systems; changes in the way in which they were viewed and used; evolution of the social, ideological, and economic systems they served; and–the crux of the matter–changes in the ways in which weapons systems served political ends, both in perception and reality; were far more complex. Regions slipped into and out of the Bronze Age as reserves of easily-worked ores were depleted and better smelting technologies were developed. Long distance trade in strategic materials became increasingly important.[5]

4 The start of Narmer's reign is generally dated from circa 3,100 BC. The palette does not tell us what the head of his mace was made of; my point is the symbolic importance attached to the mace.

5 This situation was true from a very early stage. Arsenic bronze was soon superseded by tin bronze and as McNeill, *Pursuit of Power*, 1, notes, "copper and tin ores are not found in the same regions."

"Barbarians" engaged by hydraulic civilizations as mercenary warriors overthrew their employers and established their own dynasties, merging languages, and cultures. From the military and technological perspective, the lessons are clear. Advances in technology have yielded important military advantages from the earliest times. More often than not the demands of war press technological change first and farthest. The technology of war is developed and applied within a cultural and social context of which supra-rational beliefs are an important part. Evaluation criteria revolving around tactical success are an integral part of the process of selection or rejection. Finally, the production of high technology weaponry may well depend on imported materials. This example from remote antiquity makes a two-fold point: first, that the decision-making matrix within which the technology of war is developed, tested, adopted or rejected, and applied in combat is at bottom culturally-derived; and, second, that the essential categories in that matrix have seen remarkably little change over the ages. In short, history is relevant.

After briefly describing the establishment of the geo-political, technological, and cultural context within which our military institutions presently operate, this monograph will then trace the technological developments which formed the backdrop for the supposed military technological revolution of the recent past. For a number of reasons, the story begins in 1914. First, by then the national boundaries and perceived national interests of the major economic and military players in today's international arena were in place. Second, the basic notions of national interests, national policy objectives, and the appropriate national and military strategies for pursuing them were also in place, or emerged shortly thereafter. On a technical military level, the continuing influence of Alfred Thayer Mahan's theory of sea power, promulgated in the late 1890s, substantiates this point. Finally, one can argue that our basic notions of military strategy and tactics at the national level, and professional military officers' ideas of how to go about implementing them, are firmly rooted in the experience of World War II and largely derived from World War I. The fact that the United States participated in both wars served to reinforce their lessons.

Much of our technical, military vocabulary comes to us directly from the two world wars: front lines, frontal system (as in weather forecasting), barrage, strafe, flak, spotting artillery, and calling out bandits by clock position; so do our approaches to recruiting, training, weapons system procurement, and organization of the national industrial and scientific base for war. Having briefly delineated this baseline, offered as a war paradigm, this monograph will then examine recent changes in the technology of war and address the proposition that they are of sufficient magnitude to require a paradigm shift.

When Gabrilo Princip shot Archduke Franz Ferdinand of Austria in Sarajevo to start the First World War, the essential elements of our war paradigm, though not their interpretation, were in place. Industry and, through industry, science, had been firmly and more or less formally harnessed to war, though navies were more clearly aware of the link than armies and more skilled at exploiting it.[6] Air forces existed in nascent form. Artillery technology on land and at sea had been brought to a level which differs from today's only in degree by the universal adoption of nitrocellulose-based propellants; high explosive shells with time delay fuses; hydro-pneumatic recoil systems; and, breech loading, quick-fire mechanisms. Armies throughout the world had standardized on small arms using ammunition which differs in no essential way from the predominant form in use today: a fixed round consisting of brass case, nitrocellulose propellant, impact primer, and jacketed lead bullet. The determinant of strategic capability subject to the greatest change over the next three quarters of a century lay not in weaponry, but in the industrial base. In 1914, industrialized nations with adequate reserves of coal and iron ore were essentially militarily self-sufficient, dependence on natural nitrates for high explosives production having been eliminated as a strategic consideration by the recent German discovery of atmospheric nitrogen fixation.

Transportation and steel production were the critical determinants of military, strategic power, manifested in weapon and munitions production, capital ships and railroad mobilization. All were fueled by coal. Indeed, industrial capacity, and hence military capacity, was almost congruent with coal and iron ore reserves, Japan being the only exception though, for the moment, a minor one. To be sure, the successful adaptation of the internal combustion engine to military applications was a harbinger of things to come. During the Great War, gasoline-fueled aero-engines would absorb only a tiny proportion of fuel energy expended, but aircraft would exercise a disproportionate influence on land operations by its end. The same point applies to the strategic impact of diesel-powered submarines at sea. In the run-up to war, Britain had begun construction of a new class of capital ship powered by oil-fueled steam turbines, the Queen Elizabeth class battleships, to exploit the substantial advantages of oil over coal in thermal and volumetric efficiency, consciously accepting the

6 See Jon Tetsuro Sumida, "Forging the Trident: British Naval Industrial Logistics, 1914–1918," in John A. Lynn, ed., *Feeding Mars: Logistics in Western Warfare from the Middle Ages to the Present, Boulder* (Colorado: Westview Press, 1993), 217–249. As Sumida makes clear, the Royal Navy had much closer ties with industry than did the Army and was far more experienced at negotiating its interests in the industrial/ governmental arena; in consequence, the Navy preempted industrial capacity which would have been dedicated to supporting the land war in any rationally prioritized scheme, a reality which had much to do with the munitions shortages on the Western Front. Nor was the advantage only quantitative: I would point to the generally superior quality of aircraft procured by the Royal Naval Flying Corps compared to those procured by the Royal Flying Corps in 1915–16.

strategic vulnerability implicit in dependence on imported oil–and hedging its bets by seizing the recently-discovered Iranian oil fields.[7]

Imports in certain specialized niches were important to the conduct and outcome of the First World War; significantly, these took effect at the high end of the technology spectrum: American spruce was important to French and British aircraft production. Petroleum was sufficiently important strategically–most critically for aviation gasoline though gasoline-powered trucks and tractors performed certain essential tasks beyond the capacity of horses–and the Central Powers' reserves sufficiently small that only Romania's obliging declaration for the Allies in 1916 saved Germany and Austria-Hungary significant embarrassment by placing the Ploesti oil fields in their hands. The petroleum-based lubricants of the day broke down under the temperatures produced by high performance rotary, radial aero-engines and the British blockade denied Germany supplies of the castor oil they required, effectively grounding some of Germany's best fighters of 1918. On the whole, however, the war economies of the powers whose military forces dominated the conflict were self-sufficient and coal powered. By the eve of the Second World War, oil had supplanted coal and iron as the critical determinant of military power. Coal and iron ore retained their importance to war production, to be sure; steel was vital to weapons manufacture and coking coal was needed to smelt steel. Moreover, while bunker oil had largely replaced steam coal as the fuel of navies and merchant marines, coal-fueled steam locomotives and electric power plants played a vital role in the war economies of the contending powers. Indeed, the sophistication with which German industry squeezed every possible advantage from the Third Reich's coal deposits gave the German war economy a remarkable robustness, and probably lengthened the war by six months to a year.[8] But in the final analysis, the enduring economic and industrial importance of coal notwithstanding, tactical success at the cutting edge, and thus strategic advantage, depended on machines fueled by gasoline, diesel fuel, and bunker oil.

If high performance aircraft were a critical determinant of the outcome of the First World War on land, they were an absolutely indispensable ingredient of victory on land and at sea in the Second.

Moreover, the capabilities of military aviation had increased to the point that entire air campaigns were mounted. At least one, the Battle of Britain,

7 Ian Marshall, *Armored Ships*, Charlottesville (Virginia: Howell Press, 1993), 106–110, for a concise and perceptive summation.

8 See Alfred C. Mierzejewski, *The Collapse of the German War Economy, 1944–1945: Allied Air Power and the German National Railway, Chapel Hill and* (London: University of North Carolina Press, 1988), esp. 22–34. The estimate of how much sooner Nazi Germany would have collapsed had the coal-based war economy not been so efficient is my own.

was clearly decisive and others were arguably so, though not always in a form highlighted by conventional measurements of victory and defeat. An example can be seen in the devastating effect on the German war effort of the massive losses of trained aircrew which the Luftwaffe incurred in the spring of 1943 in efforts to reinforce and then evacuate German forces in North Africa.[9]

Tanks, fueled by gasoline and diesel oil, became a critical tactical determinant of victory on land wherever terrain permitted their use. Armies spearheaded by armored fighting vehicles were supplied, at least in the attack, not by rail but by truck convoy; trucks and tractors replaced horses as field artillery prime movers. That Wehrmacht petroleum reserves were so scant that the German Army had to rely on horse traction for tactical logistics and field artillery mobility would seem, with full wisdom of hindsight, to have virtually preordained defeat. The pivotal importance of oil to the war at sea is even more apparent. The lack of petroleum reserves in the Home Islands forced the Japanese militarists to go to war or back down in December 1941. The vulnerability of Japanese merchant shipping, oilers in particular, to attack by American submarines placed the Japanese war economy in an ever-tightening noose from 1943 on, a noose made even tighter by air-dropped mines in the final year of the war. Though the American war economy was effectively self-sufficient, that of Britain was anything but, and German U Boats and aerial commerce raiders posed a similar threat to the Western Allies. From the British standpoint, the critical, strategic logistic requirement from the beginning was the provision of sufficient stocks of petroleum to support naval and air operations from British bases, a requirement which expanded to encompass the support of U.S. air forces from 1942 and the support of Allied ground forces on the continent after D Day. The war at sea intersected with that in the air and on land in the critical importance of British petroleum stocks, for those stocks had to be replenished from American and Iranian oilfields by ship.

The importance of oil to the conduct and outcome of the Second World War extended well beyond quantitative, macro-economic considerations. High-octane aviation gasoline gave British and American aircraft, particularly fighters, a critical performance boost not enjoyed by their Axis equivalents; indeed, some have gone so far as to attribute British victory in the Battle of Britain to 100-octane gasoline. Axis engineers were well aware of the performance

9 Williamson Murray, *Strategy for Defeat: The Luftwaffe, 1933–1945, Maxwell AFB* (Alabama: Air University Press, 1983), 159–156, makes this point both explicitly and implicitly; see particularly 148, Table xxx, which demonstrates graphically how the enormous German aircraft losses in the Mediterranean between March and May 1943 combined with subsequent losses incurred in Italy, at Kursk and over the Reich to break the back of the Luftwaffe.

advantages conferred by high octane, but the refining process was highly inefficient, many more barrels of crude being required per barrel of refined gasoline as octane increased. So long as their sea lanes stayed open, Britain and America could afford the inefficiency; the Germans, Japanese, and Italians could not, and their fighter pilots entered combat at a significant handicap.

Beyond the critical importance of petroleum as a fuel, the relative importance of coal and iron was further eroded by the emergence in the interwar period of new structural materials with critically important properties which could not be obtained with steel. Moreover, in most cases if not all, reserves of these structural materials lay beyond the boundaries of the major powers. Far and away the most important was aluminum, the preeminent structural material for high performance aircraft from the mid-1930s.[10] Rubber, the raw material of pneumatic tires, O rings, aircraft de-icer boots, and a host of other essential sub-technologies, was a tropical import. Rubber could be synthesized from petroleum and coal, but the product was inferior, the process inefficient, and the raw materials valuable. Tungsten carbide was required to give hardness and durability to machine tool cutting heads. Only small quantities were needed, but there was no substitute and Germany lacked tungsten deposits, a fact with implications beyond the machine shop floor. Tungsten carbide is not only extremely hard, but extremely dense and German engineers developed highly effective anti-tank ammunition using sub-caliber tungsten carbide penetrators. Guns designed to fire this ammunition were much lighter and more mobile than their Allied equivalents and were among the most powerful

10 See: Eric Schantzberg, "Ideology and Technical Choice: The Decline of the Wooden Airplane in the United States, 1920–1945," *Technology and Culture* 35, no. 1 (January 1994), 34–69. Comparisons between aluminum and other aircraft structural materials, notably wood, were by no means as lopsidedly favorable to aluminum as most general aviation histories suggest. Until the advent of jet propulsion, properly designed and fabricated wooden structures were as efficient as their aluminum equivalents if not more so, that is they represented a smaller percentage of the weight of the fully-loaded aircraft. The fastest, and arguably most successful, reciprocating-engine combat aircraft of World War II, the de Havilland Mosquito, had an entirely wooden structure except for the engine mounts, landing gear, and flight control attachment points. Wood, however, was difficult to work with and wooden structures did not hold up well under high heat and humidity. Lockheed, producer of the most efficient aircraft structures of the interwar period, converted from wood to aluminum in the mid-1930s and kept its competitive edge, but this is more a commentary on the excellence of Lockheed's design team than the inherent disadvantages of wood. It is worth noting, too, that aluminum production was extremely energy-intensive, relying on massive amounts of electrical power–mostly produced by coal-burning power plants!

anti-tank guns of the war, but shortages of tungsten forced their abandonment in 1942.[11]

In sum, the preeminence of oil as the crucial determinant of strategic capability and tactical effectiveness in World War II is clear. The result was a war in which the ability to move raw materials over extended distances by sea was both a vital determinant of victory and a potentially lethal vulnerability. At a lower level of abstraction, the ability of design engineers to harness the propulsive energy of petroleum distillates and to apply them to the propulsion of military vehicles played a major role in shaping the outcome of the war. Within this matrix, the design and production of fighter aircraft was arguably the most critical node, though the design and production of bombers, tanks, torpedoes, submarines, artillery pieces, escort and anti-submarine vessels, assault landing craft, trucks, prime movers, and so on in approximate–and debatable–descending order of importance followed closely behind. More important than their individual tactical power and strategic importance, none of these critically important technologies could be approached in isolation, either from one another or from the production and logistical apparatus which built and supported them.

Moreover, the increasing tactical and operational interdependence of the various technologies of war and their sophistication lent a new importance to the human factor. The establishments which selected and trained the crews upon whose skill the effectiveness of military hardware depended were of pivotal importance. To cite a critical and illustrative example, the ability of the U.S. and Commonwealth air forces to generate large numbers of well-trained pilots played a major role in the defeat of the German and Japanese air arms, while poor planning and worse execution in this area by the Germans and Japanese went far to ensure Axis defeat.[12] Nor was the importance of recruitment, selection, and training limited to operation of the most expensive and sophisticated weapons systems: the increasing power and diversity of infantry weapons and the increasing sophistication of infantry tactics posed serious challenges for training establishments, not least of all in the United States where a small peacetime Army and Marine Corps had to expand enormously in very short order. Raw nationalism had sufficed to mobilize and motivate

---

11 John Batchelor and Ian Hogg, *Artillery* (New York: Scribner, 1972), 90–99; my main reference is to the Krupp 75mm PAK 41.

12 While pilots were the critical node, the point applies to a crew generally and to maintenance personnel. Aircrew training was a major Allied success in both world wars, yet remarkably little has been written on the subject.

millions of recruits at the outset of war in 1914–17. By 1939–41, disillusionment with the ideologies of an earlier day called for more sophisticated appeals.

Finally, the Second World War brought home to the Americans and British, at least, the verity that science not only could contribute to the war effort but was essential to its prosecution. If the war was run and won on petroleum, it was guided, directed and shaped by electronics, optics, and acoustics. Just as coal gave way to oil, telegraph wires, and submarine cables gave way to the vacuum tube. As with the internal combustion engine, there were harbingers in the First World War: wireless telegraphy was used to control fleets, and at times armies, and was exploited for intelligence; field telephones came into common use; the artillery preparations which broke the back of the German Army in the last year of the war depended on aerial photo-mapping and photographic reconnaissance for their effectiveness; sophisticated electronic and acoustic sensors were used to direct counterbattery fire.[13]

Between the First and Second World War there was an increase in communications and sensor capability of at least an order of magnitude: radar, sonar, and electronic cryptanalysis have received the lion's share of attention; but on a day-to-day basis, aerial photo reconnaissance was of at least comparable importance, particularly to the British and their American understudies who used it most effectively; and, voice radio and intercommunications systems had enormous impact across the board.[14] But the most important change was not in capability, but in use: Sensors and electronic communications were tactically and operationally important in World War I, and at times played a vital role. In World War II, they not only totally recast the face of war tactically, they had direct strategic impact: the British, and then Anglo-American, systematic, centrally-directed collection and analysis of photographic, electronic, and human intelligence is the most dramatic example, though by no means the only one. Still, both world wars were won as much by production as technique, though the quality of what was produced was important to the outcome, particularly in the second.

In sum, both wars were won by engaging and overwhelming the enemy production and population base. In World War I, this was accomplished indirectly through attrition of manpower and material at the front and by eroding the war economy and civilian morale through naval blockade. Direct attack on allied sea lines of communication by submarine had the potential to cause

13 Shelford Bidwell and Dominick Graham, *Fire-Power: British Army Weapons and Theories of War 1904–1945 (London: George Allen and Unwin, 1982)*, 141–146; Roy M. Stanley II, *World War II Photo Intelligence* (New York: Scribner, 1981), 20–32.

14 For an insightful treatment see Stanley, *World War II Photo Intelligence*, esp. 1–14.

the collapse of the British war economy, but inadequate resource allocation doomed the German U Boat campaign to failure. In World War II, production and population bases were exposed to direct attack by aerial bombardment, and economic attrition by submarine attack came into its own, though the Germans once more put inadequate resources into the effort.

The military leaders of the United States and their civilian superiors learned the lessons of World War I as well as the leaders of any major power. Granted, the British taught the Americans some vital lessons in technique, but it is safe to say that by 1944–45, the U.S. military establishment had embraced and mastered a war paradigm which corresponded very closely to strategic reality. It is important to note that this paradigm incorporated a keen appreciation of the importance of technological advantage and of the need to systematically pursue that advantage in time of peace. This perception stemmed largely from painful memories of the unpreparedness of its air arm and field artillery when the United States entered World War I.[15] That perception found expression in the steady expansion of aerodynamic knowledge by the National Advisory Committee on Aviation (NACA) during the interwar years which yielded important dividends in World War II.[16] It also found expression in the design of a range of thoroughly capable artillery pieces, already in production when war came, and a rationalized scheme for the production of trucks and prime movers.[17] Still, at bottom, insofar as technology was concerned, the central focus of the American war paradigm was on production. The creation after World War II and continued existence of the Industrial College of the Armed Forces provide eloquent testimony to that emphasis, an emphasis which is deeply imbedded in the U.S. military services' corporate memories and thought processes. John Patrick has neatly encapsulated that reality in his perceptive description of the U.S. Army, Navy, and Air Force as "classic industrial institutions."[18]

The fundamental underpinnings of the paradigm, however, were shaken by the advent of the nuclear age. After a brief period of unexploited monopoly,

15 Attempts to produce the French '75 in the United States were a fiasco and the American Expeditionary Force was equipped almost exclusively with French artillery.

16 Notably in the empirical development of methods of reducing engine-associated drag, the most dramatic application of which was in the North American P-51, arguably the most successful fighter of the second half of the war.

17 Daniel R. Beaver, "Deuce and a Half, Selecting United States Army Trucks, 1920–1945," *Feeding Mars: Logistics in Western Warfare from the Middle Ages to the Present*, 251–270.

18 Remark from the floor, "The Revolution in Military Affairs: Defining the Army for the 21st Century," Carlisle Barracks, Pennsylvania, first afternoon session, April 27. Mr. John J. Patrick is Program Manager at Techmatics, Inc. of Arlington, Virginia.

nuclear proliferation set in. The United States and its primary allies faced the reality of the Cold War; they also faced a situation in which their industrial bases were secure from direct attack save by a massive nuclear strike which would surely have been answered in kind. In the United States, the immediate response at the national level was to embrace a new and radically different war paradigm under the rubric of massive retaliation. Confident under atomic and then thermo-nuclear umbrellas, Presidents Harry S. Truman and Dwight D. Eisenhower drew down conventional forces. To their great credit, elements within the Army and Navy challenged the validity of the new paradigm but it took the Korean War to demonstrate that defense could not be bought on the cheap with nuclear weapons. The lesson was that the United States needed deployable divisions, carrier task forces, and the draft. After the Korean War, the two paradigms uneasily coexisted side by side and the resultant inconsistencies produced problems: the Air Force, at least, continued to assume that air power should be applied as it had been in World War II. Consequently, the United States bombed North Korea as if it were Nagoya or the Ruhr Valley. Over a decade later, the essence of Air Force plans for bombing North Vietnam was not substantially different.[19]

As the U.S. defense establishment relearned the value of conventional forces, concern for the cost of national defense came to the fore, and defense dollars became tight. The attention of senior military officers and staffs was directed increasingly to the competition for funding for weapons system procurement. Secretary of Defense Robert S. McNamara's establishment of rules for the competition in the form of PPBS (Planning, Programming, and Budgeting System) only formalized and accelerated an ongoing trend. By focusing preemptively on rigorous quantitative analysis of cost benefit effectiveness, PPBS subtly inhibited technological innovation. To calculate whether or not something is worth its cost, one must know precisely what it will do. Since by definition it is impossible to precisely predict the effect of something which has never been made before, advantage in the PPBS arena accrued to systems which offered

19 This is the basic point of Earl H. Tilford, Jr., *Crosswinds: The Air Force's Setup in Vietnam*, College Station (Texas: Texas A&M University Press, 1993). In fairness to the planners who laid out the initial targeting for Rolling Thunder, interdiction of Communist manpower and supplies occupied a high priority, but this, too, was very much in the tradition of our air interdiction efforts in Korea and World War II. We should also note that the Air Force did a number of clever and successful things in Vietnam which cut against its doctrinal grain, including the widespread use of aerial FACS (forward air controllers), the development of the side-firing gunship and the creation of a specialized long-range combat rescue force. These innovations and those involved in implementing them, however, were never regarded as being in the professional mainstream.

incremental improvements of tried and true technologies whose effect could be predicted with reasonable accuracy. Should anyone be surprised, then, that an extraordinarily high proportion of weapons systems procurement programs which have yielded major advances in military capability were the product of "black" programs, originated and managed outside the PPBS mainstream? The bureaucracies that blossomed throughout the defense establishment were managed by military accountants who, in addition to confusing efficiency with effectiveness, also inhibited imagination and innovation. One should not be surprised that black programs, on the other hand, seem to have been more successful than those which have had to run the entire PPBS gauntlet.[20]

At the same time, while the awesome power of nuclear and thermonuclear weapons was monopolizing the attention of defense intellectuals and PPBS that of military professionals, emergent new technologies were beginning to gnaw away at what we might term the classical, World War-derived, war paradigm and its massive retaliation-derived supplement. Salient among these technologies in rough chronological order were: digital electronic computers; new sensor systems, particularly in the infrared spectrum; new structural materials, first simple ablatives for ICBM (intercontinental ballistic missile) warhead heat shields, and then entirely new materials such as phenolic resin honeycomb, reinforced carbon, and fiberglass boron fiber composites; and, last but most decisive in net effect, transistors and the miniaturized and more powerful analytical computers which they spawned. One could add to the list the turbojet engine, which fit the existing war paradigm very nicely, but which in the form of very small, highly fuel efficient power plants for cruise missiles and RPVs (remotely piloted vehicles) challenges it directly today. It is worth noting, however, that many of these new technologies were developed and initially exploited for explicitly military purposes during World War II.

The story of the turbojets' independent development in Germany and Britain is well known. Ablative materials were discovered more or less accidentally during the V-2 program and computer technology got an enormous boost from the demands of cryptanalysis in the United States and Britain. The guidance mechanism on the later V-2s was the first reprogrammable electronic analog computer.[21] For a time, these innovations fit our bifurcated war paradigm reasonably well. But of greater importance over the long term, the application of the paradigm continued to promote effective innovation. In the

20 The dead hand of PPBS is, of course, not the only factor involved: "black" programs are more glamorous than openly acknowledged ones and enjoy high priorities, thus presumably attracting more capable engineers and managers.

21 I am indebted to James Tomayko of Omaha State University for this information.

1950s, systematic development of jet engine technology combined with innovative airframe design and far-sighted military requirements to produce the jet bombers which made nuclear deterrence a strategic reality pending the development of ICBMs. Those same jet bombers, with the B-47 as the seminal design, spawned the Boeing 707 and its imitators, giving air transportation a whole new strategic dimension.[22] They also forced the development of air-to-air refueling which gave U.S. air operations an unprecedented flexibility, which no other nation can match.

Ablative heat shields combined with the discovery by NASA engineers of the aerodynamic and thermal advantages of detached, shock-wave reentry to make the ICBM a practical reality. Then, beginning in the 1960s, the new technologies began to combine synergistically with unprecedented and largely, unanticipated impact. Perhaps the first concrete indication of the potential of the new technologies came with the fielding by the U.S. Navy of a massive underwater passive sonar array to track Soviet ballistic missile submarines. The use of passive sonar to locate submarines was hardly new; the use of analytical computers with logic and memory circuits capable of discriminating between the whisper of a nuclear submarine's passage and the rumbles, squeaks, and whistles of the sea was. The Vietnam War produced yet other harbingers: the Army fielded the first operational stealth aircraft, the Lockheed YO-3A used for night observation.[23] Navy and Air Force airmen made effective use of electro-optical and laser-guided bombs (EOGBs and LGBs) to strike heavily-defended targets with unprecedented accuracy in the Linebacker bombing offensives.[24] These developments, however, took effect in isolation.

By the late 1980s, the emergent technologies were rapidly converging to produce an unprecedented array of new capabilities. The driving force was the transistor: transistors immediately permitted order of magnitude reductions in size for the same computing capacity, directly because transistors are much smaller than vacuum tubes which perform an equivalent function and indirectly because they require far less power and emit far less heat. Of at least equal importance transistors are infinitely more reliable than vacuum tubes. The magnitude of the difference and the speed with which it was exploited is illustrated by two examples: the SM-64 Navaho, a sophisticated vertically-launched air-breathing cruise missile of the mid- to late-1950s with a cruise vehicle about

22 Peter J. Hugill, *World Trade Since 1431: Geography, Technology and Capitalism, Baltimore* (Maryland: Johns Hopkins University Press, 1993), 290–291.

23 Bill Sweetman, *Stealth Aircraft: Invisible Warplane, Black Budget, Osceola* (Wisconsin: Motorbooks International, 1989), 53–55.

24 Tilford, *Crosswinds*, 150–151.

the size of an F-15; and the Mach 2 capable B-58, which entered service in 1957. The Navaho was an aerodynamic success, and the rocket engine, which powered its piggy-back booster, was a direct ancestor of rockets which put man on the moon, but the unreliability of its vacuum tube circuitry earned it the name "Never-go Navaho," even as a research vehicle. Like the Navaho, the B-58 was an aerodynamic success and an operational failure, though not for the same reason; fast as it was, it could not outrun Soviet air-to-air missiles. The core of the B-58 bomb-navigation system was a 1,200 pound electronic analog computer, the largest ever made, with a ticker tape readout.[25] The guidance and navigation computers of the Tomahawks and ALCMS (Air Launched Cruise Missiles) used in DESERT STORM, designed to do the same basic job as those of the Navaho and B-58, are not much larger than a carton of cigarettes.

But the importance of the size and reliability of avionics components is not the only lesson to be gleaned from the Navaho and B-58 stories. Each system exploited mature technologies which were at or near their limits of development and, barring the development of novel replacement technologies, are likely to remain so. The Navaho cruise vehicle tested the aerodynamic theories which were the basis for the development of our current generation of air-to-air fighter aircraft, pioneering the outward-splayed twin vertical stabilizers, which are a feature of every high performance multi-engine U.S. fighter since the F-14. The Navaho booster's liquid oxygen/kerosene-fueled rocket engine represented a significant improvement over earlier rockets in thrust-to-weight ratio and efficiency, but subsequent advances have been modest for reasons rooted in unalterable physical reality: the chemical energy resident in known oxidizer/ propellant combinations and the molecular weights of their decomposition products.[26]

---

25 R. Cargill Hall, "To Acquire Strategic Bombers: The Case of the B-58 Hustler," R. Cargill Hall, "To Acquire Strategic Bombers: The Case of the B-58 Hustler," *Air University Review* XXXI, no. 6 (September–October 1980): 1–20.

26 The basic measurement of rocket efficiency is specific impulse, written Is, a measurement of the number of pounds of thrust produced each second by each pound of propellant. Though expressed in seconds by mathematical reduction, Is is in fact a dimensionless ratio, the higher the value the greater the propulsive efficiency. The engine of the World War II German V-2 weighed 2,484 pounds and produced 56,000 pounds of thrust with an Is of 199 seconds. The Navaho booster weighed 1,230 pounds and produced 120,000 pounds of thrust with an Is of 230 seconds. Liquid oxygen and kerosene, the oxidizer/ fuel combination used in the Navaho and still used in the Atlas booster, has a theoretical maximum Is of 294 seconds. Liquid oxygen and liquid hydrogen, the combination used in the Space Shuttle's main engines, yields a theoretical maximum of 391 seconds, but at a considerable cost in technical complexity and risk. Liquid hydrogen is notoriously difficult to deal with and, for reasons bound up in cryogenics and the molecular physics of known structural materials, the technology for a structurally efficient, reusable liquid

As one progresses to more energetic and efficient fuel/oxidizer combinations, the difficulties of containment and handling increase exponentially, yet improvements in performance are comparatively modest.[27] For certain high priority applications, of course, the incremental increase in performance justifies the increased technical risk and cost, but it is worth noting that the Titan, a mid-1950s design, is still our workhorse booster for heavy unmanned payloads and the even older Atlas booster and Agena upper stage remain first-line equipment.[28] The same point can be made with regard to the J-79 turbojets which powered the B-58. The turbofan engines which power today's first line fighters and bombers are far more fuel efficient than the J-79, but as a practical matter few of the aircraft they drive are much faster than the B-58. USAF F-4G "Wild Weasel" anti-radiation attack aircraft powered by J-79s were first line equipment in DESERT STORM, and if their fuel-guzzling engines required more trips to the tanker than newer strike aircraft, the F-4Gs were at no great disadvantage in speed, ordnance carriage, or maneuverability. Simply put, given kerosene-based jet fuels and aluminum structures, performance improvements for manned combat aircraft have been marginal since the mid-1960s and are likely to remain so.[29] Granted, exotic engines burning exotic fuels can drive specialized air-breathing platforms made of exotic materials to much greater speeds; the largely titanium SR-71 cruised at Mach 3.5+ and its successors will be considerably faster (perhaps already are) but such vehicles are expensive,

---

hydrogen tank does not exist. Liquid fluorine and liquid hydrogen, the most efficient propellant combination known, would yield a theoretical maximum Is of only 410 seconds and the practical difficulties of containing and controlling liquid fluorine are immensely greater than those posed by liquid oxygen. Dieter K. Huzel and David H. Huang, *Design of Liquid Propellant Rocket Engines*, 2nd ed., NASA Special Pamphlet SP-125 (Washington, DC, 1971), 21–27, 36.

27 I am indebted to Jeff Cooper, Strategic Director of SRS Technologies, for pointing out to me the strikingly modest improvement in rocket engine efficiencies since the late 1950s after my spoken remarks on April 27.

28 A telling detail which indicates how quickly American engineers mastered the basic physics of rocket propulsion is the Atlas' stage-and-a-half configuration in which the vehicle leaves the pad with three engines firing and subsequently jettisons two of the three. A two-stage design, in which a first stage booster with its own tankage is jettisoned prior to second stage ignition, is inherently more efficient and was recognized as such, but the Atlas was designed so early in the game that engineers were unsure that the fuel and oxidizer would ignite spontaneously in the hard vacuum of space.

29 I do not mean that aircraft technology will remain static, or that there is no point in investing in improved piloted combat vehicles.
Clearly, such platforms as the B-2 and F-22 represent major advances in combat capability, but in terms of speed, range, and load-carrying capability and efficiency, they are not all that different from their predecessors.

highly specialized, and require enormous amounts of skill and care to operate. The one area in which the United States can anticipate major improvements in aerospace vehicle propulsion over the short term is in extremely small air-breathing engines for remotely piloted vehicles and cruise missiles, and here the improvement will be not in speed, but in compactness, fuel efficiency and range.

To return to the main argument, shrinkage of avionics packages with the advent of transistors was only the tip of the iceberg, for transistorized circuitry also permitted enormous increases in computing power. On a mundane level, the increased power of design computers turned finite element stress analysis from an exotic, time-consuming routine, applied only to the most critical of complex stress analyses, to the workhorse of structural engineering in the mid- to late-1980s. On a more exotic level, powerful design computers made feasible the complex, multivariable analyses of radar returns as function of aspect, geometry, reflectivity, and structural properties which made stealth aircraft a reality. The result was a revolution not only in the capabilities and reliability of existing categories of systems, but an ability to design entirely new kinds of weaponry. The tactical results were demonstrated dramatically in DESERT STORM. In terms of operational impact, the salient capabilities were: LGBS used in combination with night-capable target acquisition and tracking devices; sophisticated night-capable tank fire control systems; the ability to operate at night in general; the operational debut of stealth in the form of the F-117A; long range cruise missiles capable of genuine precision, notably the Tomahawk; the widespread use of secure voice communications and facsimile machines for command, control, and coordination; the massive use of air refueling, as much for operational flexibility as for simple range extension; the advent of beyond-visual-range air-to-air combat as an operational reality; and the operational debut of airborne radar systems like JSTARS, capable of monitoring the land battle in detail. The fact that they were deployed against an enemy inferior in many critical systems and training notwithstanding, the application of these capabilities in combination struck with unprecedented tactical and operational impact. They also produced unanticipated problems, notably in the difficulty of applying our intelligence systems and analytical methods to the unprecedentedly swift tempo of the war.

A handful of predictable errors notwithstanding, American ground and air forces were remarkably successful in conducting and supporting a highly mobile, highly lethal, land war without defined fronts. This was a training achievement of the first magnitude: total U.S. casualties approximated in number what a cynic might have expected to ensue from that many men and women driving that far, that fast, and at night over the same terrain without opposition. The quality of American aircrew training was equally apparent: for the Air Force, at least, the aircraft accident rate, indeed the total loss rate per flying hour, went down. One's first reaction on appreciating the full magnitude of what U.S. and allied forces had achieved in DESERT STORM was to draw a parallel with Robert Clive's victory at the battle of Plassey in 1757 in which 850 British troops and 2,100 Sepoys, trained to British standards, prevailed over a force of some 52,000 Indians and 200 French mercenaries.[30] Just as well-trained and equipped Western European armies and navies of the late 18th and 19th centuries were able to defeat forces not trained to European standards almost without regard to numerical odds, it is likely then that a balanced U.S. force of all arms, properly equipped and supported, and competently led, could defeat any number of anything else. Certain of our allies, notably the British, can achieve similar results when backed by specialized U.S. support in critical areas such as electronic warfare.

Turning from technology and tactics to strategic limits poses two questions: to what extent does the revolution in military technology expand strategic capabilities; and to what extent does the prevailing war paradigm hold true? Of the two, the second may be the more critical. Understanding the nature of war and how it is waged will directly determine not only tactical and operational effectiveness in future conflicts but also strategic success. The technological and tactical revolution witnessed over the last decade was undergirded by evolutionary adjustments in the war paradigm which have the potential to yield major strategic dividends. At the same time, the threat has changed in ways which are so fundamental as to require an extensive rethinking of the strategic equation. As suggested above, the primary technological constraints to U.S. strategic capabilities in the two world wars, in which this basic war paradigm was forged, involved productive capacity rather than tactical capability. Note, however, that the tactical capabilities of individual weapons systems were more important to the outcome of the Second World War than the First, and that the United States and its principal ally, Great Britain, were rather successful in anticipating the shift and planning for it.

In this sense, the development of the technologies fielded in DESERT STORM was more an evolutionary trend than a revolution in military thought. This encouraging thought must be balanced by a more sober one: as noted implicitly at the outset, the geopolitical and social contexts of the two world wars were strikingly similar. Indeed, some have suggested that the Second World

30 For a useful summation, see David Chandler, *Atlas of Military Strategy* (New York: Free Press, 1980), 78–79.

War was merely a continuation of the First, an argument which makes a good deal of sense. By contrast, the geographic, economic, political, and social contexts of the wars of the present differ dramatically from those of the two world wars, and it is likely that those of the 21st century will diverge even more sharply from the conflicts in which the American war paradigm was forged. Not only has the economic and diplomatic structure of the international system undergone radical change in the wake of Tiananmen Square, the collapse of the Soviet Union, and the unification of Germany; social and economic expectations around the globe have changed as well. In sum, the threat has changed substantively but the war paradigm has not. The ancient warrior, now wielding a bronze, piercing axe and wearing a helmet of Electrum, must face not only enemy champions, but a host of smaller, more subtle and more elusive foes whose objectives are not always apparent. The excellence of this hero's weapons and his strength and skill in using them still count, but he and his smith must anticipate major changes in the weaponry which he can expect to be used against him and those he is sworn to protect.

What technologies, then, should the American military expect to face in the 21st century? Beginning at the top of the spectrum of technological sophistication and working down, nuclear proliferation will remain a major concern. Indeed, sub-proliferation might be a better term, since the possession of nuclear weapons in the hands of political entities below the level of the nation-state is apt to become a reality soon, if it is not already. The bad news is the relatively easy availability of plutonium and enriched uranium on the international black market. The good news is that those most inclined to develop and use nuclear weapons lack the scientific, engineering, and technical resources to build compact weapons and sophisticated delivery systems. At least, the first wave of "outlaw" nuclear weapons will be relatively large and crude.

One terminal effect of nuclear weapons, however, which does not require the detonation of the device in close proximity to the target, demands attention. Electro-magnetic pulse, or EMP, is the wave of highly-charged electrical particles proceeding from the nuclear event like an expanding bubble at speeds approaching that of light. EMP, which takes effect as an extremely short burst of very high-voltage electrical energy, has the capacity to do enormous damage to electrical circuits. Transistorized circuits, on which the day-to-day functioning of modern society is increasingly reliant, are particularly vulnerable. A cursory review of the historical record suggests that the detonation in low earth orbit or the upper atmosphere of a relatively crude nuclear device would wreak havoc over a wide area on a host of transistor-dependent systems, including electrical power generation and transmission; telephone, radio, and television transmission and reception; information storage and retrieval systems; aircraft avionics; truck and automobile ignition and fuel control units; and, medical life-sustaining and monitoring equipment.[31] Circuits can be protected against EMP by "hardening," that is shielding, and critical military systems have been hardened on a selective basis, but hardening is an arcane science and expensive.

On a more mundane level, the allies experienced great difficulty with mobile ballistic missile systems in DESERT STORM, and the system involved was the primitive first generation Soviet SS-1 "Scud." Such systems have enormous potential for mischief, particularly if mated with chemical or bacteriological warheads. The problem is not just military, but political as well, for there are many regions of the world where the launch of such a missile could be credibly attributed to more than one entity, making the appropriateness as well as the effectiveness of the military response an issue. Fortunately, relatively long-range ballistic systems are, in principle, subject to interception and destruction by defensive missiles. Since ballistic trajectories are predictable, the interception problem is relatively straightforward, though highly demanding of sensor, avionics, booster, fusing, and warhead systems. The problem of terminal defense against ballistic missiles is solvable given the willingness to commit the necessary resources. Stealth cruise missiles present a greater challenge. While it is unlikely that any enemy will be able to approach the U.S. lead in the design and production of piloted stealth aircraft for the next few decades, the suppression of radar signatures becomes easier by a geometric ratio as the size of the vehicle is reduced. It is, thus, much easier to make a very small vehicle stealthy, a consideration which becomes more relevant in light of the advent of the extremely small, extremely efficient, air-breathing engines mentioned above. Moreover, the United States has provided potential enemies with the basis for a precision guidance system in the form of un-coded signals from Global Positioning System (GPS) satellites. The ability to develop long-range, stealthy cruise missiles with precision accuracy, at least against

31 The key event was the detonation of an American nuclear device in the upper stratosphere over Johnson Island during the Kennedy presidency. Having negotiated a nuclear test ban with the USSR, the administration came under attack by critics who argued that the Soviets were playing us for suckers and preparing a nuclear test series for immediate implementation when the ban expired. In the event the critics were right, we would be unprepared to follow suit. Embarrassed, the President ordered the AEC (Atomic Energy Commission) to conduct an immediate nuclear test. The only thing the AEC could come up with on short notice which might produce useful data was a detonation in the lower reaches of space. To the surprise of all, EMP from the test wrought electronic havoc generally and blew out street lights as far away as Honolulu.

stationary targets, is within the grasp of many national aerospace industries, and such vehicles would be exceedingly difficult to defend against.

All the above concerns address conventional munitions and delivery systems. More frightening are the less traditional systems. The design of sophisticated chemical agents does not require massive technical infrastructure and computer-assisted design capability. Moreover, such agents need not be lethal in an immediate physiological sense, and the development of agents with sophisticated psychological effects, whether transient or permanent, is a possibility that must be considered. Finally, there are nontraditional targeting and delivery methods to be considered: chemical agents can be delivered by introduction into the water supply and food chain as well as the atmosphere.

The same undemanding technical and economic constraints on design and delivery apply even more powerfully to bacteriological and viral agents than to chemical weapons. At the lower end of the technology spectrum, the use of natural selection to develop microorganisms with desired immunities and lethalities has the potential to produce highly destructive agents at extremely low cost. At the upper end, the use of genetic engineering to produce novel agents would still be relatively cheap. Disease agents offer a wide range of terminal effects which could be tailored to strategic goals. A highly virulent, fast-acting disease organism with high lethality could quickly kill off the bulk of the populace in the target area and burn itself out, leaving a depopulated area open for occupation. Conversely, moderately virulent, slow-acting bacterial or viral agents with moderate lethality, perhaps engineered to produce chronic rather than acute effects, would be at least equally difficult to counter. The longer the incubation period of the disease is, and the slower-acting it is, the greater the difficulty of detection and the greater the damage done prior to identification. In either case, the disease organism could be aimed at crops, domestic animals, or perhaps the environment, rather than directly at populations. Selective targeting could be achieved through the use of insect or animal vectors, and disease organisms biologically engineered to strike selectively according to age, gender, race, or behavior, are entirely possible. Such agents would have enormous potential for social and political disruption, particularly if used in concert with a well-crafted psychological warfare campaign.

The remarkably wide acceptance of the story that AIDS was a capitalist or racist plot hints at the potential of such a coordinated campaign. More simply, political targeting could be accomplished by a combination of biological engineering and physical distribution. It would be a relatively easy matter to develop a strain of staphylococcus immune to available antibiotics. Inadvertently, this has nearly been done through the widespread, indiscriminate use of antibiotics. Such an agent could be used to attack segments of a national health care system catering to a given socioeconomic strata or ethnic group, and the disruptive effect of such an attack could be amplified by a coordinated campaign of propaganda and disinformation designed to exploit the predictably defensive responses of public health officials.

Returning to electronics, American society, the U.S. economy, and the world trade system are increasingly dependent on the exchange and analysis of information by electronic means. Those means are vulnerable to physical attack, electronic disruption, and manipulation. As with bacteriological attack, an electronic assault could be mounted with minimal capital investment: the successes of computer hackers in gaining access to restricted data bases and planting electronic viruses in data exchange networks speak for themselves. Intrusion into classified Department of Defense and contractor databases is an obvious possibility as is compromise of militarily sensitive command and control circuits, analytical computers, and classified data bases. These circuits are presumably well-protected and monitored. However, such an attack would be the electronic equivalent of a frontal assault, likely to be detected quickly and repelled. More subtle and more likely to produce strategic results would be a sophisticated attack on selected nodes of the American social and economic fabric, again accompanied by a coordinated psychological warfare campaign. An attack on financial market and stock exchange data bases to subtly manipulate interest rates, profits, and losses so as to erode the savings–and the confidence–of selected segments of society is an obvious tactic. International credit, currency conversion, and banking systems are vulnerable as well. As with a biological attack, a slow-acting, moderately virulent "disease" would be harder to detect and counter than a fast-acting, highly virulent one. Moreover, a campaign of subtle electronic sabotage would very likely trigger a defensive reaction from governmental and semi-governmental regulatory agencies, and denials that anything was amiss would add credibility to charges that the government was deliberately inflicting injustice.[32] Intrusion into the data bases of

32 A telling and relatively recent example of such a defensive reaction can be seen in statements made by officials of the Red Cross, the American Association of Blood Banks, and the Council of Community Blood Centers in January of 1983 that there was no "hard" evidence that AIDS was being spread by blood transfusions and arguing against the use of laboratory tests to screen blood donors. At the time, the Center for Disease Control (CDC) was arguing persuasively that AIDS could, in fact, be spread by transfusions and contemporary internal Red Cross memos recently leaked to the press indicate a clear awareness that this was the case. Having drawn their line in the sand, however, the Red Cross, et. al., held their ground until March 1985 when they were forced by an overwhelming weight of evidence to adopt mandatory AIDS testing for blood donors. In fact, the circumstances surrounding the debate over mandatory donor blood testing in January of 1983 were more complicated than the above summary suggests: CDC argued for a "surrogate" test for

the Internal Revenue Service and state revenue systems to introduce injustices and inequities, perhaps by selectively targeting particular groups for audits and seizures of assets, has obvious potential as well. Any such attack would have to be undergirded not only with a deep understanding of the workings of the markets in question, but of the social, economic, and political dynamics of American society; that having been said, the potential vulnerability in this area is enormous.

Having surveyed the technological face of the threat environment, this monograph concludes with a brief examination of U.S. capabilities and posture. Here prescriptions are mainly intellectual rather than technological and tactical. Just as the Chalcolithic hero girded himself for battle and for the evaluation of new technologies with an appeal to the gods of war and his village, so, in secular guise do today's planners, programmers, and budgeters. The gods are, however, changing and mortals must take account of the change. First, there are some cautionary observations. PPBS made a good deal of sense in an era in which technological change was incremental and attacks on the enemy war economy and population base were the primary strategic focus. Those assumptions no longer hold, and the "theology" badly needs revision,–the term theology is used literally. There is no credible demonstration that the rigorous application of cost benefit effectiveness analysis to the weapons system acquisition process under the aegis of PPBS has, in fact, had a beneficial effect. PPBS has been accepted not as a methodology with demonstrable benefits, but as a verity accepted on faith, and the rational basis of that faith badly needs

hepatitis B, which it maintained could have been in place by March 1983, on the grounds nearly all AIDS patients also suffered from hepatitis B. The counter-argument, still supported in May 1994 by a senior Red Cross official privy to the internal deliberations and decision to reject testing ("Red Cross knew of AIDS risk, documents show," The Columbus [Ohio] Dispatch, Sunday, May 15, 1993), was that the hepatitis B test would have produced a large number of false positives, thus (my interpretation) excluding many potential blood donors. It was difficult at the time, and it is difficult now, not to conclude that economic imperatives outweighed humanitarian and medical concerns in the deliberations of the agencies regulating our blood supply. The point of the above example is not to portray the officials of the Red Cross, et. al., as ogres; rather it is to illustrate the deeply ingrained reluctance of official and semi-official regulatory agencies to admit error. In this instance, that reluctance was not turned against them, or the government, to major political effect, in part because the segment of society most affected by the AIDS epidemic at the time, the gay community, was ambivalent on the issue. The point is that such reluctance is real, is endemic, and is a point of potential vulnerability. The appearance of outbreaks of necrotizing fascitis, caused by a "flesh eating" strain of streptococcus, first in Britain and now in the United States, has produced a similar wave of dismissive statements by public health authorities. See: "Flesh-eating Bacteria Aren't New to This Area," The Columbus (Ohio) Dispatch, June 9, 1994.

reexamination. If PPBS is a false god, the American military may be well-advised to consider the cost of its misplaced devotion sooner rather than later.

The evidence suggests that a technological plateau has been achieved in terms of speed, firepower, armor protection, and mobility. Barring the development of revolutionary, new structural materials and technologies of propulsion which violate the laws of physics as now understood (a possibility which cannot be discounted), current state-of-the-art aircraft, ships, and land vehicles go nearly as fast and can expend about as much firepower as will be the case for the foreseeable future. The significant technological advances will come in sensors and detection systems; avionics, guidance and fire control systems; stealth technology across the board; electronic warfare systems; the biological sciences; "virtual reality" training; and, most important, analytical computers, particularly for analysis and design. Planners must be careful to avoid temptations to expend disproportionate amounts of treasure and energy to achieve incremental improvements to existing technologies. In the past, sufferers from what I term the "technological attractive nuisance syndrome" attempted to field the caseless cartridge, the diesel aircraft engine, and the atomic airplane. To cite an example of current relevance, if liquid artillery propellants offer operational advantages over those now in use which are sufficiently compelling to justify major expenditures of scarce R&D funds, the point is not readily apparent.[33]

Granted, there are areas of more of less traditional technology where we are still an appreciable distance from the edge of the performance envelope: high speed vertical takeoff and landing aircraft with sufficiently low rotor downwash velocities for general battlefield utility and very small RPVs are cases in point. There are also areas where the incremental advantage is worth the price: hypersonic atmospheric reconnaissance platforms are an obvious candidate. Cryptography would seem to have reached a plateau of security, but that is what the experts have wrongly concluded in the past on several occasions. Clearly, obsolescent equipment has to be replaced as it wears out and there are important advances to be made in reliability and stealthiness in the process, but on the whole, R&D money and energy will be far better spent on software than hardware. The United States increasingly depends on imported materials for the production and transportation of war materiel, petroleum being the most critical requirement in the short term but by no means the only one. At the same time, as in the Vietnam War, the American military is apt to go

33 I owe the addition of the diesel aircraft engine to the list to a comment in 1986 by Mr. Maxime Faget of Eagle Engineering, Houston, Texas. Faget is generally recognized as the guiding spirit behind the detailed engineering of the Space Shuttle.

up against adversaries whose war production and population base reside in other countries. The United States was not successful in dealing with this problem in Vietnam at any level: tactical, operational, or strategic. Fortunately, this problem has received objective analysis from a cadre of historians like Larry E. Cable, Mark Clodfelter, Andrew F. Krepinevich, and Earl H. Tilford, Jr. Their analyses warrant careful consideration by scholars and military professionals.

On a more positive note, there are no serious prospects of the U.S. lead in the design of high technology weapons systems being challenged in the near term, or even over next few decades–if it is careful to preserve it. The analytical and design capabilities which made possible the Tomahawk, the M1A1 fire control system, and the F-117A were the products of many decades of continuous intellectual tradition and governmental support. They could not be reconstituted overnight with the wave of a budgetary wand. Just as the Chalcolithic hero depended on his smith, so are today's progressively less numerous champions dependent on their engineers. In addition, there is a downside to these impressive technological capabilities: specifically, improved tactical capabilities breed heightened strategic expectations. The ability to take out a single building with a single bomb on a single pass fosters not only the expectation that this can be done routinely, but that to do so is strategically relevant. On a practical level, to make that capability relevant one must know with some precision what is inside the building, why it is important, and when it will be there.

Note that the answers to those questions are as likely to be political as military, posing a significant challenge for intelligence gathering and dissemination agencies. For the foreseeable future, the United States is more likely to be engaged in relatively small-scale regional conflicts where leadership, cultural symbols, and political motivation are key factors than in major conflicts where we can target the industrial base. In the past, the United States has not always been spectacularly successful in evaluating such intangible, human factors. It badly needs to improve its intelligence and analytical capabilities. The ability to minimize collateral civilian casualties, demonstrated spectacularly in DESERT STORM, the bombing of the Al Firdoz bunker notwithstanding, carries with it the assumption that any collateral civilian casualties are unacceptable, however reprehensible enemy policy goals and however vicious enemy tactics.

Similarly, the ability to minimize losses to enemy fire throws even minimal losses to "friendly fire" into stark relief. Military leaders must be forewarned not to mis-advertise their capabilities. Furthermore, they must educate their civilian superiors, all the more so because of the political attractiveness of certain of those capabilities. High technology weaponry is highly dependent on user skill and training for tactical effectiveness. DESERT STORM was as much a triumph of hard, realistic, combat-oriented training as of design engineering. Indeed, many military professionals would argue that training was more important, and some have gone so far as to speculate that we could have exchanged systems with the Iraqis and still won. That is no doubt an overstatement, but the point behind it is valid. Our military services learned the value of combat-oriented training the hard way in Vietnam; we now enjoy a huge lead in this area, but such training is expensive, and operations and maintenance budgets will remain under intense pressure. We must be extremely careful not to throw away our advantage in this critical area.

Finally, the nature of the strategic, operational, and tactical environment in which our forces will operate in the 21st century must be rethought and force structures adjusted accordingly. The recent military technological revolution has provided unprecedented, and for the moment unchallenged, tactical and operational capabilities. As was demonstrated in DESERT STORM, the American military can exploit its technological advantages when faced with a conventional threat. The challenge is to harness those capabilities to the less conventional kinds of wars which will surely arise with increasing frequency. The crux of the matter is not our ability to put bombs on target or men, munitions, and vehicles on beaches and landing zones; the pivotal issue is the identification of strategically relevant targets, beaches and landing zones, an area where improvement is imperative.

The very precision offered by technologically advanced munitions demands an equivalent precision and timeliness in targeting. That, in turn, demands timely and accurate intelligence, not only before the fact, but afterwards in accurate bomb damage assessments. The focus of the preceding observation was on air operations, but if anything the demands of ground combat for timely and accurate intelligence are even greater, if for no other reason because intelligence failure in ground combat tends to produce more friendly casualties. The need for swift, accurate intelligence is increased exponentially by the reality of smaller force structures, justified in part by the capabilities of precision-guided munitions, an observation which requires a cautionary note: certain categories of target do not lend themselves to attack by PGMs, but are best dealt with by massive, non-precision bombardment. Mobile ground formations on the move, dug in defensive lines and major storage, and production facilities are cases in point. Despite the tactical fascination with PGMs, there will always be a need for massed artillery bombardments and large numbers of "dumb" bombs dropped by small numbers of large aircraft.

In sum, intelligence must be more subtle, timely and accurate. The services can no longer afford the luxury of the compartmentalized intelligence functions which were a logical product of the old war paradigm and the cold war focus on nuclear weapons. Rather, the intelligence function must be integrated with operational planning and execution more closely than ever before. Intelligence must also expand its focus to address the sorts of nontraditional threats outlined above, a venture in which exchanges of information with civilian law enforcement agencies will become pivotal and in which the specialist noncombat arms will have an important role to play. The sticky constitutional issues raised by the prospect of close interaction between civilian law enforcement agencies and military intelligence must be resolved and indicate the changing face of strategy. The Army, by virtue of its structure and historical past, will inevitably be at the epicenter of such changes. The Corps of Engineers would seem a logical institutional locus for assessments of the vulnerability of urban areas and public utilities to attack and sabotage. Similarly, the Medical Corps should become centrally involved in assessments of hostile chemical and bacteriological capabilities and the development of appropriate countermeasures. The Judge Advocates General must develop hands-on familiarity with the nitty-gritty of operational planning and tactical execution so as to be able to provide commanders and planners with timely, relevant and accurate legal guidance.[34] The central importance of news media coverage of military operations demands that public information personnel develop a similar familiarity with operational planning and tactical execution. By the same token, commanders and operational staffs must become more sensitive to the importance of input from such specialists.

In conclusion, two basic points are clear: if strategy could ever be approached as a straightforward technical exercise in the movement of military formations across country, war on the map as Jomini put it, followed by an equally straightforward, though considerably bloodier, exercise in fire and maneuver on the battlefield, that time is long past. Similarly, if there was ever a time when war could be approached as an exercise in production line engineering, as the U.S. Army Air Forces did in preparation for World War II, that time is long past, as well.[35] The maneuver of conventional forces and industrial production will remain important integers in the strategic equation, but they are no longer preeminent.

---

34 I am indebted to W. Hays Parks, Colonel USMCR, Chief Counsel for the Secretary of the Army, for this point. An example cited by Colonel Parks of the kind of problem which can be resolved by timely legal guidance in the planning process involved the initial refusal of logistical planners during Operation DESERT SHIELD to provide Marine air units with napalm, based on the mistaken belief that the use of napalm was in violation of the laws of war.

35 See Barry D. Watts, *The Foundations of U.S. Air Doctrine: The Problem of Friction in War* (Maxwell AFB, Alabama: Air University Press, 1984).

And so this analysis of the technological limits of strategy ends on a cautionary note about the essentially human nature of war. The Chalcolithic hero understood implicitly that success in war was determined not just by strength and technological advantage alone, but by the effective application of those qualities in human context. The perceived force of his blow was increased by faith in his gods and fear of them. Just as the gods of his village presided over his swing, understanding of policy goals presides over our application of military force. Whatever the technology, war remains as Carl von Clausewitz characterized it, a test of will and faith. Do not lose sight of that reality.

# PART 1

## *The Pre-Industrial World, 1300–1800*

CHAPTER 2

# Gunpowder Artillery in Europe, 1326–1500: Innovation and Impact

*Clifford J. Rogers*

## 1 The Technical Development and Capabilities of Gunpowder Artillery: 1326–1500

In 1326–1327 an English manuscript illuminator provided us with the first depictions of guns in European art. (Illustration 2.1) In 1327, as we know from documentary evidence, the town council of the city of Florence ordered the acquisition of a number of "engines called cannon," along with the iron balls they shot, for the defense of the city.[1]

These Florentine guns of 1327 were probably similar in scale to the one found in the ruins of Monte Varmine, which was destroyed in 1341. This forged-iron weapon, meant to be mounted on a wooden stock, was 20 cm long, had a trumpet-shaped tube expanding from 4 cm diameter at the base to 6.4 cm at the mouth,

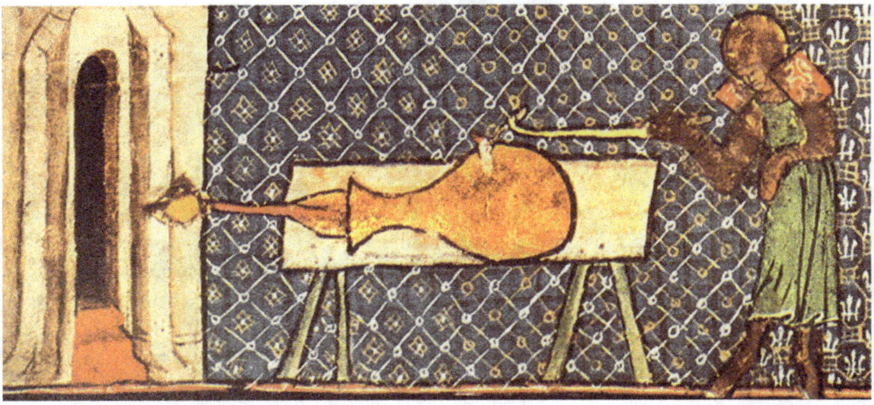

ILLUSTRATION 2.1 Depiction of Early European Cannon from *De notabilitatibus* (1327)

1 Angelo Angelucci, *Delle artiglierie da fuoco Italiane. Memorie storiche con documenti inediti* (Turin: G. Cassone, 1862), 16–17n. Note that the document is given as "die 11 februarii MCCCXXVI," and every single secondary source I have seen referring to this document dates it to 1326, but the Florentine new year did not begin until March 25, so the text actually dates to 1327.

 | DOI:10.1163/9789004393301_004

and weighed 2.7 kg. It most likely threw an iron ball of around 4.6 cm diameter weighing about 370 grams—rather larger than the acorns used as a metaphor for cannon-shot by Petrarch in the 1350s.[2] The other type of cannon in use at the time, as depicted in the 1326–1327 illuminations of Walter de Milimete's manuscripts *De secretis secretorum* and *De notabilitatibus*, was larger—a piece of artillery rather than a hand-gun—and designed to shoot either lead balls or heavy lances.

Over the next century, gunpowder artillery developed rapidly, so that the gunners of each new generation worked with weapons quite different from those their fathers had employed. Indeed, gunpowder artillery changed as much in *each* twenty-five year period between 1325 and 1500 as it did over the following three centuries. (Illustration 2.2)

But what drove this development? Some historians have suggested that the spread of and progress in gunpowder technology over its first century was

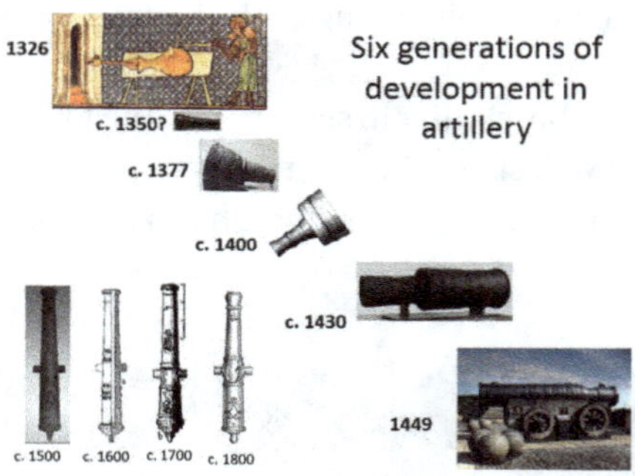


ILLUSTRATION 2.2
DEVELOPMENT OF EUROPEAN CANNON

2 Angelo Angelucci, *Documenti inediti per la storia delle armi da fuoco Italiane* (Turin: G. Cassone, 1869), 69. My estimate for the size and weight of the ball assumes it sat just inside the mouth. Angelucci's sketch shows the ball smaller and set fairly close to the touch-hole, but that would not have allowed for a reasonable weight of powder to drive an iron ball. This is similar to the bombards used by the Papal States in 1350, which fired iron balls averaging 280 grams, so about 4.2 cm in diameter, and even closer to the one-pound (339g, 4.5 cm) balls fired by four bombards purchased in 1358. By that date, the Papal States had acquired dozens of guns and thousands of iron balls. Alberto Pasquali-Lasagni and Emilio Stefanelli, "Nota di storia dell'artigleria dello Stato della Chiesa ni secoli XIV e XV," *Archivo della R. Deputazione romana di Storia patria* 60 (1937), 149–189, at 150–53. See also Gustav Köhler, *Die Entwickelung des Kriegswesens und der Kriegführung in der Ritterzeit,* 3 vols. in 5 (Breslau: W. Koebner, 1886–1887), vol. 3, pt. 1, 250, 226 (Petrarch). A better match for Petrarch's "acorns" would be the half-pound (170g, 3.6 cm) iron balls fired by certain Papal "bombarda magna" in 1358. Pasquali-Lasagni and Stefanelli, "Nota di storia dell'artigleria," 153. Angelucci also gives details of a now-lost gun of somewhat similar proportions that was cast with the date 1322; I have not referred to it in my main text because I am not confident that it is not a later fake. *Documenti*, 99.

propelled not by rational considerations of weapon efficacy, but rather by some sort of emotional reaction to the shock and awe generated by loud explosions and sheets of flame. Even Lynn White, who generally explained other technological changes in terms of practical utility, seems to have reached this conclusion about artillery: "Any rational technology assessment of the cannon in 1326 or for a hundred years later," he wrote, "would have concluded: 'Stick to the trebuchet.'"[3]

White, however, was fundamentally wrong. In fact, from the very beginning guns had important advantages that made it perfectly rational to include them in the mix of weaponry used in sieges. Then from each generation to the next the largest cannon grew heavier and stronger, capable of handling larger and larger charges of gunpowder, even while that powder was reformulated to make it stronger. Well before 1426, the power of a large bombard far surpassed that of any mechanical artillery. Other changes in gun design made it possible to fire more shots per gun per day. At the same time, the number of artillery pieces in use increased steadily. These evolutionary lines of development interacted with each other in ways that exponentially increased the destructive punch a siege train could deliver against a fortress. The rapidly growing power of gunpowder artillery mattered in qualitative ways, changing the fundamental balance between offense and defense in European warfare. This fifteenth-century Artillery Revolution was an important factor in shaping the broad sweep of Western history in the fifteenth and early sixteenth centuries, facilitating the expansion and internal consolidation of the Renaissance monarchies that became the basic patterns for the modern nation-state. I have already sketched out the technological foundations of those epochal military developments, but this essay will present fuller and more quantified detail, linking technical developments to changes in their practical consequences over time.[4] This should not only help historians better understand one particular military revolution of the past, but also provide them with insight into patterns to look for and lines of analysis to pursue in examining other exemplars of the phenomena of "revolutions in military affairs" and "military revolutions."[5]

3 Lynn White, Jr., "The Crusades and the Technological Thrust of the West," *Medieval Religion and Technology: Collected Essays* (Los Angeles: University of California Press, 1978), 269.

4 Clifford J. Rogers, "The Military Revolutions of the Hundred Years' War," in Clifford J. Rogers, ed. *The Military Revolution Debate* (Boulder: Westview, 1995), 64–76; idem, "The Artillery and Artillery Fortress Revolutions Revisited," in Nicolas Prouteau, Emmanuel de Crouy-Chanel and Nicolas Faucherre, eds., *Artillerie et Fortification, 1200–1600* (Rennes: Presses Universitaires de Rennes, 2011): 75–80.

5 For these two concepts as distinct phenomena, see Clifford J. Rogers, "'Military Revolutions' and 'Revolutions in Military Affairs': A Historian's Perspective," in Thierry Gongora and Harald von Riekhoff (eds.), *Toward a Revolution in Military Affairs? Defense and Security at the*

## 2 The Earliest Cannon

The first real description we have of a European gun in action might, at first glance, seem to support the idea that there must have been some reason other than simple usefulness for medieval governments, war-leaders, and technologists to invest in gunpowder weapons. In 1346, a seller of metal-ware from Bruges named Peter was summoned by the town council of Tournai to give a sales demonstration of one of the new weapons for them. He set up his gun to fire at a postern gate or the neighboring wall, but missed his target, even though it was much broader than the proverbial side of a barn. The projectile, a lance given extra mass by a two-pound lead ball below the point, flew out with such great velocity that the observers could not see its flight. Instead of striking the gate, it went over the both the outer wall of the town and also its inner wall (280 meters further away) to the plaza in front of a church, where it happened to kill an extraordinarily unlucky passer-by.

As a seller's demonstration of a new technology, that was a pretty epic failure.

But one of the first things I learned from Joe Guilmartin was that medieval people were not stupid, and that—especially in the realms of war and technology—if they did something that seems stupid to you, there is a good chance that the problem is in your lack of understanding rather than theirs. Consider again Peter of Bruges's demonstration. His weapon may not have done what it was supposed to do, but in the process it demonstrated remarkable potential. Since the plaza where the victim was killed was about 440 meters from the outer gate and 160 meters from the inner wall, the total horizontal flight must have been over 600 meters[6]—and since Peter was targeting

---

*Dawn of the 21st Century* (Westport: Greenwood Press, 2000): 21–36, and Williamson Murray and MacGregor Knox, "Thinking About Revolutions in Warfare," in *The Dynamics of Military Revolution, 1300–2050*, ed. MacGregor Knox and Williamson Murray (Cambridge: Cambridge U.P., 2001). For the long tradition of looking at military developments as actually or potentially "revolutionary," see Clifford J. Rogers, "The Idea of Military Revolutions in Eighteenth and Nineteenth Century Texts," *Revista de História das Ideias* 30 (2009): 395–415.

6 Text in L. Lacabane, "De la poudre à canon et de son introduction en France," *Bibliothèque de l'école des chartes*, 2e ser., I (1844), 45–46n. The total flight of the missile cannot be determined, because there is no clear indication of how far from the wall the gun was initially sited, but the distance between the Morel gate (the target) and the plaza of St. Brice (where the victim was killed) is about 440 meters, and the distance from St. Brice to the inner wall is about 160m, based on Isaac van der Kloot's map of the 1709 siege (https://upload.wikimedia.org/wikipedia/commons/f/f1/Siege_of_Tournai_%28Doornik%29_in_1709_%28Isaac_van_der_Kloot%29.jpg) and the 1649 van Loon atlas (https://upload.wikimedia.org/wikipedia/commons/0/02/Tornacum_-_Tournai_-_Doornik_(Atlas_van_Loon).jpg). Since Peter was aiming for the wall, it is likely that he was firing at a relatively flat trajectory. The projectile

the gate, he was likely using a low trajectory (as we'd expect from the Milemete illustration), which would mean the cannon-shot could have carried much further still if it had been elevated to a higher launch angle.[7]

On the other hand, despite its relatively high velocity—probably at least triple the speed of a crossbow-bolt[8]—the projectile would probably not have been able to smash open the gate even if it had hit, and even much larger projectiles were of little use against stone walls. Hence, for the first seventy-five years of their use in Europe, the purpose of guns used in offensive warfare was mostly what Peter of Bruges had said to the town council in describing his gun—that it was an engine for "shooting *into* a walled town when it is besieged"[9]—rather than what, presumably at their insistence, he tried to do with it. Traditional engines such as traction trebuchets and springalds (roughly, giant tension- or torsion-powered crossbows) were also used for this same role alongside the cannon.

Each technology had advantages and disadvantages. Traction trebuchets, lever-action engines that used the force of a team of pullers hauling on ropes to launch their projectiles, did not require well-rounded stones, specially-made quarrels, or gunpowder, so they were cheap to operate.[10] They could sustain a

---

must still have been on a rising trajectory when it passed the first wall, or it would not have made it over the second wall. Assuming the two walls were the same height, and recognizing that in a ballistic trajectory through air the projectile falls more steeply than it rises, the launch distance must therefore have been more than 160 m from the outer wall, for a total horizontal flight of more than 600 m.

7 For example, not accounting for air resistance, a missile launched at 10 degrees would need an initial velocity of 131 m/sec to travel 600 m. With the same initial velocity, fired at 45 degrees the projectile would have an *in vacuo* range of over 1,750 m. These numbers and similar calculations reported in subsequent notes were calculated using the online tools at http://hyperphysics.phy-astr.gsu.edu/hbase/traj.html.

8 Assuming a total flight of 600m and 10 degree elevation, the initial velocity would be over 130 m/sec, which is a close match for the median results achieved in modern testing with a gun of probably similar design, firing a lead ball and using dry-mixed fourteenth-century powder formulations (133–142 m/sec, though ranging up to 210 m/sec), though much higher than the maximum 87 m/sec achieved with an (unweighted) arrow. Medieval Gunpowder Research Group, "The Ho Experiments. Report Number 1: The Firing Trials," (Nykøbring. F., Denmark: Middelaldercentret, 2002), 10. A contemporary crossbow for battlefield use, by comparison, would have launched bolts with velocities somewhere around 39–50 m/sec. Andreas Bichler, "Testing Reconstructed Medieval Crossbows Featuring Composite Bows," http://www.historiavivens1300.at/biblio/beschuss/beschuss1-e.htm, accessed 2 September 2012; note also Stephen Grancsay, "Just How Good Was Armor?" *True* (April 1954), 90, for 42.16 m/sec from a 336 kg draw weight crannequin crossbow (with a light bolt).

9 Lacabane, "Poudre," 20n: "pour traire en une boine ville quand elle seroit assise."

10 For examples of the very rough rounding of 4.5 kg trebuchet stones, see Laure Barthet, "La prise de la barbacane de Montségur (Ariège) en février 1244: une introduction à

very high rate of fire and could be surprisingly accurate. Springalds may have been able to outrange bows and crossbows, and were definitely effective antipersonnel weapons even against armored combatants. Cannon, aside from the psychological value of their noise and smoke in frightening the enemy (which must have been short-lived as a major effect), were equally effective against armored men-at-arms, much cheaper to buy than springalds,[11] and easier to transport than either trebuchets or springalds. They also had the advantage of suffering far less deterioration over time than wooden siege engines.[12] On the other hand, they were at first terribly inaccurate and, relatively, very expensive to operate, because the cost of powder was much higher in the fourteenth century than in the fifteenth or later.[13]

---

l'archéologie de la poliorcétique," in Nicolas Prouteau, Emmanuel de Crouy-Chanel and Nicolas Faucherre, eds., *Artillerie et Fortification, 1200–1600* (Rennes: Presses Universitaires de Rennes, 2011), 44; note also Reynaud Beffeyte, *War Machines in the Middle Ages* (Rennes: Éditions Ouest-France, 2004), 28. The excavations of Burg Tannenberg (destroyed in 1399) found clearly distinguishable rough-hewn stones for trebuchets and smooth-rounded stones (for guns). J. von Hefner and J.W. Wolf, *Die Burg Tannenberg und ihre Ausgrabungen* (Frankfurt am Main: Verlag der S. Schmerber'schen Buchhandlung, 1850), 77. Stones that were not very carefully rounded could cause guns to burst. J. Bertin, "Le siège du château de Vellexon en 1409," *Bulletin de la société d'agriculture, sciences et arts du département de la Haute-Saône*, 3d Ser., 29 (1898), 120, 170.

11 Rogers, "The Military Revolutions of the Hundred Years' War," 64, for an example of a springald costing five times as much as a small cannon.

12 Edward I of England spent a small fortune on the giant trebuchets he had built at the Tower of London in 1278, but within 35 years they were unusable and valuable only for salvage. Michael Prestwich, "The Trebuchets of the Tower," in Gregory I. Halfond, ed., *The Medieval Way of War: Studies in Medieval Military History in Honor of Bernard S. Bachrach* (Farnham: Ashgate, 2015), 284–290. By contrast, the bronze bombard *Faule Mette,* cast in 1411, was used for the defense of Brunswick as late as 1550, and fired for ceremonial purposes as late as 1728. C.W. Sack, Emil Ferdinand Vogel, and Ferdinand Spehr, *Alterthümer der Stadt und des Landes Braunschweig,* 2d. ed. (Brunswick: Friedrich Otte, 1861), 77–82.

13 Inaccuracy: the experience of Peter of Bruges is reflected in modern tests with a replica of the Loshult gun: five shots all failed to hit a two-meter-square target at 200 paces. Medieval Gunpowder Research Group, "The Ho Experiments. Report Number 2" (Nykøbring. F., Denmark: Middelaldercentret, 2003), 4. Costs: Powder costs per kilogram were especially high early on—for example, powder purchased by the town of Arnhem in 1354 cost 13 *livres* for 11.7 kg, so each kilogram cost the equivalent of over 20 days' wages. Karl Jacobs, *Das Aufkommen der Feuerwaffen am Niederrheine bis zum Jahre 1400* (Bonn: Peter Hanstein, 1910), 33. During the fifteenth century the cost per kilogram fell dramatically, but the quantities of powder used increased even more dramatically. In France, for example, the price of gunpowder fell by around 80% from the late fourteenth century to the late fifteenth, due to a variety of factors including importation of mined saltpeter and economies of scale in production. See Bert S. Hall, *Weapons and Warfare in Renaissance Europe* (Baltimore: Johns Hopkins U.P., 1997), 58. However, this was more than offset by the need to purchase more and more gunpowder resulting from the use of increasingly

Although contemporary sources surprisingly do not emphasize the fact, it does seem from scattered references and modern testing that the biggest advantage of early guns, relative to mechanical artillery, was that they could impart much greater velocity to their projectiles, allowing for shots that carried much further and still retained enough power to kill, even through armor. One recent analysis suggests springalds had maximum ranges similar to those of crossbows, at around 300 meters.[14] Both traction and counterweight trebuchets generally operated at shorter distances. In one fourteenth-century case, two large counterweight trebuchets were set up at around 155 and 240 meters from their target, which is a good match for the range of distances achieved with modern replicas.[15] By contrast, recent tests demonstrate that a small (9 kg)

---

powerful charges for increasingly large numbers of guns. For example, in 1346, for the Crécy-Calais campaign, Edward III of England made what was for the time an extremely large purchase of saltpeter and sulphur, enough to make around 865 kg of gunpowder. French plans for the campaigns of 1406 and 1442, however, envisaged the provision of 2,940 and 9,800 kg of powder, respectively. By 1453, a well-informed Burgundian author reported that the Turks besieging Constantinople were using 490 kg of powder per day. By 1513, Henry VIII's siege train could use as much as 29,000 kg *per day*. T.F. Tout, "Firearms in England in the Fourteenth Century," *English Historical Review* 26 (1911), 690; Napoléon III and I. Favé, *Études sur le passé et l'avenir de l'artillerie,* 6 vols. (Paris: J. Dumaine, 1846–1871), 3:121; Hall, *Weapons and Warfare,* 62, and Philippe Contamine, "Les industries de guerre dans la France de la Renaissance: l'exemple de l'artillerie," *Revue historique* 271 (1984), 278; Jean de Waurin, *Croniques et anchiennes istories de la Grant Bretaigne,* vol. 5, ed. William and Edward L.C.P. Hardy (London: Rolls Series, 1891), 252; Michael Mallett, *Mercenaries and their Masters* (Totowa: Towman and Littlefield, 1974), 163; cf. also Philippe Contamine, *War in the Middle Ages* tr. Michael C. Jones (Oxford: Basil Blackwood Ltd., 1987), 48–49, and for 1567 Thomas F. Arnold, *The Renaissance at War* (London: Cassell, 2001), 33.

14 Peter Purton, *A History of the Late Medieval Siege* (Woodbridge: Boydell, 2010), 397, for J. Liebel's estimate of 180 m at fifteen degrees elevation, shooting from the top of a tower, which can be calculated to equate to about 300 meters on level ground and using a 45 degree angle. This is based on a torsion-model springald; springalds seem to usually have been tension-powered devices, and those might have had greater or lesser ranges. See the drawing of a springald in the 1315 charter of the town of Carlisle (https://upload.wikimedia.org/wikipedia/commons/4/48/Siege_of_Carlisle_1315.jpg); the drawing in Österreichisches Nazionalbiliothek manuscript CPV 3069, fo. 22v; and the mention of an *archum* for a springald in David S. Bachrach, "English Artillery 1189–1307 : The Implications of Terminology," in *English Historical Review* 121 (2006), 1414. Alain Salamagne, "L'Artillerie de la ville d'Arras en 1369," in Nicolas Prouteau, Emmanuel de Crouy-Chanel and Nicolas Faucherre, eds., *Artillerie et Fortification, 1200–1600* (Rennes: Presses Universitaires de Rennes, 2011), 52, gives a higher estimate of 500 m based on a rather unclear statement by Marino Sanudo that he interprets as meaning that springalds outranged crossbows.

15 *Ville de Périgueux. Inventaire sommaire des archives communales antérieures à 1790*, ed. Michel Hardy (Périgueux: R. Delage & D. Joucla, 1894), 92; my thanks to Dr. Nicolas Savy for help identifying the locations. In 1320, the Dinanters besieging Bouvignes apparently

cannon of a very early style can send a lead ball over 1,000 meters.[16] An iron or stone ball, being lighter, would, with the same powder charge, have carried much further still.[17] The results of this increase in range were demonstrated as early as the siege of Algeciras in 1342. Siege camps were usually positioned to be out of range of fire from the besieged fortification's walls, but in this case the defenders used cannon to shoot iron balls the size of apples at the besiegers, and some not only reached the camp, but passed right over and landed on its far side.[18]

The high velocity of the projectiles not only allowed for this long-range shooting, it also meant, at shorter ranges, very high impact energy compared to muscle-powered projectile weapons. Even an extraordinarily strong English longbow or a windlass-drawn, composite-bowed crossbow would rarely impart

---

had to set their trebuchet up near the bank of the Meuse River in order to throw house-smashing stones into the town, which, given the width of the river there, suggests a range of 180 m or less. Jean de Hocsem, *Chronique,* ed. Godefroid Kurth (Brussels: Commission royale d'histoire, 1927), 167. At the thirteenth-century sieges of Dryslwyn and Newcastle Emlyn, the operators of a large English trebuchet needed a protective shelter, suggesting they were operating at less than the range of the Welsh shortbow, which was probably not more than 200 m, if that much. John France, *Western Warfare in the Age of the Crusades,* 1000–1300 (Ithaca: Cornell U.P., 1999), 123–124. For modern reconstructions, see the NOVA television show "Secrets of Lost Empires- Medieval Siege" [114 kg stones for 183 m]; Renaud Beffeyte, *L'art de la guerre au Moyen Âge* (Rennes: Éditions Ouest-France, 2005), 81; Beffeyte, *War Machines*, 13 [125 kg over 170m]; Peter Vemming Hansen, "Experimental Reconstruction of a Medieval Trébuchet," *Acta Archaeologica* 63 (1992), accessed at http://web.archive.org/web/20070105190530/http://www.middelaldercentret.dk/acta.html; and (for shorter ranges) cf. Tanel Saimire, "Trebuchet—A Gravity-Operated Siege Engine. A Study in Experimental Archaeology," *Estonian Journal of Archaeology,* 10 (2006), 75–78.

16 Medieval Gunpowder Research Group, "Report 2." New tests with various powder formulations, using both the *Feuerwerkbuch* loading method (which leaves an empty space in the powder chamber behind the tampon, and will be described more fully below) are much needed to improve our understanding of early artillery.

17 $Ke=1/2\ M\ v^2$ so as mass falls, for a given amount of Ke, $v^2$ increases in inverse proportion. Horizontal range increases in proportion with $v^2$ (though this does not account for air resistance). Hence, for example, cutting mass to 1/3 with unchanged Ke allows for triple range. The specific gravities of lead, iron, and marble are around 11.35, 7.21, and approximately 2.56, so *in vacuo* ranges should be roughly 1.57 and 4.29 times as far for iron or marble vs. lead.

18 Köhler, *Entwickelung,* 3:1:223; José Antonio Conde, *Historia de la dominacion de los Arabes en España, sacada de varios manuscritos y memorias Arabigas* (Paris: Baudry, 1840), 604. Although the source is late, this is not the sort of story that mere anachronism would lead to, since a late writer transposing the warfare of his own time back to an earlier period would expect the camps to be further back in proportion with the increased range of the defenders' weapons.

more than 150 joules of energy to its missile, whereas a small fourteenth-century gun firing a 1.5 cm lead ball could easily launch its projectile with *ten times* that amount of punch, such that "no armor can withstand it," as Jean Buridan wrote as early as 1358.[19]

## 3 Growing Size and Power

Thanks to these advantages in range and penetrating power, between 1351 and 1375 cannon went from being something of a novelty to an absolute commonplace in siege operations. The largest ones also grew much bigger. As late as 1372, an English clerk could still describe a gun weighing 55 kg as "large,"[20] but just three years later an iron gun weighing around a tonne was forged at Cahors for the siege of St.-Sauveur-le-Vicomte.[21] From that point, cannon continued

19 Kinetic energy for gun: Medieval Gunpowder Research Group, "Report 1," table 5, shot 2. Buridan: Hall, *Weapons and Warfare,* 45. The lethality of these early guns is also emphasized in Spanish sources. The description in the *Cronica de Alfonso XI* of iron balls able to pass entirely through an armored soldier (*pasaba un ome con todas sus armas*) at the siege of Algeciras in 1342 might perhaps be questioned, since the text dates to the early fifteenth century, but the poem *Alfonso el Onceno*, composed before 1348, also refers to cannon-shot killing men and beasts in the besiegers' camp. Francisco Javier López Martín, "La evolución de la artillería entre los siglos XIV y XVI, con especial atención a los manuscritos de Walter de Milemete y los primeros usos de la artillería en Europa," *Fortificações e Território na Península Ibérica e no Magreb (séculos VI a VXI)*, vol. 2, ed. Isabel Cristina F. Fernandes (Lisbon: Edições Colibri, 2013), 605–606. The validity of Buridan's statement (at least at short range) is, moreover, confirmed by the results of tests with the replica Loshult gun. Robert Douglas Smith, *Rewriting the History of Gunpowder* (Nykøbing Falster: Middelaldercentret, 2010), 117–118. My thanks to the author for sending me a copy of this publication. Arrows: Clifford J. Rogers, "The Battle of Agincourt," *The Hundred Years War (Part II): Different Vistas*, ed. L.J. Andrew Villalon and Donald J. Kagay (Leiden: Brill, 2008), Appendix 1. Crossbow: Alan R. Williams, *The Knight* and the *Blast Furnace: A History of the Metallurgy of Armour in the Middle Ages & the Early Modern Period* (Leiden: Brill, 2003), 919–920, suggests 200J, but modern testing indicates much lower energies. (See n. 8, above.) 150J may therefore be generous for the crossbow, but various indications suggest that windlass- or crannequin-drawn crossbows were comparable to longbow arrows in range and penetrating power, and therefore (since crossbow bolts and longbow arrows were comparable in mass) also in velocity and kinetic energy. Clifford J. Rogers, "The Longbow, the Infantry Revolution, and Technological Determinism," *The Journal of Medieval History* 37 (2011), 332 n. 73.

20 T.F. Tout, "Firearms in England in the Fourteenth Century" *English Historical Review* 26 (1911), 693: "pro ij gunnes grossis feri, precia pecia xl. s." Each would be a gun of around 55 kg given the normal pricing of 4 d./pound (454 g). Ibid., 682–683.

21 Peter J. Burkholder, "The Manufacture and Use of Cannons at the Siege of St-Sauveur-le-Vicomte, 1375," (MA Thesis, U. of Toronto, 1992); Napoléon III and Favé, *Études,* 4:p.j.

to get steadily bigger and more strongly built, so that they could handle stronger charges of powder and impart more energy to their projectiles. In 1394, a gunfounder in Frankfurt cast a bombard more than three times as massive as the largest St.-Sauveur gun. *Faule Mette,* cast in 1411, was twice again as heavy as that, and *Luxembourg* (cast before 1447), was more than twice the weight of *Faule Mette.*[22]

The destructive capacity of the weapons increased in close parallel with their weight. (Table 2.1) Already by 1378, cannon had surpassed the strongest mechanical artillery in power.[23] A master engineer asked in 1297 to build the largest trebuchet possible constructed one that threw stones of 200 pounds.[24] A bigger weapon casting 400-pound stones had been built for the siege of Damietta in 1218-19, and this seems to have been a plateau level for the normal range of counterweight trebuchets: two hundred years later, French and Burgundian examples lobbed stones of 100, 300, and 400 pounds.[25] A few were even bigger:

xvii–xlii. There is also a reference to a 660 kg Pisan bombard as early as 1362 in the *Cronica di Pisa,* though that is obviously not as reliable a source as a financial account (and is not included in Table 2.1 below). In Muratori, *Rerum italicarum scriptores* XV, col. 1037.

22 The weights were respectively 3,995; 9,583; and at least 23,520 kg. Bernhard Rathgen, *Das Geschütz im Mittelalter* (Berlin: V.D.I. Verlag, 1928), 256–257; Sack, et al., *Alterthümer... Braunschweig,* 77; Joseph Garnier, *L'Artillerie des Ducs de Bourgogne, d'après les documents conservés aux archives de la Côte-d'Or* (Paris: Honoré Champion, 1895), 172.

23 Already in 1378, bombards could shoot 1,000 *Schritt* (below, note 29), so around four times as far as a trebuchet throwing large stones (above, note 15), indicating double the velocity and quadruple the energy for a given mass. Also in 1378, we have records of a "grooten donrebusse" of the count of Holland able to throw 400-pound (196 kg) stones, and a bombard made at Chalon firing 450-pound (221 kg) stones. However, it is likely that the bombards of 1378 with the thousand-Schritt range (which was presumably noted in the chronicle because it was an increase over previous experience) used the roughly 5% powder:ball ratio of the 1394 Frankfurt bombard and several 1397 *Centnerbüchse,* while the earlier big guns used only about half that proportional charge. (Below, notes 72, 71, 59). If so, that suggests the Chalon bombard of 1378 would have imparted about 1.5 times the kinetic energy of the 600-pound stones of the near-contemporary Périgueux trebuchet described below. Jacobs, *Aufkommen der Feuerwaffen,* 56; Napoléon III and Favé, *Études,* 3:103; J.R. Partington, *A History of Greek Fire and Gunpowder* (Cambridge: W. Heffer & Sons, 1960), 114, 122; Garnier, *Artillerie,* 12.

24 *Istore et croniqes de Flandres,* ed. Kervyn de Lettenhove, 2 vols. (Brussels: F. Hayez, 1879–80), 1:213. The trebuchet stones excavated at Burg Tannenberg (from 1399) ranged from 24–36 cm, so up to a maximum of about 63 kg. Hefner, *Burg Tannenberg,* 77.

25 Paul Chevedden, "The Hybrid Trebuchet: The Halfway Step to the Counterweight Trebuchet," in Donald J. Kagay and Theresa M. Vann, eds., *On the Social Origins of Medieval Institutions: Essays in Honor of Joseph O'Callaghan* (Leiden: Brill, 1998), 187; Contamine, *War,* 195; Ch. L. Grandmaison, ed., *Documents inédits pour servir à l'histoire des arts en Touraine* (Tours: Soc. Arch. de Touraine, 1870), 124. In between, for example, a French trebuchet at the siege of la Roche Derrien in 1347 cast stones of 300 pounds (147 kg?), while the engines

an engine owned by the town of Périgueux in 1398 launched 600-pound projectiles.[26] But in the period 1399 - c. 1415, as we know from both archival evidence and surviving weapons, bombards could throw gigantic stones of 700; 800; 1,100; even 1,700 pounds.[27]

---

of the Count of Hainault at a siege of 1345 threw projectiles of "200 pounds (94 kg?) or more." *Grandes chroniques de France*, ed. Jules Viard, 9 vols. (Paris: SHF, 1920–53), 9:300; Hocsem, *Chronique*, 336.

The claim that two Venetian engines threw stones of three thousand pounds (*libbre tre mille*) (one tonne if the Venetian *libbra sottile* is meant) is not credible; wooden throwing arms simply cannot be made strong enough for that. *Istoria dell'assedio e della ricupera di Zara fata da'Veneziani nell'anno MCCCXLVI, scritta da autore contemporaneo,* in Jacopo Morelli, ed. *Monumenta Veneziani di varia Letteratura* (Venice: Carlo Palese, 1796), ix. The same goes for the stones of 2,700 pounds (*cantariorum...decem octo*) supposedly thrown by the Genoese engine *Troia* at Cyprus in 1373 (even if the *libra peso sottile* of 317 grams is meant). Giorgio Stella, *Annales Genuenses,* in *Rerum italicarum scriptores,* ed. Muratori, vol. 17 (Milan: n.p., 1730), 1104–05. Roland of Padua is a better source, but his claim that at the siege of Este in 1249 the Paduans and their allies had fourteen trebuchets casting stones of 1,200 pounds (408 kg) and more (*rotabant hedificia quedam lapides...ponderis librarum 1200 et ultra*) should be viewed with skepticism. *Rolandini Patavini Chronica,* ed. Philippo Jaffé in MGH, *Scriptores,* ed. G.H. Pertz, 19 (Hannover: MGH, 1866), 90. It is not, however, beyond possibility. Guillaume Henri Dufour, *Mémoire sur l'artillerie des anciens et sur celle du moyen âge,* vol. 1 (Paris: A. Cherbuliez, 1840), 104, calculated that a 500 kg stone could be thrown by an engine with a 15,000 kg counterweight, and Edward I in 1278 had built a trebuchet with a lead counterweight of over 17,000 kg. By comparison, the Urquhart fixed-counterweight trebuchet built for the NOVA television show "Secrets of Lost Empires- Medieval Siege" was able to throw 250-pound stones for 200 yards with a counterweight of only eight tons (about 7,250 kg). "Secrets of Lost Empires- Medieval Siege"; Ed Levin, "The Highland Fling," *Timber Framing: Journal of the Timber Framers Guild* 50 (1998): 14–19.

26 *Petit Livre Noir* (Registre consulaire de Périgueux. 1360 - 1449. Archives départementales: BB. 13), fo. 47v, transcription by Jean Roux online at https://web.archive.org/web/20150405145020/http://www.perigordoccitan.fr/Paleographie/Livre_Noir/LNtexte.htm, accessed 24 Jan. 2017.

27 The last-mentioned figure is for the still-extant *Pumhart von Steyr,* forged 1414–1424 (probably at the start of that period), now in the Heeresgeschictliches Museum in Vienna. It has a conical bore, to allow for stones of different sizes. Since it has a caliber at the mouth of 88 cm (tapering down to 76 cm), it could have handled a ball of at least 83 cm, which would weigh around 772 kg, depending on the density of the stone used. For the measurements, see Volker Schmidtchen, *Bombarden, Befestigungen, Büchsenmeister* (Düsseldorf: Droste Verlag, 1977), 32–34. In this article I have made calculations based on a specific gravity of 2.56 for stone, though in the majority of cases extant stone cannonballs that have been weighed have lower specific gravity, ranging from 2.02–2.39. Rathgen, *Geschütz,* 49, 203; a personal communication from Sascha Pries of the Kölnisches Stadtmuseum providing the diameter and weight of the ball associated with their "Steinmörser" indicates a s.g. of 2.34. These data, however, are all from Germany and mostly from the fourteenth century. English documents of 1428 and 1440 show a preference for cannonballs

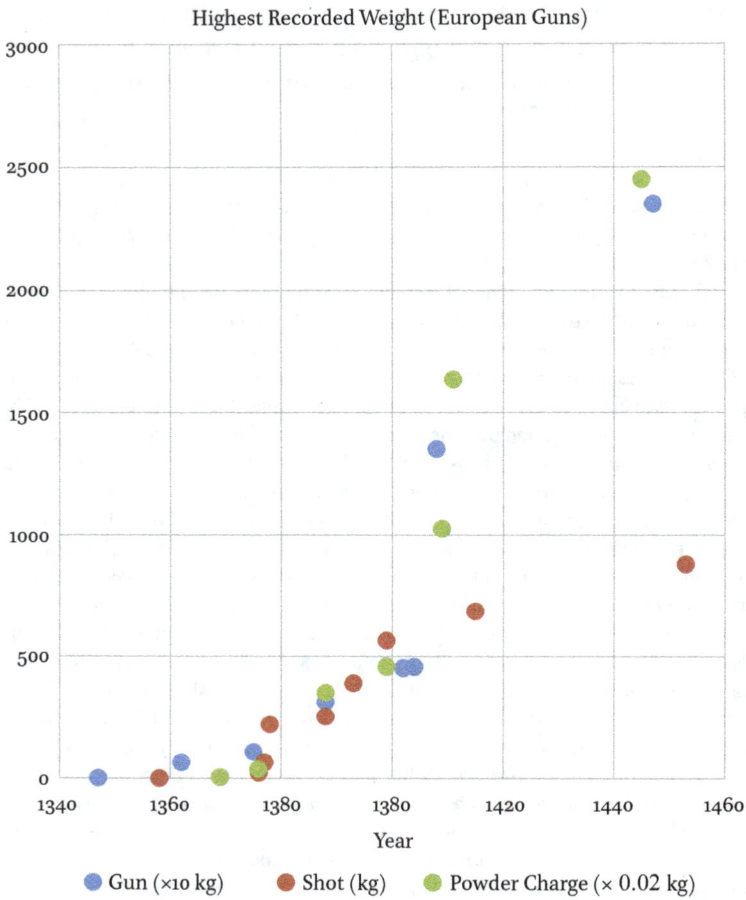


TABLE 2.1 From around 1380–1405, the highest recorded gun weights (blue dots), shot weights (red dots) and powder charge weights (green dots) rose in tandem: rapidly, but in a linear fashion. Thereafter maximum shot weight reached a plateau (with only one somewhat doubtful report of a bombard firing an eighteen hundred pound stone), but increases in maximum gun weights and powder charges accelerated asymptotically until they peaked around end of the Hundred Years War.

---

from Maidstone, indicating Kentish ragstone (s.g. ca. 2.7) and fifteenth-century Burgundian accounts describe bombard stones as marble (s.g. 2.4–2.7) or more rarely sandstone (commonly s.g. ca. 2.6). The National Archives, Kew, E101/51/27; Joseph Stevenson, (ed.) *Letters and Papers Illustrative of the Wars of the English in France during the Reign of Henry the Sixth, King of England*, 2 vols. (London: RS, 1861–4), 2:2:587; Garnier, *Artillerie*, 122, 127, 134, 173, 175, 87. Two 44 cm. cannonballs recovered from the siege of Bolesławiec in 1396 have an average s.g. of 2.38. Piotr Strzyz, "Archaeological Traces of Use of Firearms During Sieges. Selected Examples from the Territory of Poland," *Fasisculi Archaeologiae Historicae*

Moreover, even before 1400, bombard stones had substantially greater velocity than trebuchet projectiles—and increasingly so over time, as shown by increases in range. In 1377, the Büchsenmeistser Walter Judenkind of Arles contracted to forge an iron *Centnerbüchsen* (gun throwing hundred-pound stones) with a range of at least 300 *Schritt* (215 m)—about the same as a large trebuchet.[28] But three bombards cast in Frankfurt just one year later could shoot large projectiles 1,000 paces (715 m), indicating an initial velocity about 1.8 times that imparted by a large trebuchet.[29] Just one decade later, in 1388, a gunfounder in the same town offered to cast a cannon able to shoot 1,000 *Klaster* (1,716 meters).[30] That seems to have been exceptional for the time, but was quite normal by around 1405-10. A passage in the *Feuerwerkbuch* that probably dates to that period says that bombards should be able to shoot up to 2,500 *Schritt* (around 1,850 meters).[31] A slightly later text increases that to 3,000

28 (2015), 112. Granite and related stone (s.g. ca. 2.5) is the most common material for the stones at Olszytyn. Piotr Strzyz, Piotr Czubla, and Adam Mackiewicz, "Cannonballs from the Olsztyn Turret," *Fasisculi Archaeologiae Historicae* 28 (2015), 123–124. My thanks to Grzegorz Żabiński for pointing me to this article. For a Turkish bombard reportedly firing an 1,800-pound stone in 1453, see Waurin, *Croniques*, 5:252.

28 A pace (*Schritt*) of 2.5 feet varied in length from place to place, as did other traditional measurements of weight and distance. Throughout this article, I have made use primarily of the following references to determine the appropriate metric equivalencies for traditional measures: Johann Leuchs, *Der Contorwissenschaft*, pt. 3, 4th ed. (Nuremberg: C. Leuchs und Comp., 1834); M.R.B. Gerhardt, *Allgemeiner Contorist*, pt. 1 (Berlin: Arnold Wever, 1791); P. Kelly, *The Universal Cambist*, 2nd ed., vol. 2 (London: Printed for the Author, 1821); Franz Mozhnik, *Lehrbuch des gesammten Rechnens für die vierte Classe der Hauptschulen in den k.k. Staaten* (Vienna: Verlage der k.k. Schulbücher Verschleiß-Administration, 1848). Note that traditional forms of measurement such as "miles" and "pounds" are sometimes retained instead of (rather than used alongside) metric linear measures, particularly when there is doubt about appropriate local usage.

29 Martin Crusis, *Annalium Sveveicorum* (Frankfurt: Nicolai Bassaei, 1596), 291: three bronze bombards cast in 1378 which reportedly threw iron balls (*globum ferreum*—which is likely a misreading of an original document that described the *bombard* as iron, though the rest of the passage certainly seems to be based on a contemporary municipal fiscal account or gun-maker's contract) of 127, 70, and 50 pounds to a distance of a thousand paces (*ad spatium mille passuum*). The *passus* here is probably a translation of a German *Schritt* of 2.5 feet, rather than the proper Latin meaning of two *gradi* (five feet), because bombards of this period generally had a maximum powder charge of 5–8% of stone weight and used un-corned powder, whereas the later *Feuerwerkbuch,* which indicates a range of about a mile for bombards, also envisions a use of an 11% charge of corned powder, 50% more powerful than un-corned powder. (See below.) Initial velocity vs. trebuchet: 4x the range (taking average trebuchet range as 180m) indicates double the initial velocity.

30 Rathgen, *Geschütz,* 249–251.

31 From a manuscript dated 1432, but incorporating various earlier strata of text: Gerhard W. Kramer, ed., *The Fireworkbook: Gunpowder in Medieval Germany*, tr. Klaus Leibnitz

*Schritt* (2,220 meters) or more.[32] Recorded shots of around 1,600 meters in 1430 and in 1452, and instances when large medieval bombards threw stones 3,200 meters or more during the sixteenth century, show these generalizations are perfectly credible.[33] Guns manufactured at the very end of the Middle Ages

(London: Arms and Armour Society, 2001), 56, 58, gives the range of a cannon-shot as 1,500 *Schritt* to (with stronger powder) 2,500 paces [1,850 meters or 1.15 English miles], from a gun with a powder charge of one ninth the weight of the stone, and a ball weighing a Venetian hundredweight (30kg). This version of the text probably dates to c. 1405–10 as it recommends a 1:9 powder charge ratio (which matches a Burgundian gun cast in 1409: Garnier, *Artillerie,* 25–26) and a barrel length of 5 *Klotz*, which would be 2 calibers (not 5 calibers as usually interpreted based on an unsustainable assertion that in this one instance *Klotz* means ball rather than the chamber plug), which matches the gun design shown in Conrad Kyeser's 1405 *Bellisfortis*. For the ratio between Klotz (=tampon = powder chamber diameter) and barrel caliber, see the document (Nürnberg HS 1481a, from ca. 1405–1410) quoted in Max Jähns, *Geschichte der Kriegswissenschaften, vornemlich in Deutschland,* pt. 1 (Munich: R. Oldenbourg, 1889), 391. For barrel length, see University of Freiburg Handschrift 362, fo. 87, which is the basis for the Kramer edition; it does not, however, match the translation given there. Instead, it is close to the text in the printed 1529 version of the *Feuerwerkbuch*. Kramer, *Fireworkbook,* 56; Wilhelm Hassenstein, *Das Feuerwerkbuch von 1420. 600 Jahre Deutsche Pulverwaffen und Büchsenmeisterei* (Munich: Verlag der Deutschen Technik, 1941), 33, 71. A closely related French version of the *Feuerwerkbuch* makes it up to 2,000 *pas* (1,624 m) or more, but does (probably because the translator was thrown off by the practice of his own day, c. 1430) recommend barrels of five calibers (rather than five tampons). Napoléon III and Favé, *Études*, 3:154, 153.

32 Jähns, *Geschichte*, 1:389. This is from the Ambraser Sammlung codex 67, dating to 1408 or shortly thereafter.

33 1430: *A Parisian Journal, 1405–1449*, tr. Janet Shirley (Oxford: Clarendon Press, 1968), 241 (from the Porte St. Denis of Paris to well beyond St. Lazare); Purton, *Late Medieval Siege,* 325 (Ottoman bombard cast in 1452 for the siege of Constantinople firing a test shot one mile). The Boxted bombard, which probably dates to the 1440s-50s, was used into the eighteenth century to throw balls into a hill "about a mile distant" for the amusement of fair-goers. Robert D. Smith and Ruth Rhynas Brown, *Bombards: Mons Meg and her Sisters* (London: Royal Armouries, 1989), 52. In 1558, the medieval bombard Mons Meg fired a stone from Edinburgh castle to Wardie Muir, 3.2 km distant (though also about 75 meters lower). J. Hewitt, "Mons Meg, the Ancient Bombard Preserved at Edinburgh Castle," *Archaeological Journal* 10 (1853), 28. Judging by the style of gun-carriage shown in a c. 1600 stone-carving of the gun—reproduced on the back cover of Smith and Brown, *Bombards*—this was likely at a low angle (perhaps 15–20 degrees), not 45 degrees, suggesting a maximum potential range of three miles or more (albeit with Early Modern gunpowder). Even *Faule Mette,* cast in 1411, was recorded as firing from Brunswick to and over a besiegers' camp near Melverode, indicating a flight of approaching 5 km—though this was in 1550 and using a 70-pound charge rather than the c. 55-pound charge employed in earlier times. A 1717 shot at 45 degrees elevation with a 52-pound charge shot a 341 kg ball 660 *Ruthe* (around 3 km). C.W. Sack, Emil Ferdinand Vogel, and Ferdinand Spehr, *Alterthümer der Stadt und des Landes Braunschweig,* 2d. ed. (Brunswick: Friedrich Otte, 1861), 77–81.

were designed to have ranges as great 3,200 meters, roughly thirteen to sixteen times what a heavy trebuchet could manage.[34]

Thanks to their high velocity, already by the turn of the fifteenth century gunstones could hit with significantly more energy than even the largest projectiles from counterweight trebuchets. For example, the second-largest bombard used at the siege of Burg Tannenberg in 1399 already would likely have thrown its 54 cm stones with more than double the kinetic energy of the somewhat heavier stones cast by Périgueux's great trebuchet.[35]

These giant bombards could sometimes make a big difference in siege warfare. In 1405, the captain who held Warkworth castle for the rebellious earl of Northumberland at first refused to surrender the place to King Henry IV of England, because his garrison was well provided with men and supplies. Hearing the captain's final response, as Henry himself wrote shortly afterwards to his council, "We immediately directed Our cannon against the castle, and they did such service that within seven shots the captain and all the others of his company, crying for mercy, submitted themselves to Our mercy."[36] It made a big impression when a bombard knocked a hole in a fortification's wall with a single shot, as happened at a castle besieged by the bishop of Minden in 1408.[37] An English translator of Vegetius commented that same year that "great guns

34 In 1453, the Venentians employed a German gunfounder to cast bombards capable of throwing 400-pound stone balls with a two-mile range. Michael E. Mallett and J.R. Hale, *The Military Organisation of a Renaissance State: Venice c. 1400 to 1617* (Cambridge: Cambridge U.P., 1984), 83. In 1451, Stralsund purchased a bombard supposed to throw a stone a half of a *Meile,* so either somewhat under or somewhat over two English miles (3.2 km), depending on the *Meile* meant. Otto Piper, *Burgenkunde* (Munich: R. Piper, 1905), 372n.

35 As noted above (note 29), already in 1378 some bombards could shoot a thousand *Schritt,* so around 4x as far as a trebuchet throwing large stones (above, note 15), indicating double the velocity and quadruple the energy for a given mass, * 250 kg /293kg for the difference in weight, = 3.41 ke. However, it is likely that the half-mile range was achieved by bombards with powder charges of around 8% of ball weight, rather than the 5% ratio that seems to have been more common in the 1370s–90s, both for *Centnerbüchsen* throwing hundred-pound stones and for larger bombards. (Below, notes 69–70.) So, we can fairly conservatively estimate the power of the Tannenburg gun as 3.41 * 5/8 = 2.13 times that of the Périgueux trebuchet. The weight for the stone is calculated based on its diameter, for which see Hefner and Wolf, *Tannenberg,* 77.

36 *Proceedings and Ordinances of the Privy Council of England,* ed. N.H. Nicolas, 7 vols. (London: Record Commission, 1834–7), 1:275.

37 Minden: *Scriptorum rerum Brunsvicensia* (Hanover: Nicolas Förster, 1710), 2:203: "Vicit et recuperavit potenter *castrum Wedegonis*; Nam habuerunt ibi de *grotten Metten* & cum ea uno ictu jactaverunt intra muros unum magnum foramen & foderunt subtus terram, ita ut accepterint urnam de fonte, & aliqui de castro suspense sunt."

nowadays shoot stones of so great weight that no wall may withstand them, as has been well shown both in the north country and in the wars of Wales."[38]

## 4 Practical Capabilities in the Early Fifteenth Century

That last comment, however, was an exaggeration, more reflective of the psychological impact of the new bombards than their actual power. It has often been noted that Berwick town and castle were captured by Henry IV in 1405 using his bombards, but the rarely-mentioned details are revealing. The town's walls were low, thin, and in such poor repair that they were on the verge of collapse from their own weight. The castle, however, was much stronger, and initially the stones from the king's guns simply shattered against its solid walls. It was only when a lucky shot struck a window and passed through to kill several men that the demoralized garrison surrendered.[39]

A campaign conducted in 1411 by the young duke of Guienne (the heir of the king of France) against the rebellious duke of Berry provides another excellent case-study for delimiting what gunpowder artillery both could and could not do in this period.[40] Guienne's army aimed to capture the town of Étampes. To clear his lines of communication, he needed to capture the castle of Bretonnière. The captain of the place refused to surrender initially, but then fled with his garrison by night once he saw artillery emplaced for battery. The duke's army then pushed on to Étampes. The townsmen had no interest in facing a siege for the sake of the duke of Berry and surrendered while Guienne's army was still approaching. Berry's garrison, however, held out in the castle of the town. This fortification was old but strong, with large towers and substantial walls, and on high ground. Since its situation made it practically impossible

38 Malcolm Hebron, *The Medieval Siege: Theme and Image in Middle English Romance* (Oxford: Clarendon Press, 1997), 31. Similarly, Thomas Walsingham reports that Henry IV took to Berwick in 1405 several guns, one of which was so large that, "it was believed" (*ut creditur*), no wall could withstand it. Thomas *Walsingham, Annales Henrici Quarti, in Chronica et annales regnantibus Henrico Tertio, Edwardo Primo, Edwardo Segundo, Ricardo Segundo, et Henrico Quarto,* ed. Henry Thomas Riley (London: Rolls Series, 1868), 411.

39 Town: Clifford J. Rogers, "The Military Revolutions of the Hundred Years' War," repr. with revisions in Clifford J. Rogers, ed. *The Military Revolution Debate* (Boulder: Westview, 1995), 88. Castle: *Eulogium historiarum,* ed. F.S. Haydon, 3 vols. (London: Rolls Series, 1858–63), 3:408. At Plaue in 1414 cannon destroyed walls so thick that a wagon could be driven along the ramparts—but they were only made of brick, not stone. Dorothea Goetz, *Die Anfänge der Artillerie* (Berlin: Militärverlag der DDR, 1985), 22.

40 This paragraph and the next are based on Louis Bellaguet, ed., *Chronique du religeux de St. Denys (1380–1422)*, 6 vols. (Paris: Crapelet, 1839–52), 4:572–577.

to undermine, it was considered almost impregnable. Despite the presence of the artillery train that had successfully intimidated the captain of Bretonnière, the duke's initial attempts to assault the place, mounted over several days, seemed to bear out this conclusion. Each attack was thrown back with serious losses to the besiegers. The situation changed, however, when the great bombards of Paris arrived.[41] "The wall was broken down in several places by the enormous, massive stones—as big as millstones—that these engines cast without ceasing," writes the chronicler of St.-Denys. "When the shot had at last broken a way through, the French consigned all the houses next to the walls to the voracious flames," compelling the defenders to retreat to the extremely strong twelfth-century donjon. It is not clear whether the cannon had opened a breach that French soldiers stormed through, or merely exposed flammable structures inside the walls to an attack with incendiaries, but either way the bombards had managed to do enough damage to solid masonry walls to make a major difference in the conduct of the siege.[42]

The next phase of the operation, however, showed the limits of the new bombards. Several assaults failed badly, and against the sixteen-foot-thick walls[43] of the donjon the gun-stones accomplished nothing but to inspire the "derision and mockery" of the ladies inside the stronghold, and the great displeasure of the leading besiegers. They saw the ineffectiveness of the missiles and worried that their efforts were being wasted and that the lengthening siege might end in failure. They even seriously considered breaking off the operation, though the failure of his first campaign would have brought "eternal shame" on the duke, and on the royal house of France. This disastrous result

41 These are described in the *Chronique de St. Denys* as "machinas jaculatorias" [throwing machines], and this has often been taken to mean trebuchets. It can, however, be taken as fairly certain that bombards are meant, because in the chronicler's account of the siege of Bourges the following year the author also uses the same noun to refer to what is unquestionably a bombard, named *Griete*, likewise firing "immense stones," but specified as using a large charge of powder and making a thunderous noise. Ibid., 4:652.

42 The French translation that accompanies the Latin text of the *Chon. St. Denys* has the French entering by the breach, but the Latin does not actually say that. For another very early example of the power of artillery at the turn of the century, see Denis Lalande, ed. *Le livre des fais du Mareschal Bouciquaut* (Geneva: Droz, 1985), 131, for a claim that Marshal Boucicaut's engines (including at least one huge trebuchet firing 600-pound stones) and cannon, firing constantly, were able to thoroughly demolish large sections of the walls of the "seemingly impregnable" stronghold of Montignac in 1398—albeit over the course of two months.

43 The St. Denys chronicler says the walls, where they were eventually broken through, were ten feet (3m) thick, but that would have been at the entrance, where internal guard-chambers left a void in the middle. The gunstones would have hit higher up where the wall was presumably solid.

was avoided when a Parisian engineer organized the construction of a giant mantlet of extremely thick oak, which was emplaced against the walls. Under its shelter, teams of workers with steel picks worked away at the masonry. After five days, they had so undermined the base that they could threaten to bring down the whole structure by firing the props they had emplaced. The captain of the garrison "suppressed his stubborn pride" and surrendered to the duke.[44] Ancient techniques and technology, more than new ones, were the key to the success of the capture of the stronghold.

In this period, the failure of the guns against the donjon of Étampes was still more the norm than artillery's success against the outer castle walls. The French offensives against English Aquitaine in 1404–1408 illustrate the situation well. In 1404, even the little towns of the Chalosse region in the Landes were still so confident of the superiority of the defense in siege warfare that when threatened by a substantial French army they declared they would not pay so much as fifty pennies to secure a truce. In 1405, the French required six months to capture the minor fortification of Chalais, and three months to take Mortagne in Saintonge.[45] In 1405–6, the town of Bourg was unsuccessfully besieged by a large army under the French king's brother, the duke of Orléans. The duke set up three bombards "in order to weaken the walls by the continuous blows of large stones," mounted frequent assaults, and rained a steady hail of arrows and bolts down on the defenders. But from the end of October to mid-January all his efforts failed. With his army suffering from miserable conditions and his finances in disarray, Orléans admitted failure and retired into France "in great dishonor."[46] Similarly, in Italy in 1405–6, the castle of Vicopisano was besieged for nine months by the Florentines, and although the interior of the place was completely demolished by bombardment (by both trebuchets and large bombards), the attackers' assaults on the walls were always repulsed "with shame and damage" until hunger forced the defenders to surrender.[47]

Up to this point, the spread of gunpowder artillery had actually probably favored defenders more than attackers. Either stone walls or earth-and-timber bulwarks were good protection against all but the largest cannon-shot. But during an assault the attackers had to leave their shelters and expose themselves to the lethal shot of guns, whereas defenders did not. Indeed, the

44 *Chronique de St. Denys*, 4:576.

45 Guilhem Pépin, "The French Offensives of 1404–1407 against Anglo-Gascon Aquitaine," *Journal of Medieval Military History* 9 (2011): 1–40.

46 Ibid. and *Chron. St. Denys*, 3:452–61; Germaine Lefevre-Pontalis, "Petit chronique de Guyenne jusqu'à l'an 1442," *Bibliothèque de l'école des chartes* 47 (1886), 64 (quotation).

47 *L'Assiedo di Pisa (1405–1406). Scritti e documenti inediti*, ed. Giuseppe Odoardo Corazzini (Florence: Ulisse Diligenti, 1885), 66–67; *Purton, Late Medieval Siege*, 225–227.

besieging gunners manning large bombards were themselves vulnerable to accurate counter-fire from the smaller cannon of the defenders (which shot more accurately, and when fired from city walls had an advantageous falling trajectory). In 1419, for example, a Moroccan army with gunners on loan from Granada aimed to recapture Ceuta in North Africa from the Portuguese. They hoped

> to wear them down with that violent invention—the bombard—like the ones we [Portuguese had earlier] brought to conquer this place. The Moors had two, a lot for them, the mainstays of this diabolical offensive. They shot at the walls but with no success because we brought up our own, which inflicted real damage. Then, as the enemy was repairing, a wily engineer shot off his piece so well that he bested the enemy's main gun, slew the gunner, and then so adroitly suppressed the fire of the other weapons that the Moors lost all their guns.[48]

Thus, observed a German master-gunner of around 1420, the defender of a strong fortress provided with sufficient artillery ought to be able to repulse any attacks "until he is relieved, or the enemy is given a good thrashing and departs the siege."[49] As we have already seen, that is what happened at Bourg in 1406. Because of the great kinetic energy they could impart to their projectiles, which made them capable of lethal hits even against well-armored attackers, small-to-medium guns were very effective in defeating assaults. At Bourges in 1412, some of the Burgundian soldiers besieging the town tried to pursue a defeated Armagnac sally force back through the open town gates. The Armagnac defenders on the walls began to shoot so strongly with cannon that the Burgundians had to fall back. Although the besiegers were provided with "many large bombards for laying siege," some ten weeks later they still had accomplished nothing and decided to starve the defenders out with a distant blockade.[50]

Because guns were so effective in the defense of town walls, they became increasingly popular acquisitions for municipal governments and individual bourgeois. In 1406, the town of Mons, for example, possessed 66 guns.[51] In the

48 Zurara, quoted in Weston Cook, "Warfare and Firearms in Fifteenth Century Morocco, 1400–1492," *War & Society* 11 (1993), 28.

49 Hassenstein, *Feuerwerkbuch von 1420,* 31.

50 Pierre de Fenin, *Mémoires de Pierre de Fenin,* ed. Mlle. Dupont (Paris: SHF, 1837), 27.

51 Depreter, "Le prince," 299, which also notes 95 cannon and 38 ribaudequins at Bruges in 1411, fifty more guns purchased by the same city in 1416, and a large iron bombard acquired the next year.

same year Lille had an arsenal of at least 37 guns.[52] On top of that stock, in response to the siege of the nearby city of Arras in 1414, the latter town acquired an additional 63 cannon, almost all *veuglaires* or "fowlers," a recently-developed type of longer-barreled gun shooting stone or lead shot. These weapons were breech-loaders provided with two or three beer-stein-shaped powder chambers each, and typically weighed around 50 kg per piece.[53] The large bombards of this era, firing stones each of which weighed as much as several *veuglaires*, were obviously far more expensive, and less useful (due to their slow rate of fire) in defeating an assault. However, as noted above, they did have occasional dramatic successes in offensive siege warfare. Such weapons were, therefore, sometimes acquired not only by territorial princes, but also by wealthy urban governments, both as status symbols and for their practical value.[54] In 1420, for example, the city of Bordeaux commissioned a great bombard to throw 346 kg stones.[55] The big guns could be used in enforcing a city's authority in its district, and towns that owned them gained opportunities to be useful to allies or lords, or to the king, by loaning them out.[56]

52 Alixandre de la Fons-Mélicocq, *De l'artillerie de la ville de Lille aux XIVe, XVe et XVIe siècles: archers, arbalétriers, canonniers* (Lille: Lefebvre-Ducrocq, 1854), 15. In 1412, there were four large cannon called bombards, four other large cannon, and 24 small iron guns, and the town councilors commissioned the casting of two more "portable" bronze guns, weighing 44 pounds combined, "considering that there are none of that style, and that all are of iron." Ibid.

53 Ibid., 16. Compare to the 11 guns (probably the town's first) purchased by Regensburg in 1379, which collectively weighed only 120 Pfund (67 kg). Köhler, *Entwickelung*, 3:1:247.

54 The inscription on the "large and noble" bombard *Redoubtée* (meaning "dreaded, feared") nicely reflects the typical motivations of townsmen in acquiring the big guns, pitching the purpose as defensive but implying also offensive use: "The year 1436/ I was made, in order to spend my time / on guard and for the defense / against those who against Metz make offense / to punish them and bring them to justice. / Suitable I am for this profession; / and for those who would know my name / *Rebdoubtée* is what I'm called." Jean-François Huguenin, ed. *Les chroniques de la Ville de Metz, 900–1552* (Metz: S. Lamort, 1838), 199.

55 Seven *quintaus* of 49.402 kg. *Archives Municipales de Bordeaux*, IV: *Registres de la Jurade. Déliberations de 1414 à 1416 et de 1420 à 1422* (Bordeaux: Imprimerie G. Gounouilhou, 1883), 426; for the weight of the quintal of Bordeaux, see Jean Cavignac, *Jean Pellet, commercant de gros, 1694–1772* (Paris: S.E.V.P.E.N., 1967), 35.

56 Although the theme of his article is how territorial princes worked to *reduce* their dependence on artillery loaned by the towns, the importance of such guns is clear in Michael Depreter, "La prince face à l'artillerie communale dans les Pays-Bas au début du XVe siècle: Flandre, Brabant, Hainault," in *Revue du Nord,* collection histoire (hors série) no. 35, *Autour d'Azincourt: Une société face à la guerre (v. 1370- v. 1420)*, ed. Alain Marchandisse and Bertrand Schnerb (2017), 285–304, especially 289–298. Middling nobles could also use their guns both to pursue their own purposes and to aid their lords. Among the guns used by the Duke of Burgundy's army for the siege of Vellexon in 1409–10, for example, were a large bombard borrowed from the lord of Villars, three small bombards

## 5 Quantitative Evolutions and the Artillery Revolution: More Guns

The increased breadth and pace of acquisition of cannon meant that the number of guns available for deployment at major sieges—both wall-breaking bombards and smaller supporting weapons—increased sharply. In the sieges of 1398–1406 mentioned above, even armies led by the French king's brother or by his marshal (one of the three highest-ranking military officers of France) had only two or three big guns.[57] In 1405, the wealthy count of Holland (who was also count of Zeeland and Hainault and duke of Bavaria-Straubing) had at least four big bombards (and around twenty fowlers) for the siege of Hagestein.[58] In 1407, King Juan II of Aragon had five large bombards for the unsuccessful siege of Setenil; his eldest son had only three for the prior siege of Zahara.[59] John the Fearless, Duke of Burgundy, who may have had the best artillery train in the world at the time, had at first two and ultimately five at Vellexon in 1409–10.[60] In 1415, when Henry V deployed nine large bombards

---

belonging to the town of Besançon, and two medium bombards belonging to the lord of Montaigu. Below, note 59, and see also Depreter, "La prince," 294.

57 As an additional example, a major Polish royal expedition in 1396 probably had just three bombards of 26–50 cm caliber. Strzyz, "Archaeological Traces," 111.

58 M.J. Waale, *De Arkelse oorlog, 1401–1412. Een politieke, krijgskundige en economische analyse* (Den Haag: Koninklikje Bibliotheek, 1990), 114 (*grote bussen* and *vogelaars*). The defenders made use of two *bussen* and some *vogelaars* as well; this is by several years the earliest reference to *veuglaires* I have found, though the *Bochsen* cast in four parts [i.e. with three chambers] at Marienburg in 1403 may be *veuglaires* without the name. *Das Marienburger Tresslerbuch der Jahre 1399–1409,* ed. Erich Joachim (Königsberg: Thomas & Opperman, 1896), 217; Grzegorz Żabiński, "Technology of Manufacture of Firearms in the Teutonic Order's State in Prussia—Gun Barrels and Metal Projectiles," *Fasisculi Archaeologiae Historicae* 28 (2015), 102. Over the course of the siege of Hagestein, according to financial accounts, the besiegers' *grote bussen* used some 15,480 kg of powder, though one wonders if this actually covered the powder expended on the smaller guns also. Waale, *De Arkelse oorlog,* 254, 260.

59 Fernán Pérez de Guzmán, *Cronica del serenissimo rey don Juan el segundo deste nombre* (Valencia: Benito Monfort, 1779), cap. xli, xxxvi.

60 Initially two large bombards (one of bronze, one of iron) belonging to the marshal of Burgundy and three small bombards belonging to the town of Besançon; then the large bombard *Dijon,* a bronze gun weighing 833 kg; then an additional large bombard of iron (the bombard of "M. de Chalon" or of "M. d'Arlay" = belonging to the lord of Châlon Arlay); plus two more "medium" bombards ("à M. de Montagu"). On 17 December, the Burgundians commissioned a new large bronze bombard at Auxonne, designed to cast a stone of 157 kg and weighing 3,381 kg, including the bronze of the *Dijon,* which was melted down after bursting. The bombard of Châlon Arlay also burst and was rebuilt with a new chamber made of cast bronze, weighing 833 kg. This was a huge increase from the old chamber, which only weighed 147kg. The lords conducting the siege were clearly aware that larger powder charges made for more powerful shots, but also were dismayed by the frequency

of "unheard-of size" for "playing tennis" with the defenders of Harfleur, it was considered quite a remarkable show of force. But thirty-four years later Charles VII emplaced almost twice as many (sixteen) for the siege of the same town.[61] At Caen in 1450, King Charles had twenty-four great bombards, some of which were reportedly so large that a man could sit erect inside the barrel.[62] At Belgrade in 1456, Mehmet the Conqueror had twenty-eight newly cast bombards of roughly 55 cm caliber.[63] So, the number of big bombards at a major siege increased by a factor of around ten over the course of a half-century.

This quantitative increase mattered in qualitative ways. What two bombards could do in two months, eight could do in two weeks, and twenty-four could do in five days. Indeed, the practical effect of larger numbers of big guns was even greater than that, because the large bombards fired so slowly that in between shots defenders could work to reinforce threatened sectors of walls, repair breaches, or build new defenses behind them.[64] Thus, slower progress in the work of demolition actually increased the total amount of demolition

---

with which their guns were bursting, so they evidently were erring very much on the side of strengthening the guns. The new chamber was designed to support a 36-pound powder charge for a 320-pound stone (11.25%)—the ratio recommended in the *Feuerwerkbuch*—suggesting that the old chamber must have been intended for a much smaller charge of weaker powder, perhaps around 2.5% of "meal" powder. Bertin, "Le siège du château de Vellexon," 101, 90, 108.

Late in the siege two more bombards arrived; the large bombard the lord of Villars and another sent by the duchess of Austria (the Duke of Burgundy's sister) from the castle of Maison-Vault. Villars' gun, however, burst at its first shot, so I have not counted it.

Another very large iron bombard was purchased for the siege, but at 3,773 kg it was so heavy that 31 horses and 32 oxen could only pull it around 5 kilometers per day, so that it did not arrive at the siege in time. It was designed (with a conical barrel able to accommodate stones of different sizes) for shot of 343–416.5 kg. Yet another large bombard, shooting 600-pound (294 kg) stones, was not brought because it burst during testing. All these details are from Garnier, *Artillerie*, 22–30; see also Bertin, "Siège."

61 1415: *Chron. St. Denys*, 5:536 (*inaudite*); *Ingulph's Chronicle of the Abbey of Croyland: With the Continuations by Peter of Blois and Anonymous Writers*, tr. Henry Thomas Riley (London: Bohn's, 1854), 365 (tennis). 1449: Jean Chartier, *Chronique de Charles VII*, ed. Vallet de Viriville (Paris: P. Jannet, 1858), 2:178.

62 Thomas Basin, *Histoire de Charles VII et Louis XI*, ed. J. Quicherat, vol. 1 (Paris: SHF, 1855), 240.

63 Gábor Ágoston "Firearms and Military Adaptation: The Ottomans and the European Military Revolution, 1450–1800," *Journal of World History*, 25 (2014), 92; Purton, *Late Medieval Siege*, 331; *Magyar történelmi tár: a történelmi kútfők ismeretének előmozdítására*, vol. 9 (Pest: Eggenberger Ferdinánd M. Akad. Könyvárusnál, 1861), 66.

64 Smaller guns were used precisely to hinder these efforts, but they could not stop reinforcement of a wall before it was breached, or the construction of back-up retrenchments behind a weakened wall.

that needed to be done.[65] That logic would apply even if the guns themselves had not changed, but in fact—for reasons that will be laid out in the next two sections—the bombards of the 1430s–1440s were much more powerful than those of the period up to 1420, and could be fired significantly more rapidly.

## 6 Quantitative Evolutions and the Artillery Revolution: More Shots Per Day

Up to around 1400, bombard barrels were very short, like early-modern mortars. Powder chambers were of smaller diameter and relatively long. Working from the mouth of the weapon, the gunner had to ladle and pack in enough powder to fill 3/5 of the chamber, and then plug the opening tightly with a wooden tampon that filled another fifth of the space, leaving an equal volume of empty air between the powder and the wood. The ball was then loaded in, wedged in place with triangular pieces of soft wood, and covered with some mixture of mud, loam and hay, which then had to be allowed to dry. All of this served to keep the ball in place as a plug for the powder chamber until the pressure of the gas generated by the burning powder built to a high level, which was necessary because all the explosive force of the small charges that were then used had to be applied in a single spitting push, as the shortness of the barrel allowed for relatively little application of pressure after the ball started to move. Because of this complicated procedure (which of course was especially difficult and time-consuming when it involved manipulating balls of 300 or 700 kg), rates of fire were very slow, typically 5–6 shots per day.[66] This was a poor showing in comparison to trebuchets—of which even the largest

65 To cite one example of many: at Melun in 1420, for some eighteen weeks, "as fast as their walls were broken down by the engines of their adversaries, [the defenders] strengthened them with hogsheads full of earth, straw, wool," and other suitable materials. John de Waurin, *A Collection of the Chronicles and Ancient Histories of Great Britain, Now Called England (1422–31)* tr. William and Edward L.C.P. Hardy (London: Rolls Series, 1887), 311.

66 Johann Ottsen, "Über den derzeitigen Stand unserer Kentnisse von den Anfängen der Pulverwaffen," *Beiträge zur Geschichte der Technik und Industrie* 13 (1923), 8 [40 shots in 7 days at Tannenberg in 1399]; Pérez de Guzmán, *Cronica Juan el Segundo,* cap. xli, xliij [8 shots per gun, firing all day and part of the night, at Setenil in 1407]; Wenceslai Hagecii, *Böhmische Chronica,* vol. 2 (Cadan: J.S. Zluticensem, 1596), fo. 114v; Thomas Fudge, ed. and tr., *The Crusade against Heretics in Bohemia, 1418–1437: Sources and Documents for the Hussite Crusades* (Aldershot: Ashgate, 2002), 155–6 [largest guns only six or seven times per day]. Note also: *Chroniques de la ville de Metz,* ed. J.F. Hugenin (Metz: S. Lamort, 1838), 200 [3 shots in a day, 1437].

could be fired as many as 89 times in 24 hours, though the practical norm was more like 10–18 shots per day.[67]

But the situation changed as gunners improved their art and as changes in gun design and loading technique favored a higher rate of fire. Gun barrels grew steadily longer, up to around 5 calibers by 1440s. Longer barrels allowed gunners to dispense with the wedges holding the ball in place, and to do without the loam seal at the mouth used in earlier guns.[68] By around 1410, gunners were also using "loading irons" that enabled them to consistently and rapidly set in place the right amount of powder to load a gun, without needing to measure the distance filled.[69] The rate of fire increased correspondingly. It is difficult to say how fast fifteenth-century gunners normally served their weapons, but during a particularly intense period of firing at the siege of Rheinfelden

67 *Petit livre Noire* of Périgueux, fo. 47v: "hi tiret en I jorn XLV peyras, en Iª nuech XLIIII." For an average of only 18 per day at Karlstein in 1422, see Rathgen, *Geschütz*, 369. In the secondary literature, I have often read of the siege of Edinburgh in 1296 as an example of a rate of fire of around 18 stones per day, based on 158 stones thrown by three engines in three days. However, although the figure of 158 stones in three days and nights is given by the Lanercost chronicle, the number of engines is not. It was probably well more than three, since the same source describes "immense machines for casting stones having been set *all around* [*circumquaque*]" the castle, and Hemingbugh's chronicle says the English had "plentiful" [*copiosas*] large machines. Hence, this suggests an average of at *most* 10 shots per 24 hours per machine, assuming "plentiful" and "all around" imply at least five engines. Joseph Stevenson, ed., *Chronicon de Lanercost, MCCI-MCCCXLVI* (Edinburgh: The Bannatyne Club, 1839), 178; Walter of Hemingburgh, *Chronicon Domini Walteri de Hemingburgh* ed. H.C. Hamilton. (London: English Historical Society, 1849), 2:105. For 10–12 heavy stones thrown per day in the 1380s, see Jean Froissart, *Chronicles*, tr. Thomas Johnes, vol. 2 (London: Routledge, 1874), 254. During a large siege by the count of Hainault in 1345, 10 trebuchets ("mangonalibus") threw 1,100 large stones over seven weeks—an average of less than three per day per machine. Jean de Hocsem, *La Chronique de Jean de Hocsem* ed. Godefroid Kurth (Brussels: Commission Royale d'Histoire, 1927), 336. We get almost the same low figure for the siege of Ypres in 1383, where two trebuchets together threw 400 or 450 stones over 8 weeks: 3.5–4/machine/day. Kervyn de Lettenhove, ed., *Istore et croniques de Flandres*, vol. 2 (Brussels: F. Hayez, 1880), 296, 311–312. This last example is often mistakenly cited in secondary literature as indicative of the rate of fire of *cannon* at the time, but the chronicle is clearly referring to trebuchets, not guns.

68 This was clearly expressed in a 1445 manuscript of the *Feuerwerkbuch* quoted by Max Jähns: "Diewyle die Buchssen vor dem pulversak als kurtz waren, wenn der stein dar in geladen wart, dass er ein wenig für die Buchs gieng, zu den zyten und zu denselben Buchssen, was bedurfft, das man den stein verbisset. Aber zu den Buchssen, die man yetzunt hat, die langen Ror haben vor dem pulversack, so die Buchs eingeladen wirt mit pulver und mit stein, da bedarff der stein nichts denn usschoppens." Jähns, *Geschichte*, 397.

69 Quoted in Jähns, *Geschichte*, 387.

in 1445, four large guns managed 74 shots in 8 hours: two per gun per hour.[70] Four Milanese and Genoese bombards in 1464 each managed ten shots in just a very few hours, so around two to three per hour per gun.[71] When *Mons Meg*—forged in 1449—was put into use in 1571, its crew fired four times per hour.[72] But, as with the trebuchet, the practical norm seems to have been much lower than the theoretical maximum, perhaps never passing 12–19 shots per day—still a large increase over the standard of the turn of the century.[73]

## 7 Quantitative Evolutions and the Artillery Revolution: More Kinetic Energy Per Shot

The first great bombards fired slowly in the other sense as well: the muzzle velocity of their projectiles, though higher than that of mechanical artillery, was quite low in relation to cannon of the mid-fifteenth century. Compared to later guns, the earliest bombards used very small amounts of powder relative to the weight of their shot, so the amount of potential energy available to be converted to kinetic energy was low, and so was the velocity of the projectile. When the first bombards throwing large stones were built in the 1370s, they had very short barrels (just long enough to hold the ball, or a bit longer) and small powder chambers designed to hold about 2.5% of their projectiles' weight in dry-mixed powder.[74] A generation later the design was similar, though powder

70 "Am mendag, als ich und Schaltenbrant in der forstat uff wacht giengend, von 8 zaltend wir 74 hŏptchútz bisz früge uff 4." Ed. A. Gessler, "Basler Geschütznamen," *Basler Zeitschrift für Geschichte und Altertumskunde* 14 (1915), 87.

71 Luca Beltrami "Le bombarde milanese a Genova nel 1464," *Archivo storico lombardo,* 2nd ser., 5 (1888), 802.

72 I.e. ten times in two and a half hours (not 24 times, as I have read in a number of modern works), according to an eye-witness. Richard Bannatyne, *Journal of the Transactions in Scotland during the Contest between the Adherents of Queen Mary and those of Her Son* (Edinburgh: Constable, 1806), 153. For a similar rate of 140 shots in a day from two guns in 1484, see Prescott, *History of the Reigns of Ferdinand and Isabella,* 1:268.

73 At Rheinfelden in 1445 over two days, the average was 18.75 shots for each of four guns (including the period of more intense fire just noted). Gessler, "Basler Geschütznamen," 87. For 14 shots/day in 1474: ibid., 89, and Vaughan, *Charles the Bold,* 295. Note also: *Chronique de St. Denys,* 652 (*Griete* at Bourges in 1412) [12 shots in a day]; Napoléon III and Favé, *Études,* 2:96) [guns with 1,200-pound stones once per two hours by the Turks in 1453].

74 For a record-based example of 2.5% powder ratios for a 1376–7 gun firing a 60-lb. stone with a 1.5 lb. charge, see Contamine, *War in the Middle Ages,* 141. The Nürnberger bombard *Kriemhilt* or *Chriemhilde* is often given in secondary literature as firing a 560-pound stone with a 14-pound charge (2.5%), but actually the source does not give the weight of the

chambers were now constructed more solidly and powder charges had increased to about 5% of ball weight.[75]

From then until the 1430s, gun design changed in ways that allowed bombards to fire more rapidly in both senses. The most obvious change was that (as already noted) barrels grew proportionally longer. At the turn of the century, barrels were still commonly about 1–1.5 calibers long: that is, with an internal length 1–1.5 times their bore width. A manuscript probably written shortly after 1408, but incorporating an earlier text, implies a norm of barrels 1.6 calibers long.[76] *Faule Mette*, a bronze bombard cast in 1411, had a barrel 2.5 calibers long.[77] A manuscript of about the same date says bombards with barrels 2.8

---

stone. *Chroniken der deutschen Städte. Nürnberg*, vol. 1 (Leipzig: S. Hirzel, 1862), Beilage IV.B., 177. The Bodiam Mortar, which can be dated c. 1405 based on its very close similarity to an illustration in Conrad Kyeser's *Bellifortis* (see Schmidtchen, *Bombarden*, 15), has been estimated to have used a 2.2% charge, apparently based on corned powder loaded up to the base of the ball. Brig. Gen. J.H. Lefroy, *Official Catalogue of the Museum of Artillery in the Rotunda, Woolich* (London: HMSO, 1864), no. 1. It would have been correspondingly lower if used with a 3/5 chamber fill. The Amsterdam bombard, which may date to the 1370s, would have used a similarly small charge, though it is difficult to estimate due to the uncertainty of loading method and powder type that early. In any case, it has less than one-third the ratio of powder chamber volume:stone volume of the otherwise fairly similar in form (though more sophisticated in manufacture) *Centnerbüchse* of the Kölnisches Museum, which can be estimated to have held a powder charge of about 4.8%, based on a specific gravity of .88 and a 3/5 load. I have determined to use that figure (which falls between the .80 used by Angelucci and the .94 used by Rathgen) for specific gravity because it is the number that would make the charge weight ratio match the volume ratios given in the early fifteenth century text of HS 1481 a. of the Germansiches Nationalmuseum in Nürnberg (in Köhler, *Entwickelung*, 3:1:291-2n), assuming a specific gravity for stone of 2.56. My thanks to Sascha Pries of the city museum for providing me with measurements of the Köln gun.

75 Financial records show that the bombard of Frankfurt at the siege of Tannenberg in 1399 used 7 Centner and 33 pounds of powder to fire forty stone balls weighing 3.5 Centner each (5.2%). Rathgen, *Geschütz*, 6; Schmidtchen, *Bombarden*, 15. In 1397, an Imperial campaign against a robber baron included three contingents each bringing a *Centnerbüchsen* along with 20 hundred-pound stones and a hundred pounds of powder, indicating a 5% powder:ball ratio. Karl Heinrich Lang and Max von Freyberg eds., *Regesta sive rerum boicarum*, vol. 11 (Munich: Royal Press, 1847), 108.

76 It calls for filling a gun to a maximum of 1/3 with powder. One third of total length would equate to 3/5 of the powder chamber length given an overall design of 3 (powder) + 1 (space) +1 (tampon) + 4 (barrel). Given that the ratio of powder chamber bore (= Klotz thickness and diameter) to barrel caliber was 2/5 (per Jähns, *Geschichte*, 391), a barrel 4 Klotz long = 8/5 calibers long.

77 Sack, et al., *Alterthümer...Braunschweig*, 76–82. However, because it had a conical barrel, the barrel length was 3 rather than 2.5 times the length of a ball that would fit at the base of the barrel.

calibers long shoot the furthest.[78] A manuscript drawing from the 1420s shows a bombard designed based on a barrel length of 3 calibers.[79] The contract for the casting of the bombard *Bourgogne* specified a barrel the length of four stones (just slightly below four calibers) in 1430, and when a new barrel was made for the gun in 1437–8 the same ratio was retained.[80] Large late-medieval bombards like the two "Michelettes" and *Mons Meg* had barrels of from 5 to 5.6 calibers.[81]

These longer barrels meant that the ball remained subject to the impulsion of exploding powder for a longer period of time, making them more suitable for larger charges of powder that could give balls more velocity and kinetic energy. *Faule Mette* (1411) used a powder charge of around 7–8% of the weight of its balls—about a 50% increase from what was typical at the turn of the century, but already on the low side for its own day.[82] A text of about that same date or a little earlier calls for a 10% powder load.[83] Burgundian bombards of 1409 already were designed for even larger powder charges: around 1:9, as recommended in the *Feuerwerkbuch*.[84] The more than fourfold increase in proportional powder charge since the first substantial stone-throwing cannon required greatly increasing the quantities of metal used in manufacturing guns, to make them strong enough to contain and direct more powerful explosions. Events at the siege of Vellexon in 1409 illustrated this dramatically. After an old bombard burst in operation, a new powder chamber was made to go with the old barrel. The new piece was made with almost *seven times* the weight of bronze used for the part it replaced.[85]

In 1430, Burgundian authorities commissioned a bombard that was to fire a 600-pound (294 kg) stone using 70 pounds (34.3 kg) of powder—so still using a

---

78 I.e. 7 Klotz, the Klotz (equal to the powder chamber diameter) being 2/5 calibers at that time. Jähns, *Geschichte*, 388–391.

79 Hassentstein, *Feuerwerkbuch*, 101; Köhler, *Entwickelung*, 3:1:293.

80 Garnier, *Artillerie*, 70; Monique Sommé, "Les Mesures dans l'artillerie bourguignonne au xve siècle," *Le médiéviste et la métrologie historique: Table ronde Université Charles de Gaulle, Lille III, 9 Mai 1989*, 1st part (Lille: Université Charles de Gaulle, 1989), 49.

81 Brown and Smith, *Bombards*. The *Mons Meg* group of bombards have barrels of from 5.1 (the Basel bombard) to 5.6 (*Mons Meg*) calibers.

82 *Faule Mette:* Sack, et al., *Alterthümer...Braunschweig*, 76–81.

83 Text in Köhler, *Entwickelung*, 3:1:291–2n.

84 Sack, et al., *Alterthümer...Braunschweig*, 76–82. *Feuerwerkbuch*, ed. Kramer, 25 (with this part of the text probably dating to around 1410, rather than to c. 1380 as Kramer thinks, since records show powder loads of only 5% in 1379–99 (above, note 72), and around 11% only around 1409). For three guns with powder:ball ratios of 10.42% to 12.15% in 1409, see Garnier, *Artillerie*, 25–26.

85 Garnier, *Artillerie*, 26, 265; Bertin, "Siege," 100.

powder-to-ball ratio of 11.6%, similar to the one for the bombard commissioned in 1409 and still aligning closely with the *Feuerwerkbuch*'s recommendation for a 1:9 powder-to-ball ratio.[86] However, there was not enough metal available to make the new gun as planned, so its barrel was redesigned for a 400-pound (196 kg) stone, allowing for a charge of 17.5% of the ball weight. It is quite possible this accident led to an appreciation of the value of larger powder charges, since in 1446 another Burgundian bombard was designed from the start to fire a 350-pound (172 kg) stone with 72 pounds (35.3 kg) of powder (a 20.6% charge), and the year before that *Bourgogne* was re-built to use a 100-pound (49 kg) charge for a stone of about 174 kg—a charge of about 28%.[87] Thus, the amount of powder per pound of gunstone, for the largest guns, had increased around ten-fold over 75 years, and almost six-fold in just 50 years. The increasing powder charges made guns substantially more powerful even though smaller stones were being used. Even if both had used the same formulation of gunpowder, for example, a 174 kg stone thrown from the remade *Bourgogne* (1445) would have hit with about five times the kinetic energy of the 190 kg stones from the large bombard cast at Frankfurt in 1394, and more than triple the punch of the much larger stone of the *Pumhart von Steyr* (c. 1415).[88]

But the guns of the 1430s–1440s were not only using much more powder per pound of projectile than had been the case a century or a half-century earlier, they were also using powder made in a different way, so that it was in practice significantly stronger. In the mid-fourteenth century, before the first substantial stone-throwers came into use, the ingredients of gunpowder were normally finely powdered and mixed together dry. Probably in the 1370s, many master gunners had learned the "secret" of making wet-mixed powder. When the main ingredients were moistened with brandy or other liquids during the

86 Burgundian: Garnier, *Artillerie*, 70. In 1428, three English bombards (two firing 16" and one firing 18" stones) were tested using 80 pounds of powder, indicating a similar charge of around 11.8%. The National Archives, Kew, E101/51/27 (Receipts and other documents relating to the account of John Parker, master of the ordnance under the earl of Salisbury.)

87 Garnier, *Artillerie*, 70, 112, 109. For the Burgundian pound for artillery (490 grams), see Sommé, "Mésures," 46.

88 If the same formulation of powder was used in each case and the gun conveyed its potential energy into kinetic energy with equal efficiency, the muzzle energy would increase in proportion with the powder charge, so that the 1446 bombard would impart 4.95 times the muzzle energy of the 1394 Frankfurt gun, and 3.15 the muzzle energy of the Pumhart (assuming a 3/5 load of 0.88 s.g. powder). However, the later gun would have a substantially bigger advantage in *impact* energy, because its smaller stone would lose less energy to air resistance. Moreover, because of its longer barrel, the 1446 bombard would have done a better job capturing the potential energy in the powder and transferring it to the shot, for a higher muzzle energy from the same powder charge.

manufacturing process, the sulfur and saltpeter were absorbed into the fibrous structure of the charcoal. Modern tests using the 3/5 loading method have not yet been performed, but such tests as have been done do confirm that wet-mixed powder is significantly more powerful than dry-mixed.[89] The earliest large stone-throwers probably already used this type of powder. The next stage was the development of "corned" powder, in which the gunpowder was formed into larger or smaller balls or grains. The interstices between the powder grains provided the air needed for an explosion to progress by the more rapid inter-grain burn rate, rather than the slower intra-grain rate. The *Feuerwerkbuch* manuscripts consistently say that such powder was fifty percent more powerful than the older style of meal powder.[90]

Thus, relative to a major siege of around 1400, one conducted around 1450 might involve something on the order of 8 times as many great bombards, each delivering 7.5 times as much kinetic energy per shot, and shooting 3 times as frequently, for a total of around *one hundred and eighty times* as much

89 The earliest text on the subject says that wet-mixed powder with a 4:1 saltpeter:sulfur ratio, reinforced with small quantities of salarmoniac and camphor, was three times as powerful as earlier formulations. Österreichische Nationalbibliothek Wien, Codex 3069, fo. 2 (1411); cf. Rainer Leng, ed. *Anleitung Schiesspulver zu bereiten, Büchsen zu laden und zu beschiessen: Eine kriegtechnische Bilderhandscrhift im cgm 600 der Bayrischen Staatsbibliothek München* (Wiesbaden, Reichert Verlag 2000), 72–73. That ratio happens to almost exactly match the results of modern testing with a replica fourteenth-century gun (not employing the tampon loading method), which found that two shots with dry-mixed "Marcus Graecus" (6/1/1) powder averaged muzzle velocities of 101 m/sec, whereas two shots using the same formulation but wet-mixed resulted in average muzzle velocities of 176 m/sec—i.e. triple (3.03 times) the kinetic energy. However, this may be due to an improper loading for the first dry-mixed powder shot; discarding that as an outlier, the wet-mixed powder was about 76% more powerful than the dry-mixed. Medieval Gunpowder Research Group, "Report 1," table 5. This is another area where more testing needs to be done, using the loading methods and powder recipes specified in same text.

90 The various German manuscripts of the *Feuerwerkbuch* all agree that two pounds of powder wet-mixed and rolled into balls (*Knollenpulver*) does more than three pounds of "meal" powders, which latter are called either (depending on the manuscript) *gestossen* [meaning either "stamped" or "rammed"] or "sieved" powders. E.g. Royal Armouries *Feuerwerkbuch* (MS I-34), fo. 4v: "knollen pulfer zway pfunt mer tun dann gestossen pulfer drey pfunt"; Staatsbibliothek zu Berlin, Ms. germ. qu. 1187, fo. 15v; Kramer, *Feuerwerkbuch,* 25; Hassenstein, *Feuerwerkbuch,* 45; Heidelberg Universitätsbibliothek *Feuerwerkbuch,* Cod. Pal. germ. 502, fo. 4. Cf. the French translation of c. 1430 in Napoléon III and Favé, *Études,* 3:146, which clarifies the meaning of *Knollen* and makes the ratio 3:1 rather than 3:2: "La question est que si les pouldres ainsi meslées et faites par boulles et pelottes sont meilleurs…les dictes pouldres accumulées en monceaux en façon de boulles et pelottes vallent et font mieux et plus une seule livre que trois d'autres pouldres criblées."

destructive power deliverable per day.[91] This offers an excellent example of how the multiplicative interaction of several more linear or "evolutionary" developments can cause exponential or "revolutionary" change.

## 8 The Effects of the Artillery Revolution: Siege Warfare and Strategy

Fortification design responded in a variety of ways to the increasing power of gunpowder artillery, but the strength of defenses rose nowhere near as rapidly as the strength of artillery siege trains, and so for the first time in history it became possible to rapidly breach strong masonry fortresses by means of bombardment. This changed the dynamic of siege warfare dramatically, with tremendous consequences for the art of war and indeed for the course of European and World history.

Although, as we have seen, the development of efficient wall-breaking artillery resulted from the interaction of several strands of evolutionary change, the real tipping point seems to have come in the mid-to-late 1420s and the 1430s, at least in France. Before that time, as noted above, guns aided defenders as much as attackers, and only relatively weak walls like those of Berwick in 1405 could be quickly breached. Major sieges—including sieges of well-built but small strongholds—continued to last, typically, from two to six months, and to end by starvation rather than assault. Long sieges were then still the expectation as well as the reality. In 1400, the indenture between the King of England and the castellan of Roxburgh forbade the castellan, if besieged, to surrender until three months had passed—the same standard that had been applied in the case of Berwick back in 1356. In 1409, Christine de Pisan's essay on modern siege warfare, which called for besiegers to deploy numerous heavy guns, advised defenders to provide themselves with six months' worth of food supplies, which shows how little she expected the attackers' bombards to actually accomplish.[92] Those expectations not only reflected but also shaped reality, since defenders who expected to be able to sustain long sieges would lay in food supplies accordingly, and since garrison commanders and soldiers would be highly motivated to hold out for a respectable amount of time.

91 Reckoning 3 bombards firing 5 times per day with 5% charges of wet-mixed "meal" powder vs. 24 bombards firing 15 times per day using 25% charges of 50% more powerful corned powder.

92 Michael Prestwich, *Armies and Warfare in the Middle Ages* (New Haven: Yale U.P., 1999), 241; Christine de Pizan, *The Book of Deeds of Arms and of Chivalry*, tr. Sumner and Charity C. Willard (University Park: Penn. State U.P., 1999), 110–111.

Long sieges meant that there was a huge amount of "strategic inertia": The normal advantage of the defensive in warfare was magnified, and it was far easier to retain a region than to capture it. This encouraged commanders on the offensive to avoid efforts at gradual conquest and to rely instead on devastating expeditions (*chevauchées*) that bypassed strong fortifications and took the war to the relatively vulnerable agricultural population. For those on the strategic defensive, however, the tendency was to prefer a Fabian or "Vegetian" style of defense, which relied on stone walls to provide a secure "reservoir of lordship."[93] Small powers (little states or sub-sovereign territorial lords) could use the leverage of their fortresses effectively to resist strong neighbors or their own superior lords. All this tended to eternalize war and to make it especially damaging to civilians and to the European economy.[94]

By 1425, however, it took only eight days for a large Burgundian siege train to leave the walls of Braine-le-Comte so "ripped apart and broken down" that the place would soon have been taken by assault if the garrison had not surrendered.[95] Stronger places could still hold out longer, but by 1449–1451 even the formidably defended town of Harfleur was captured after just seventeen days, and the walls of Blaye "were completely thrown down in many places" after only five days.[96] The full extent of the "Artillery Revolution's" impact on the conduct of war really became clear in those years, when Charles VII of France was able to overrun first Normandy and then Aquitaine each within a single year. "Never," wrote the Berry Herald, "was so large a country conquered in such a short time."[97] At the other end of the continent, Mehmet the Conqueror used many massive bombards in his successful siege of Constantinople in 1453. His siege of Belgrade in 1456 did fail, but only because of the extraordinarily

93 On the Vegetian defensive strategy see John Gillingham, "Richard I and the Science of War in the Middle Ages," in John Gillingham & J.C. Holt, eds., *War and Government in the Middle Ages* (Boydell Press, Woodbridge 1984); on its limits, see Clifford J. Rogers, "The Vegetian 'Science of Warfare' in the Middle Ages," *Journal of Medieval Military History* 1 (2003): 1–20.

94 On the destructiveness of warfare in this period, see Clifford J. Rogers, "By Fire and Sword: *Bellum Hostile* and 'Civilians' in the Hundred Years War," in *Civilians in the Path of War*, ed. Mark Grimsley and Clifford J. Rogers (Lincoln: University of Nebraska Press, 2002): 33–78.

95 Jean de Waurin, *Recueil des Croniques et anchiennes istories de la Grant Bretaigne, a present nomme Engleterre*, 5 vols. (London: Rolls Series, 1864–1891), 5:165 ("deschirez et desrompus"); similarly at Jargeau in 1428, ibid., 5:241.

96 Rogers, "Military Revolutions," 66–67, and Clifford J. Rogers, "The Artillery and Artillery Fortress Revolutions Revisited," in Nicolas Prouteau, Emmanuel de Crouy-Chanel and Nicolas Faucherre, eds., *Artillerie et Fortification, 1200–1600* (Rennes: Presses Universitaires de Rennes, 2011), 77–78, for these and other examples.

97 Rogers, "Military Revolutions," 67, 89n. 110.

vigorous defense mounted by Jan Hunyadi. Although the city was then "probably one of the strongest fortresses in Europe," Mehmet's numerous great bombards were so effective that, according to an eyewitness, "within the space of ten days almost all the walls of the citadel were leveled to the ground." A few days later, Belgrade's defenses were so thoroughly demolished that, according to Hunyadi himself, the place "could be called a field, rather than a fortress": *non castrum, sed campum.*[98]

Although the new artillery contributed greatly to the rapid conquests of the mid-century, there was of course much more to the story. A siege train, no matter how powerful, is of little use if you cannot bring it to bear on the enemy's fortresses, and it is difficult to do that if your army cannot hold its own in open battle. For most of the Hundred Years War, the French had failed signally to convert their superior resources into tactically superior field armies, but by the 1440s English field forces had grown weaker, while their enemies had experienced a large improvement in their qualitative effectiveness. The latter development was largely the result of a sustained and energetic program of military reform that began to be implemented in 1444–45, after the French monarchy broke noble resistance to the increased taxes necessary to sustain a substantial body of soldiers during a period of extended truce with England. Charles VII broke up the private mercenary bands that had formed the bulk of his army to that point, then commissioned a carefully selected sub-set of their leaders to head twenty new regular or "ordinance" companies, each composed of 100 veteran men-at-arms (armored cavalry), 100 light cavalry, 200 mounted archers, and 200 support personnel. All these soldiers, who were enlisted for long-term, year-round service, were selected on the basis of their strength, their fighting prowess, and the quality of their mounts, armor, and arms. It was the ordinance companies that provided the hard edge of the French armies that conquered Normandy and Aquitaine in 1450–51 and that—along with the impressive French artillery train—made France's army by far the most powerful in Christendom.[99]

---

98 The number of large bombards is given in various places as 12, 22, or 28. Gábor Ágoston, *Guns for the Sultan: Military Power and the Weapons Industry in the Ottoman Empire* (Cambridge: Cambridge U.P., 2005), 67; Kenneth M. Setton, *The Papacy and the Levant, 1204–1571*, vol. 2 (Philadelphia: American Philosophical Society, 1978), 174–181, quotations pp. 176, 177–178. Hunyadi, in *Monumenta Hungariae historica,* 33 (Budapest: Kiadja a Magyrar Tudományos Akadémia, 1907), 208: "In tantum enim castrum per ictus bombardarum destruxit, quod ipsum castrum non castrum sed campum dicere possimus, quod usque ad terram murus castri est destructus."

99 Paul Solon, "Charles VII and the *Compagnies d'Ordonnance*, 1445–61: A Study in Medieval Reform" (Ph.D. Dissertation, Brown University, 1970); idem, "Valois Military Administration

## 9 Conclusion

Over the course of their first century of use European gunpowder weapons had come a very long way. By the end of the fourteenth century, and even more so by 1410, big bombards far surpassed the largest trebuchets in range and hitting power. They then continued to see a rapid growth in the strength of powder charge they could handle—more than doubling again the punch of the largest guns, even though the stones they used fell to "only" a few hundred pounds. Meanwhile the number of wall-breaking bombards employed rose steadily, so that by the 1440s there had been a true quantum leap in the total amount of destructive power a king or great noble could bring to bear against a besieged stronghold. The multiplicative interaction of the effects of more guns shooting more rapidly with more power meant that it was possible for a royal siege train of 1450 to do as much damage to a fortification in one day of firing as could have been accomplished in a half-year of bombardment in 1400. This completely changed the balance between offense and defense in siege warfare and therefore in warfare as a whole, in ways that are important for explaining the rise of the expansive empires of the second half of the fifteenth century, including France, Spain, and the Ottomans. The consequent military revolution would not have been possible using mechanical artillery, which had already reached more-or-less its maximum potential by the end of the thirteenth century. Thus, despite Lynn White's suggestion to the contrary, there is no need to look to non-rational factors to explain why medieval rulers and artillerists turned to gunpowder artillery first to supplement the trebuchet and then—well before 1426—to replace it as the principal technology for bombarding fortified positions.

---

on the Norman Frontier, 1445–61: A Study in Medieval Reform," *Speculum* 51 (1976): 91–111; Philippe Contamine, *Guerre, état et société à la fin du moyen âge. Etudes sur les armées des rois de France, 1337–1494* (Paris: Mouton, 1972), 277–530.

CHAPTER 3

# To the Seas: The Genesis of Ship Rigging in the Medieval and Early Modern Worlds

*Sarah K. Douglas*

The genesis for this chapter can be found in an excellent article written by Dr. John F. Guilmartin, Jr. for the *Journal for Maritime Research*, published in 2011.[1] Titled "The military revolution in warfare at sea during the early modern era: technological origins, operational outcomes, and strategic consequences," the paper was originally a presentation for a conference roundtable, and, despite being only 10 pages long, it is one of the more important pieces of military revolution historiography written in the last decade.

Admittedly the term "military revolution" is a contentious one, and can bring forth quite avid debate from all quarters of the historical profession. The term refers to a paradigm shift in the conduct of military affairs on land and sea c.1450 to c.1650 that generated new methods of warfare unique to Western Europe. The reciprocal relationship between increasing political centralization, economic growth, and both military technology and doctrine dictated changes which transformed Western European society on the grand scale. Europeans then projected this military supremacy beyond its shores, and grew from controlling only 7% of the global inhabitable landmass in 1500 to approximately 34% by 1750.[2]

Although these developments were first alluded to by C.W.C. Oman, the theory of a western European Military Revolution was first formally put forward by Michael Roberts in a 1955 lecture at Queen's University of Belfast that

1 John F. Guilmartin, Jr., "The military revolution in warfare at sea during the early modern era: technological origins, operational outcomes and strategic consequences," *Journal for Maritime Research* 13 no. 2 (November 2011), 129–137. Also: John F. Guilmartin, "The Military Revolution in Warfare at Sea During the Early Modern Era: Technological Origins, Operational Outcomes and Strategic Consequences," in *Revue d'Histoire Meritime: Marine, État et Politique*, no. 14 (Paris: Presses de l'université, Paris Sorbonne, 2011), 223–237.

2 A personal global survey of European empires as they existed in 1500 and 1750 using global land surveys generated 31% control of the global landmass with a 34% control of the global inhabitable landmass in 1750. This survey was done by estimating tangible control of claimed territories as opposed to European mapped territories, which often claimed land that was untouched by European presence or not realistically controlled given terrain or experienced local resistance measures.

 | DOI:10.1163/9789004393301_005

became the article "The Military Revolution, 1560–1660."[3] Roberts postulated that the changes instituted by Maurice of Orange and Gustavus Adolfus led to the rise of innovative tactics such as linear formations. Drill was emphasized and light field guns were adopted to boost mobility. Strategically, the scope of warfare increased with the growth of army size, which in turn fueled the centralization of state authority due in large part to the fiscal burdens necessary to sustain such unprecedented growth.

The first major challenge to Roberts' theory came from Geoffrey Parker, initially in the 1976 article "The Military Revolution, 1560–1660—A Myth?" and then in his 1988 book titled *The Military Revolution: Military innovation and the rise of the West, 1500–1800*.[4] While Parker agreed with Roberts insofar that there was an increasing impact of warfare upon society and that an augmentation of army size is a key factor in this debate, Parker pointed out that this surge in manpower occurred *before* the reforms of Maurice and Gustav Adolph, notably with the Spanish before 1560, and that their tactical innovations were not unprecedented. Their changes, therefore, could not have led to the Military Revolution. According to Parker, it was in fact the development of the *trace italienne* in response to the creation of effective cannon that became a driving force of change. The amassing of large armies and the construction of these complex fortresses not only led to the dominance of the siege over open battle and the ascendancy of infantry over cavalry, but, just as Roberts stated, it depended upon the growth of centralized government power capable of supporting the financial burden of these new circumstances. With the growth of this centralized military-fiscal state, technological innovation on both land and sea generated a feedback loop with social change that constituted the Military Revolution.

Subsequent historians have either debated with or built upon this basic premise. In 1991 Jeremy Black published *A Military Revolution? Military Change and European Society, 1550–1800*, arguing that the most important developments vis-à-vis European global expansion occurred after Parker's period of

3 Oman makes no overt claims but the groundwork for the later debate is present. C.W.C. Oman, *The Art of War in the Middle Ages* (Ithaca, NY: Cornell University Press, 1953). Michael Roberts, "The Military Revolution, 1560–1660," *Essays in Swedish History* (Minneapolis: University of Minnesota Press, 1967), Ch.1.

4 Geoffrey Parker, "The Military Revolution, 1560–1660—A Myth?," *Journal of Modern History*, 48 (1976). Geoffrey Parker, *The Military Revolution: Military Innovation and the Rise of the West* (Cambridge: Cambridge University Press, 1996).

consideration.[5] In Clifford J. Rogers' edited volume *The Military Revolution Debate*, Parker and Black reiterate their arguments in contributor chapters, while Rogers puts forth a new concept, that of punctuated equilibrium.[6] Borrowing from biological theories of species evolution, Rogers argued that such changes were not due to one specific revolution but rather a *series* of sequential military revolutions, each attempting to reverse a disequilibrium introduced by the previous one. Two revolutions, for example, occurred during the Hundred Years War (1340–1453)—an artillery revolution and an infantry revolution—and these two movements in concert laid the groundwork for subsequent early modern military changes. Historians MacGregor Knox and Williamson Murray followed this publication with their 2001 *The Dynamics of Military Revolution, 1300–2050*, which by contrast focused on the RMA—the revolution in military affairs—which involves changes in military affairs that exist as subsets of broader military revolutions.[7]

Certainly there are those who disagree with the existence of either one or multiple revolutions. Historians Kelly DeVries and Bert Hall were notable for their critique of Parker's *Military Revolution* in their 1990 review of the work for *Technology and Culture*.[8] DeVries' subsequent work in particular has focused on the very real and sometimes-problematic nature of the military revolution concept, given that it lasted for such a prolonged period and assumes a certain amount of stasis in the medieval era prior to this period of great change that in fact did not exist. Historians in non-western fields have also downplayed the idea of any western exceptionalism the military revolution concept might engender. Those such as Peter Lorge, Andew de la Garza, and Tonio Andrade have made arguments for Asian military revolutions prior to and/or as successful as what occurred in Europe.[9] This situation is particularly true of the Chinese Empire, Lorge and Andrade argue, as they were the first to successfully develop effective gunpowder weapons and seafaring vessels.

5 Jeremy Black, *A Military Revolution? Military Change and European Society, 1550–1800* (Atlantic Highlands, NJ : Humanities Press, 1991).

6 Clifford J. Rogers, ed., *The Military Revolution Debate: readings on the military transformation of early modern Europe* (Boulder: Westview Press, 1995).

7 William H. McNeill, *The Pursuit of Power: Technology, Armed Force and Society since AD 1000* (Chicago: University of Chicago Press, 1982).

8 Bert S. Hall and and DeVries, Kelly, "Essay Review – the 'Military Revolution' Revisited," *Technology and Culture*, 31 (1990), 500–507.

9 Peter A. Lorge, *The Asian Military Revolution: From Gunpowder to the Bomb* (Cambridge: Cambridge University Press, 2008); Andrew de la Garza, *The Mughal Empire at war: Babur, Akbar, and the Indian military revolution, 1500–1605* (London: Routledge, 2015). Tonio Andrade, *The Gunpowder Age: China, Military Innovation and the Rise of the West in World History* (Princeton: Princeton University Press, 2016).

By and large, however, most of the existing historiography is focused on military changes that occurred on land. Historians like Jan Glete, Carl Chipolla, Nicholas Rogers, and Guilmartin have addressed how important the changes in maritime technology were to this greater revolution, but their work has only scratched the surface of what should be a key component of the military revolution debate.[10] This point is the thrust of Guilmartin's article.

As he rightly states, despite the plethora of works on the military revolution on land, by and large, historians have ignored revolutionary changes happening at sea. Anyone who knew Guilmartin (Dr. G to his graduate students), however, knew that he was dedicated to the idea that pivotal changes had occurred on sea during this "revolutionary" period. He felt, moreover, that it was our job as historians of the debate to "explore the causal mechanisms behind developments in warfare and commerce at sea that had profoundly important consequences."[11] As he argued,

> between the turn of the sixteenth century and the final decades of the seventeenth, use of the sea for commerce and warfare by certain of the nations of Europe underwent a series of changes so profound in their effects and as far-reaching in their consequences as to merit labeling the process a military revolution at sea. The term is not a perfect fit, for changes in commerce and war were inextricably intertwined, but by and large military factors led the way, so military revolution will do. The changes in question were driven and shaped by a variety of forces and impulses ranging from economic through political and cultural and religious. That said, the impetus for change more often than not revolved around technological considerations, defining technology broadly as the application of an idea to achieve physical effect by means of an object, artifact, or thing.[12]

He goes on to state that Europe experienced four discreet military revolutions at sea that occurred either sequentially or in parallel, and "in the final analysis,

10 Jan Glete, *War and the State in Early Modern Europe: Spain, the Dutch Republic and Sweden as fiscal-military states, 1500–1660* (New York: Routledge Publishing, 2002). Carl Cipolla, *Guns, Sails, and Empires: Technological Innovation and the Early Phases of European Expansion 1400–1700* (Minerva Press, 1965). Nicholas A.M. Rodgers, *The Safeguard of the Sea: A Naval History of Britain, 660–1649* (New York: W.W. Norton and Co., 1997). John F. Guilmartin, Jr., *Gunpowder & Galleys: changing technology & Mediterranean warfare at sea in the 16th century* (Annapolis, MD: Naval Institute Press, 2003).

11 Guilmartin, "The military revolution in warfare at sea," 129.

12 Guilmartin, "The military revolution in warfare at sea," 130.

these combined in their effects to create the first global maritime empires, to remake the political and economic map of Europe, both in terms of national boundaries and in terms of international organization, and to create a global maritime economy run along European lines."[13]

Although the bulk of the paper goes on to discuss the impact of mounting gunpowder weapons on vessels of various types, he notes that the first of these revolutions involved the development of the fully-rigged ship. The transition in Europe from single-mast, coastally-bound ships with limited range to ocean-fairing, multi-masted, complex-rigged sailing ships is an essential one to understand, and yet, like the wider debate about the revolution at sea, has not received the attention that it is due. More specifically, if historians can understand why this change happened, it can, perhaps, illuminate the cause for the wider military revolution.

Technology, as Guilmartin argued in his contributing chapter to this volume and throughout his career, is essential to contemplate when trying to understand military affairs in any era. Moreover, technology is a product of the society that creates it, and functions according to that society's outlooks, aims, and practices. Therefore, technology changes as society changes, and vice versa. Given that technology provides this reflection, to understand why naval technology underwent a massive change in the 15th century—so that the single-mast ships of limited capability seen in the mid-1300s gave way to the oceanic vessels Christopher Columbus sailed to central America in 1492—one must look at equally massive changes to society in that era. Upon such examination, one potential catalyst stands out: the Black Death.

It should come as no surprise that the Black Death has been a favorite topic of medieval historians for decades. Not only have historians investigated every aspect of the plague itself, but researchers have contemplated its influence upon nearly every aspect of medieval European society. Contemplating its social impact, Jim Bolton in 1996 felt the Black Death created a "mass realignment within the peasantry" and shifted wealth to urban centers rather than rural estates.[14] Colin Platt argued that the plague was responsible for an increased concern with death and the afterlife, while Joseph P. Byrne believed that the plague broke down any remaining vestiges of feudalism and lowered standards for entry into the Church due to the high death rate of Latin-educated

13 Guilmartin, "The military revolution in warfare at sea," 130.

14 Jim Bolton, "'The World Upside Down.' Plague as an Agent of Economic and Social Change," in *The Black Death in England*, E.M. Ormrod and P.G. Lindley, eds. (Stamford: Paul Watkins, 1996), 45.

clergy.[15] David Herlihy postulated that the plague improved women's rights, increased the living standards of the lower classes, and ushered in bettered health standards and practices.[16] John Aberth in *The First Horseman* asserted that the Black Death led to the development of the printing press, promoted secular patronage in Renaissance art, increased wages for agricultural workers, and may have even laid the groundwork for the Reformation.[17] As he wrote, "no longer can we afford to write off Europe at the end of the Middle Ages as a wasted society waiting for the modern era to begin. Rather, Europe's rebirth was forged in the crucible of its terrible yet transcendent ordeal with the Black Death."[18]

Beyond my doctoral dissertation on this topic, virtually no historians of the Black Death yet contemplate the disease's impact upon warfare. Herlihy did, however, postulate that the pandemic revolutionized maritime travel.[19] As he argued, in order to maintain the economic boom taking place at the time, a labor-saving technology had to be invented that would allow bigger vessels to be sailed with less manpower. In other words, because of the demographic pressure imposed by the Black Death, the labor-saving device that is ship rigging was invented so that fewer sailors could operate the much-needed bigger vessels. This intriguing argument is supported by existing, although at the current time limited, evidence.

The first hurdle to studying a topic like this is terminology. We do know that medieval vessels can be grouped into roughly four developmental categories: "Viking," the "cog," the "hulc," and the "punt."[20] As Joe Flatman discusses in *Ships and Shipping in Medieval Manuscripts*, the differences in these types generally fall around hull style and structure, vessel propulsion, and steering.

15 Colin Platt, *King Death: the Black Death and its aftermath in late-medieval England* (London: University College London Press, 1996). Joseph P. Byrne, *The Black Death* (Westport, CT: Greenwood Press, 2004).

16 David Herlihy, *The Black Death and the Transformation of the West* (Cambridge: Harvard University Press, 1997).

17 John Aberth, *The First Horseman: Disease in Human History* (Upper Saddle River, NJ: Pearson Education, 2007).

18 Aberth, *The First Horseman*, 25.

19 Herlihy, *The Black Death and the Transformation of the West*, 51. My own work begins to do this, however: Sarah K Douglas, "The Price of Pestilence: England's response to the Black Death in the face of the Hundred Years War," (Unpublished PhD thesis, The Ohio State University, 2015). Sarah K Douglas, "The Black Death and Nation-State Wars of the Fourteenth Century: Environment, Epigenetics, Excess and Expiation, 1346–1450," *Epidemics and War: The Impact of Disease on Major Conflicts in History* (Prager: 2018).

20 Joe Flatman, *Ships and Shipping in Medieval Manuscripts* (London: The British Library, 2009), 69.

Yet it is difficult to tease out specific times and places where each of these construction traditions were employed, because development did not follow particularly rational paths, and one of the seemingly simplest aspects of ship identification is actually one of the most difficult: vocabulary. In medieval English records alone, over 60 terms are used to refer to various types of ships, and each region or maritime tradition would use similar, or conversely different, terms to describe either the same or different vessels, respectively.[21] The *Mary Rose*, for example, was an English royal vessel built in 1511 and sunk in 1545. Detailed records exist of its construction and maintenance, and the vessel itself was excavated from the bottom of the English Channel in 1982. Yet despite having records of its rigging components such as the "poleankers," "stryks," "trepgatelynes," "tragetes," "dryngs," and "sherwynes" naval archeologists still do not know what these items actually *were*.[22]

This issue brings us to the next problem, namely the information that can be gleaned from archeological discoveries of sunken ships. Such evidence is problematic, because in the medieval period vessels employing different levels of technology coexisted without much problem.[23] Some merchants might build new ships once developments emerged, others would just superimpose new technologies onto older vessel models, and still others would stick to more traditional designs for (most likely) cost reasons. Ships from different centuries could therefore coexist without much problem; other ships could even function as veritable floating museums of maritime technology over a given period because it was far easier to maintain an existing ship than undertake the immense cost of building a new one. In 1335, for example, King Edward III could only afford to maintain three royal ships, one of which he had bought from an English merchant for the enormous sum of £450.[24] There is a reason

21 Flatman, *Ships and Shipping*, 69–70.

22 Damien Sanders, "Knowing the Ropes: The Need to Record Ropes and Rigging on Wreck-Sites and Some Techniques for Doing So," *The International Journal of Nautical Archaeology* 39, no. 1 (2010): 4.

23 Flatman, *Ship and Shipping*, 70. Damien Sanders, "Knowing the Ropes: The Need to Record Ropes and Rigging on Wreck-Sites and Some Techniques for Doing So," 2–26. Although the article mostly focuses on methods of ship excavation, Sanders also emphasizes the importance of ship rigging and hull structure when attempting to identify sunken ships, even though the problems with this method, as explained, abound.

24 Jonathan Sumption, *The Hundred Years War I: Trial by Battle* (Philadelphia: University of Pennsylvania Press, 1990), 175. To provide scale, in 1346 nobleman Edmund de Bereford paid the English Crown a sum of £100 to compensate for not providing any soldiers for any of Edward III's campaign (which lasted 13 months) from any of his lands in Oxford, Berkshire, Warwick, Narthampton, Derby, Middlesex, and Cambridge. To add insult to

that in naval warfare, seizing enemy ships intact has generally been considered a favored outcome as opposed to sinking by ramming, burning, or shelling.

Of course, there are physical descriptions of vessels and records of tonnage and crew size, but even distinguishing specific tonnage of ships is problematic because there are six different ways to measure it in the modern context. There's gross registered tonnage, net registered tonnage, displacement tonnage, deadweight tonnage, measurement freight tonnage, and old registered tonnage. In the medieval world, however, we do not know how they calculated tonnage. As Susan Rose points out, in 15th century Venice they used tonnage formula that may have also been used in England (keel length x bredth x depth / 6). Yet in England, many records show that tonnage was often recorded *per estimacionem* (or by estimation).[25] As Rose also notes, a further problem was that tonnage was not objective, as at times merchants wished to either be lighter or heavier, bigger or smaller, than they actually were.[26] As such, tonnage must be considered carefully and comparatively.

Given these elementary difficulties, what can we say for certain? We know that during the largest pre-Black Death campaign undertaken by the English—the Crécy-Calais campaign in 1346–47 against the French in the opening phase of the 100 Years War—the army of 14,000 men, 5,200 horses, carts, weapons, and victuals was transported to France by 747 ships of 90 ton average, manned by an average of 36 crewmen per ship.[27] We know that all 747 ships sailed roughly at the same time and that the majority of them were merchant vessels impressed by the Crown into service. We know too that the campaign, which ended with the fall of Calais in August 1347, was a veritable success for the English, depending upon your view of their grand strategic design for the operation.

Then came the Great Mortality. Epidemiologists and biologists have confirmed that the Black Death was the work of the soil bacterium *Yersinia pestis*, typically carried by the fleas upon various rodents in areas like Uganda, Western Arabia, Kurdistan, Northern India, and the Gobi Desert. This "Great Pestilence" broke out in central Asia in the 1330s, made its way west into the Crimea in

---

injury, perhaps, two of these three ships were captured by the French in 1338, only to be recaptured by the English in 1340 (*Calendar of Fine Rolls, Edward III, 1337–1347* (London: His Majesty's Stationary Office, 1913–1923), 486).

25 Susan Rose, *Medieval Ships and Warfare* (Burlington, VT: Ashgate Publishing, Routledge Press, 2008), 16.

26 Rose, *Medieval Ships and Warfare*, 16.

27 Douglas, "The Price of Pestilence," 146–159. Craig L. Lambert, *Shipping the Medieval Military: English Maritime Logistics in the Fourteenth Century* (Rochester, NY: The Boydell Press, 2011), 139.

1346, Constantinople in early 1347, Cyprus and Sicily by the Autumn, Marseilles by January 1348, Paris in spring, followed by Germany, the Low Countries and Great Britain by the end of the year. This massive biological disaster spread at a pace of over 4 km/day, and by 1351 an estimated 50% of the European population was dead.[28] As one medieval chronicler in England noted, the plague left "not a city, a town, a village, or even, except rarely, a house, without killing most or all of the people there…As a result, there was such a shortage of people that there were hardly enough living to look after the sick and bury the dead."[29]

Surprisingly though, recovery was swift. Within 18 months most abandoned homes in many port towns in England were reoccupied, the King's Bench was hearing cases, and the Exchequer was back up and running. Trade companies resumed their commerce, churches resumed their practicing, and notably, violence against minority groups (most commonly Jews but not exclusively so) ceased to occur.[30] Overall, things seemed to be settling down. That is, until the Black Death returned in 1362, and over a dozen more times before 1415. Although some historians argue that European populations had begun a swift recovery in the decade after the Black Death, with a 20% average mortality rate during each successive plague outbreak, European populations stagnated for over 70 years.[31] As historian Elisabeth Carpentier notes, "the plague's impact can be seen properly only by viewing the disaster as a recurrent phenomenon, to be seen as a totality, and not by confining our attention to the first great epidemic of the mid-fourteenth century."[32]

That this severe and prolonged manpower shortage undoubtedly affected every sphere of life goes without saying. Not only did it make manual labor occupations highly lucrative, but it made labor in general a far scarcer commodity than it had been previously.[33] It also, I argue, contributed to the emergence of the fully rigged ship.

We can see proof of the pressure this labor shortage caused when examining the largest English military campaign in the decade after the Black Death: the 1359–60 Reims Campaign. King Edward III's goals for this campaign were

28 Douglas, "The Price of Pestilence," 2.

29 F.S. Haydon, ed., *Eulogium Historiarum sive Temporis*, Vol. 3 (London: 1863), 214.

30 Douglas, "The Price of Pestilence," 200–204.

31 Elisabeth Carpentier, "The Plague as a Recurrent Phenomenon," in *The Black Death: A Turning Point in History?*, William M. Bowsky, ed. (New York: Holt, Rinehart and Winston, 1971), 35–37.

32 Carpentier, "The Plague," 37.

33 David Farmer, "Prices and Wages, 1041–1350," in *The Agrarian History of England and Wales* Vol. I, II, Joan Thirsk, ed. (Cambridge: Cambridge University Press, 1988). Douglas, "The Price of Pestilence," 203.

nearly identical to that in 1346–47, although he employed 1/3 of the manpower, horses, and supplies he had in 1346–47. Moreover, because he could not simply raise soldiers from county arrays as he had done before the plague, this army was manned almost entirely by contracted retinues and fed with supplies purchased almost entirely in France.[34]

Yet, despite this smaller army and equally smaller supply load, transporting this army to the Continent proved a monumental challenge. Indeed, Edward III had to ship his army to France in six separate crossings between August and October 1359.[35] The medieval chronicle account of the campaign, the *Scalachronica*, noted that this was due to a lack of available ships, and historian Craig Lambert adds that the Crown also had difficulty finding sailors to man the ships they did have.[36] It is even argued that the time it took to get the English army to France caused operational delays—namely pushing the campaign into the winter months, which would prove quite harsh for the English—that in turn caused a lack of success that had been seen 13 years before.

More generally, research shows that while English merchant shipments in the early 14th century were usually done in ships of 150–250 tonnage, after the Black Death, ships even in the 150 ton range were rather rare. Although in the early 15th century the number of ships used by the English increased, and they certainly increased in size (200+ tons on average), their crews remained comparatively small.[37] In essence, because of the continued demographic pressure posed by the plague outbreaks, the maritime community was feeling the pressure to produce bigger ships that could be manned by smaller crews.

It seems no coincidence then that the earliest known record of a two-masted ship dates from 1410, which was a Genoese vessel captured by the English King Henry IV. It also seems no coincidence that it was found on a new type of vessel called the "carrack," *caravel*, or *karvel*, which was a mixture of all the effective elements of the previously-existing ship types. The Portuguese would then use the caravel structure to begin their forays to the western coast of Africa, and Portugal would eventually become the base for the three-masted,

34 Douglas, "The Price of Pestilence," 220.

35 TNA E101/393/11, mm. 81, 82d, 84, 85–85d. Lambert, *Shipping the Medieval Military*, 148.

36 Thomas Gray, "Scalacronica 1271–1363," Andy King, ed., trans., *The Publications of the Surtees Society* 209 (Rochester, NY: The Boydell Press, 2005), 171. Lambert, *Shipping the Medieval Military*.

37 Ian Friel, "Oars, Sails and Guns: The English and War at Sea, c.1200–c.1500," in *War at Sea in the Middle Ages and the Renaissance*, John B. Hattendorf and Richard W. Unger, eds. (Rochester, NY: The Boydell Press, 2003), 70.

full-rigged, carvel planked, skeleton-first built sailing ship capable of intercontinental voyages.[38]

One must ask, however: is this an example of correlation or causation? Did military and commercial demands in the face of demographic shortage caused by the 1348–1415 Black Death era lead to the development of ship rigging? Or perhaps did ship rigging emerge as a technology at the same time, and is correlation the true answer? Although certainly not monocausal, this author argues that it is indeed the former. Black Death historians have long supported the causal link between the pandemic and the emergence of labor-saving devices, and given both Europe's commercial and military expansion in the late 14th and 15th centuries, the emergence of a more efficient oceanic vessel seems inevitable.

It was this vessel that moved European goods around and beyond the Continent, and it was this vessel that exported European militarism across the globe. The Black Death era, therefore, laid the groundwork, not just for the emergence of the fully-rigged ship, but for the broader Military Revolution.

38 Friel, "Oars, Sails and Guns," 72.

CHAPTER 4

# The Lost World: Change and Continuity in Mughal Military Technology

*Andrew de la Garza*

Scholars who discuss the concept of a 'Military Revolution' in Europe—a process in which drastic changes in military technology and systems eventually led to a reorganization of the state and society—often use Asia and the Muslim world as counter-examples. They describe these regions as places where no comparable advances took place and where that absence would have lasting and negative consequences. There are several recurring arguments made against the existence of any comparable Asian military revolution. Many of these assertions involve technology and its implementation. Not without reason, technology and its change over time are central themes in the study of military history. The tools used to enable and wage war are integral to the success of armies and states. They represent not just creative or destructive power but the ability of leaders and nations to manage resources, pursue knowledge and better understand the world around them.

In studies of early modern warfare, Asian states were often found wanting in all of these areas, starting with the fundamental assumption that they were bound by tradition and slow to adopt new technology. Even when new technologies were adopted in the East, the prevailing explanation was that Asian governments and armies did not completely understand their utility and implications and thus failed to use them to their full potential. The persistence of old and 'outdated' technologies in Asia served as further evidence to support these claims.

In early modern India, however, the reality was much different. The founders of the Mughal Empire were not myopic traditionalists, but a product of the dynamic cultural and intellectual environment of the Timurid Renaissance in Central Asia. After the foundation of the Empire by Babur and his introduction of new military technology and tactics to India these tools and methods were soon adopted by its rivals, spreading virally, to the point where they became the standard in the region within two decades. The Mughals devoted a great deal of attention and study to firearms, artillery, and other devices. They embraced new technology instead of simply accepting it as a necessary evil. Babur, Akbar,

 | DOI:10.1163/9789004393301_006

and other early Mughal leaders were amateur engineers and scientists, and their hunger for new and better implements of war promoted the growth of a novel military system.

The resulting application of technology and tactics was often different than in Europe, but this situation was a result not of ignorance but of necessary adaptations to the unique military environment of India. 'Old' technology like bows and war elephants did persist in the Mughal military—because they were still useful and effective. These items were also used in new and different ways to complement more recent weapons and methods. The armies built from these components were an effective synthesis of technology and tactics from all over the world—Central and South Asia, the Middle East and Europe.

While the Mughals and their rivals adopted many inventions made by others, they were not simple imitators. They were informed and astute consumers of new technology who adapted these tools to the specific needs and challenges of their environment. Early modern South Asia was often described as a military 'Lost World,' like Arthur Conan Doyle's forgotten continent still inhabited by dinosaurs and other prehistoric creatures, where the ways of a bygone era persisted long past their natural lifespan.

In reality, the Mughal military system was the complete opposite. In many ways—an emphasis on small unit and combined arms tactics, the application of field fortifications, the use of camouflage and military rocketry, and an advanced understanding of logistics and support personnel including their integration into the regular army—the Mughals were anachronistic in the opposite direction, foreshadowing many developments from warfare in the 19th and 20th centuries.

In *Carnage and Culture*, his extended explication of the theory of a superior 'Western Way of War,' Victor Davis Hanson described an inherent strength in European armies that frustrated attempts by non-Western powers to gain and to maintain military supremacy. 'The Western way of war is so lethal precisely because it is so amoral—shackled rarely by concerns of ritual, tradition, religion or ethics, by anything other than military necessity.' By contrast he describes Muslim states, both their rulers and subjects, as slaves to tradition:

> The political power of the state was never separate from Islamic control. This general ubiquity of Muslim ideology had the effect of placing most commercial and intellectual life ostensibly under the auspices of the Koran. While Muslim scholars were able to create centers of religious

> teaching ... no real research in universities that might lead to military innovation, technological progress or an economic renaissance was possible.[1]

In early modern Central Asia and the surrounding areas influenced by military, political, and cultural movements coming out of that region—including India—the reality could not have been more different. Ziauddin Muhammad Babur, the founder of the Mughal Empire, was a Timurid, a descendant of the famous Central Asian conqueror Timur Leng. This lineage, however, implied more than simple family ties or presumed right to rule. Timur, despite his Turkish ancestry, was in many ways an exemplar of the theoretical 'Western Way of War' described by Hanson, John Keegan, William McNeill and others. He was notorious for his ruthlessness and his single-minded pursuit of military and political power, regardless of the material, moral, or human cost.

Yet, Timur's influence extended far beyond battles, campaigns and conquests. In an attempt to assert his legitimacy and to be taken seriously as a statesman as well as a military commander, he engaged in patronage of the arts and sciences on a massive scale. He acquired and commissioned works of art, built monuments, and retained a literal army of artists, scholars and scientists—as much by coercion and by theft as by free exchange. It is arguable just how much Timur internalized concepts of high art and culture, or whether he could ever truly be considered a man of learning and sophistication. But there could be no doubt that his descendants—who grew up surrounded by scholars and their works and who were taught their childhood lessons by acknowledged masters in many fields—were truly civilized, becoming not a new generation of nomadic chieftains, but the aristocracy of sedentary, civilized, highly urbanized states built around the Silk Road trading and manufacturing centers of Central Asia.

Those leaders would oversee the so-called 'Timurid Renaissance' in the 15th and early 16th centuries, a period of intellectual ferment and great advances in art, scholarship and science. Ironically it was the political failure of Timur's empire that facilitated this process. After the empire collapsed in a series of succession crises, a number of Timurid rulers were left in control of smaller, but still very wealthy, city states. These leaders competed with each other economically, politically, and militarily, but they also competed for status and prestige. This conflict led to an intellectual arms race in which kings and princes strove to hire the brightest minds and to create the most remarkable works

1 Victor Davis Hanson, *Carnage and Culture: Landmark Battles in the Rise of Western Culture* (New York: Anchor Books, 2001), 21 and 226.

of literature, art, and architecture. The essential context—a politically divided region at a nerve center of world trade—and results closely paralleled those of the Renaissance in progress at the same time in Italy and southern Europe.

Babur and the first Mughals were part of an extended family that included figures such as Ulugh Beg, the mathematician and astronomer whose work would become part of the foundation of the European Scientific Revolution, and Hussein Bayqara, famous for his patronage of literature in both Persian and his native Turkish. Like other Timurid aristocrats, Babur and his descendants received extensive education in the arts and sciences, including canonical works from both the Persian and Greco-Roman traditions.

The first Mughals, however, were also notable as 'Renaissance men' who produced groundbreaking work of their own.[2] Babur was a highly esteemed poet in Persian and Turkish. Humayun was a natural scientist as well as an amateur engineer. Akbar was apparently handicapped by dyslexia or a similar 'learning disability,' and could only read and write with great difficulty. Yet this did not stop him from becoming both an engineer and inventor, and an avant garde philosopher. Far from being a slave to Islamic tradition, Akbar often received harsh criticism from the Muslim clergy for his eclectic beliefs and his imposition of religious tolerance and harmony as state policy. Babur and his heirs also broke the mold in other ways—especially in their embrace of new forms of technology and organization. Nowhere was this more apparent than in their evolving understanding of warfare.

The military state of the art in northern India at the time of Babur's invasion emphasized traditional Central Asian cavalry, including mounted archers, supplemented by war elephants and rudimentary gunpowder weapons. Infantry was essential, for its cost effectiveness, if nothing else, but foot soldiers were more often used to occupy territory, and to consolidate gains than to hold the line in battle. Babur was successful in his campaigns because, with the help of Ottoman Turk and Persian advisors, he had developed a new military system that combined cavalry with effective infantry, field fortifications, and more powerful artillery and small arms fueled by corned gunpowder.

The Mughal plan of attack combined the strategic and operational offensive with the tactical defensive. Raids by Babur's Turkish and Mongol cavalry placed enough pressure on the enemy to make them offer battle, but when the fighting started most of the Mughal army retreated to the safety of entrenchments and wagons assembled as a mobile fortress. When the enemy attempted

2 For a more detailed exploration of the intellectual life of Babur and other Timurids, see Stephen Dale, *The Garden of the Eight Paradises: Babur and the Culture of Empire in Central Asia, Afghanistan and India, 1483–1530* (Boston: Brill, 2004), especially Chapters 3 and 5.

to dislodge them, they were punished by sustained artillery and small arms fire. Once the attack stalled, infantry and cavalry reserves counterattacked from the center, while light cavalry and horse archers on the flanks conducted the traditional Central Asian double envelopment maneuver, disrupting and then routing the enemy army. This basic plan allowed Babur to win decisive victories at Panipat and Khanua despite being outnumbered by a factor of 3 to 1 or greater.

Yet in less than two decades, all that he had accomplished was nearly lost. This reverse was not due to any inherent flaw in the new system or a revival of the old order. Instead, the Mughals were nearly overwhelmed by Sher Shah Suri, another aspiring warlord, who adopted and refined many elements of Babur's military system. When the Emperor Humayun and the Afghan usurper clashed, they both fielded thousands of musket-armed infantry and artillery parks with hundreds of cannon.

The new technology and tactics had already become the status quo in South Asia, adopted by the Mughals, their allies and their enemies alike. Groups as diverse as the Afghans, the Rajputs, the Deccani sultanates, and the petty kingdoms of central and eastern India quickly built armies based around the combined arms system of cavalry, archers—both mounted and on foot—heavy infantry, musketeers, and artillery. A new style of naval warfare based on cannon and purpose-built warships became the standard almost as quickly. The spread of new military concepts in the region was neither halting nor hindered by tradition. It was viral.[3]

While guns and other new technology spread far and wide in early modern Asia, some historians argue that non-Western leaders and states did not fully understand the tools that they were adopting, realize their true implications, or grant high esteem to the weapons and the people who made and used them. The resulting polities, using popular terminology coined by scholars like Marshall G.S. Hodgson and William McNeill, are often described as incurious and inefficient 'Gunpowder Empires.' As Thomas F. Arnold explains, 'what was critical was not just the physical presence of firearms and cannon, but their mental presence in the minds of Europe's ruling class.'[4]

Yet, in Mughal India gunpowder weapons were also part of the mental landscape, not just for emperors and generals, but for ordinary citizens as well.

---

3 The development of Mughal tactics and their impact in the early battles of the Empire are described in Chapter 2 and, in more detail, in Chapter 4 of Andrew de la Garza, *The Mughal Empire at War: Babur, Akbar and the Indian Military Revolution, 1500–1605* (London: Routledge, 2015).

4 Thomas F. Arnold, "War in Sixteenth Century Europe: Revolution and Renaissance," in *European Warfare 1453–1815*, ed, Jeremy Black (New York: St. Martin's, 1999), 38.

The first Mughal emperors, especially Akbar and his son Jahangir, were connoisseurs and obsessive collectors of muskets and of other small arms; people who would, in a later era, be referred to as 'gun nuts.' Gun culture and participation in hunting and other shooting sports soon became an essential part of the socialization of elites, both within the Empire and among its rivals. Guns were appreciated as weapons, as tools, and as works of art. Cannon cast in fanciful shapes to resemble lions or dragons were often the subject of ridicule by later Western observers, but the status of gunpowder weapons as objects of ostentatious display, pomp and circumstance indicates just how deeply both the Mughals and their Indian rivals appreciated these new devices.

This appreciation of guns was accompanied by a keen understanding of their practical value. As Akbar's chronicler Abu al-Fazl famously stated, firearms were 'wonderful locks for protecting the august edifice of the state; and befitting keys for the door of conquest.'[5]

> The Mughal army maintained rigorous standards for the manufacturing, quality and specifications of artillery and small arms. The construction and preparation of weapons for Akbar's elite troops shows an almost fanatical attention to detail: "When a (musket) barrel is completed lengthways, before the transverse bottom piece is fixed to it, they engrave on it the quantity of its iron and the length, both being expressed in numerals. In this imperfect state they are sent to His Majesty, and delivered, in proper order… At the same time, the weight of the ball (caliber) is fixed, and the order is given for the transverse section of the matchlock… When the barrels are polished, they are again sent … and preserved in proper order. They are afterwards taken out, and closed by the order of His Majesty with a transverse bottom piece. Having been put to an old stock, they are filled to one-third of the barrel with powder, and fired off. If no *tarawish* (failure) takes place, and the trial is satisfactory, they take the barrels again to His Majesty, who gives the order to finish the mouth piece of the barrel. After this the gun is again placed on the stock, and subjected to a trial. If the ball issues in a crooked line, the barrel is heated, and straightened by means of a rod introduced into it, and, in the presence of His Majesty, handed over to a filer. He adorns the outside of the barrel in various ways, according to orders… The wood and the shape of the stock are then determined … Several things are marked on every matchlock… the weight of the raw and the manufactured iron, the former marks being

5 Abu al-Fazl. *Ain-i-Akbari*, Trans. H. Blochmann (Los Altos, CA: Packard Humanities Institute, 2006. Online Edition), Volume 1, Chpt. 88.

> now removed, the place where the iron is taken from, the workman, the place where the gun is made, the date, its number."[6]

Soldiers at all ranks, from commissioned officers to privates, were expected to train in marksmanship and the handling of firearms. The growing role of muskets in military service led to a more egalitarian gun culture in which the musket became a symbol of self-sufficiency and personal freedom. Despite the government's efforts to limit their proliferation, the affordability and mechanical simplicity of firearms made them widely available to civilians. Musket-armed commoners defended their homes and communities as members of local militias and built careers and status for themselves as mercenaries or members of state armies.

From the perspective of many outsiders, however, all of this attention to firearms and their uses appeared misguided. The tactics for fighting with artillery and small arms in South Asia were often very different from the state of the art in Europe. Most notably, linear infantry tactics never developed in any meaningful way. Instead of using 'pike and shot' formations like the *tercio* or similar to later formations such as lines and columns, Indian soldiers moved across the battlefield in what appeared to be scattered, disorganized swarms. Some European accounts compared their movements to those of roaming animals or flocks of migratory birds. One French observer explained,

> These immense armies frequently perform great feats, but when thrown into confusion it is impossible to restore them to discipline… I could never see these soldiers, destitute of order and marching with the irregularity of a herd of animals, without reflecting upon the ease with which five and twenty thousand of our veterans from the Army of Flanders, commanded by Prince Conde or Marshall Turenne would overcome these armies, however numerous.[7]

Such behavior was understood to be the result of indiscipline or of lack of training, but most foreign observers did not fully understand the context of these actions. Compared to their Western counterparts, Indian battlefields were saturated by fire.

In addition to muskets and varied types of light and heavy cannon similar to those used in the West, there were many other hazards—archers, both mounted and on foot, as well as extremely mobile and dangerous light artillery like

6 *Ain-i-Akbari*, Volume 1, Chpt. 89.

7 Jean-Baptiste Tavernier, *Travels in India*, trans. V. Ball (Macmillan: London, 1889), 391.

rockets and 'camel guns.' The latter were small swivel guns or oversized muskets that were mounted on draft animals—camels or elephants, which could each carry two or more such weapons. These guns could fire projectiles the size of a golf or a tennis ball at nearly the same rate as ordinary muskets with far greater range and striking power. A sizeable field army might field several thousand camel guns and tens of thousands of rockets. Carlo Cipolla's assertion that 'Moslems never developed artillery into field weapons' does not apply to Mughal warfare, even if the devices involved did not always resemble their European counterparts.[8]

The resulting volume of fire made linear formations unsafe, despite their inherent advantages of coordination and concentrated firepower. Instead infantry was forced to disperse, using skirmishing tactics or fighting from within field fortifications. That led to the 'flocking' behavior described in contemporary accounts, with soldiers scattering under heavy fire and then gathering again to counterattack:

> Here in skirmishes and battles, one makes no use of pikes to prevent the cavalry from breaking in as usual in European warfare. In this country they do not keep ranks when marching, skirmishing or retreating as they do in the European quarters, but everyone attacks and strikes wherever one feels fit or best to harm the enemy… although they keep no fixed rule in their ranks, they take care that their troops or regiments, be it horse or foot, are not mixed up, and in attacking everyone keeps to his duty, taking little notice of one being in the advanced or rear guard.[9]

Shock action by concentrated heavy infantry or cavalry did take place, but only at opportune moments when an enemy formation had been disrupted by fire, or maneuvered into a vulnerable position. This combined action was a style of warfare that emphasized small unit tactics and initiative on the part of junior officers and individual soldiers. Contrary to conventional wisdom, these tactics demanded a greater degree of discipline and of initiative than fighting from within the shelter of dense formations, where combatants were always in the view of comrades, sergeants, and officers.

The persistence of supposedly outdated technology in Indian armies was also interpreted as evidence of their inferiority, but older weapons could still

---

8 Carlo Cipolla, *Guns Sails and Empires: Technological Innovation and the Early Phases of European Expansion, 1400–1700* (New York: Pantheon, 1966), 92.

9 From the journal of Geleynssen de Jongh, translated and excerpted by Jos Gommans in *Mughal Warfare* (New York: Routledge, 2002), 159–160.

be very effective in a new setting, even if their applications were different. One weapon that remained important in India until the start of the colonial era was the bow. In the West, bows and muskets were often seen as two species competing for the same ecological niche, with one destined to supplant the other. In Central and South Asia, however, bows and muskets evolved to complement each other.

Muskets came into wide use, because they were cheap to manufacture and usable even with limited training. They were also valuable because they could be used from behind cover, and from almost any stance, including a prone position. Muskets were ideal for sneaking and sniping, as a soldier could prime the weapon, wait indefinitely for the right moment, and then fire instantly when a target was in sight—something that could not be done with a bow held at full draw. A round musket ball lacked the inherent accuracy of an arrow, but Indian muskets were designed around this limitation, with a number of features that promoted accuracy. These enhancements included iron sights, integral bipods, recoil pads to reduce flinching, and slings that were used for steadying the weapon as well as carrying it (Illustration 4.1).

Muskets also had a great advantage in penetrating and stopping power. Musket balls were heavier than arrows, had a greater initial velocity, and deformed on impact, transferring much more energy to the target. They could easily pierce body armor at close range; they were much more likely to immediately kill or incapacitate enemy soldiers;, and, they could easily bring down larger

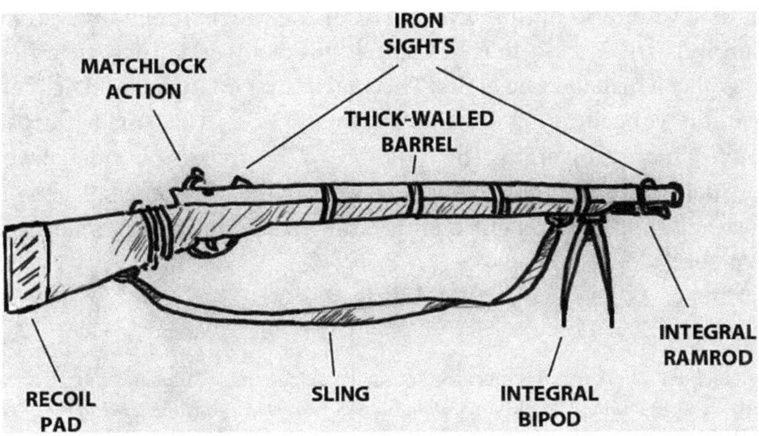


ILLUSTRATION 4.1 This diagram shows some of the design features of Indian muskets (although not all of them may have been present on a particular weapon). They indicate how these small arms were used differently than in Europe, with applications for fighting in open order or behind cover and an emphasis on marksmanship over rate of fire.

animals as well. For these reasons, they became the weapon of choice for hunting dangerous game—whether the targets were lions and tigers or warhorses and war elephants.[10]

Bows had one great disadvantage: unlike muskets, they required years of practice to use effectively in combat. Trained archers were a rare and valuable commodity in Europe, but they were omnipresent in South Asia, where both the traditions of Central Asian mounted bowmen and native Indian yeoman archers had shaped the military landscape for centuries. This relative glut of trained archers was a defining characteristic of Indian warfare and a primary reason why tactics there evolved so differently than in the West.

The bow did have one important advantage over the musket—rate of fire. A well trained musketeer could fire only two to three shots per minute, and this rate slowed even further if he was prone or inside restrictive cover. A trained archer could fire six or seven aimed shots every minute, and if needed he could sacrifice accuracy for speed and fire at the rate of almost an arrow every second. The bow was in effect the only 'select fire' weapon on a premodern battlefield, invaluable for delivering suppressive fire and for hindering the movement of enemy troops. Unlike a musket, with sufficient training, a bow could be readily used from horseback. Until the advent of rifle muskets and more effective anti-personnel artillery, the mounted archer was the most dangerous man on the battlefield not only due to his ability to move quickly, to fire aimed shots, or to deliver rapid fire—but to do those things simultaneously. The more limited early modern firearms did little to limit the power and utility of horse archers.

Foot archers were also an important part of the Mughal military system. It was in an infantry context that bows and muskets worked best together. Archers provided a high volume of fire. This meant that infantry could fight effectively without concentrating into linear formations. This also made the use of linear tactics more dangerous. The rapid fire of the archers provided cover for the more deliberate, but more punishing, work of the musketeers. The typical composition of a Mughal light infantry unit was about three archers for each musketeer.[11]

---

10 For a detailed discussion of the specifications and applications of Mughal-era small arms and artillery, see Iqtidar Alam Khan, *Gunpowder and Firearms: Warfare in Medieval India* (New Delhi: Oxford, 2004), especially Chapters 4 and 5.

11 Iqtidar Alam Khan argues that the Mughals used mixed formations of foot archers and musketeers. See Chapter 7 of *Gunpowder and Firearms*. According to the *Ain-i-Akbari*, only a quarter of each senior officer's infantry complement was actually composed of musketeers. Town militias and other irregular forces clearly used mixed formations. There are contemporary accounts of men and women fighting together during village uprisings,

Such a combined arms approach at the small unit level persisted in Central Asian armies even after the Mughal era. While 18th and 19th century European and North American armies focused on the standardization of small arms, with every soldier in each unit using the same model, the Ottomans and others utilized a combination of smaller caliber repeating rifles and longer ranged, harder hitting single shot models. Instead of engaging in the endless arguments over range and accuracy versus rate of fire, they, like the Mughals, adopted the best of both approaches.

Another ancient weapon, the war elephant, remained in use after the rise of the Mughals. Its application on the battlefield, however, changed drastically. In ancient and medieval combat, elephants were used in the line of battle. Their mass and thick hides—especially when reinforced with additional armor—made them highly resistant to attacks by edged weapons and arrows. In India, war elephants were often used in place of heavy infantry, to deliver shock action or to stiffen the line against enemy attacks. No cavalry could withstand an elephant charge, and only the most disciplined of infantry could hold their ground. Elephants were in many ways better adapted to working in tropical climates than heavily armored humans on foot, who had obvious difficulty enduring extended maneuvers and combat in hot weather.

Gunpowder weapons countered the elephant's natural advantages. Cannon and rocket fire could literally tear an elephant apart. Small arms were also lethal. The penetrating and stopping power of musket balls meant that a single musketeer could easily inflict fatal injury to an elephant. A squad of infantry with muskets could drop one in its tracks. Cavalry and human heavy infantry, who could more effectively use cover, concealment, and defilade while approaching the enemy, became a more practical means of delivering shock action.

For all of the above reasons, elephants lost their place on the front line, but they remained valuable in different roles. The development of 'camel guns' turned elephants into standoff weapons platforms, each capable of carrying several small cannon that had several times the effective range of muskets. Their size provided gunners with elevated positions from which they could easily see and engage distant targets. This advantage in visibility also made elephants popular as mobile observation platforms for commanders and senior officers. Elephants still played a limited role as direct combatants, but this usually happened when an army attempted to contain a break in their lines—or to exploit a breakthrough of their own. They were also used as living

the men using muskets and the women carrying bows. See Niccolo Manucci, *Storia do Mogor, or Mogul India*, trans. William Irvine (London: Asiatic Society of Bengal, 1907), 134.

tractors and bulldozers, transporting heavy equipment and breaking down obstacles and enemy fortifications.

The development of the *trace italienne* or 'star' fort was a distinguishing feature of the European Military Revolution. The distinctive geometric outline of these fortresses allowed for interlocking fields of fire by defensive guns, while their concentric earthworks and low profile made them much more resistant to enemy artillery than the relatively fragile medieval castles with their higher, thinner stone walls.

The near total absence of such structures in early modern India is often interpreted as yet another example of the backwardness of military practice in the region. In reality, there had already been a revolution in the design of South Asian fortresses. That development took place in the 14th century during the later dynasties of the Delhi Sultanate, when fortifications were reshaped in response to the introduction of primitive gunpowder weapons and the more powerful *maghribi*, or 'western,' trebuchet from the Middle East and Europe. The resulting structures were massively over-engineered, with concentric layers of lower, thicker stone walls that each often exceeded 15 meters in width (Illustration 4.2). This architecture was a brute force solution, not as elegant as and more expensive than *trace italienne* forts. Yet Indian states and rulers were wealthy enough to afford these fortresses, and such defenses were adequate for their needs. These buildings proved remarkably resistant to artillery, even the more advanced weapons introduced during the Mughal era, and they could be easily retro-fitted to mount their own defensive guns. Indian forts remained

ILLUSTRATION 4.2 The fortress at Tughlaqabad (left) built in 1321 is an example of what was then a new style of fortification based on concentric layers of lower, thicker stone walls. These forts were developed in response to the introduction of more powerful trebuchets and primitive gunpowder weapons, but the design proved quite resistant to cannon and was a model for Mughal-era artillery fortresses like the Red Fort in Agra, completed in 1573 (right). This limited the incentive to adopt the newer *trace italienne* style.

serious obstacles well into the colonial era, withstanding sustained fire from even the artillery of 19th century armies.[12]

Other technologies more commonly associated with the Middle Ages persisted in early modern South Asia. Body armor remained in wide use into the 18th century, long after it had largely disappeared in Europe, outside of a few specialized cavalry applications. While armor had limited utility against firearms, it was still very important as protection from arrows. The persistence of archers as a threat on Indian battlefields meant that armor was still a valuable and practical defense. Medieval siege artillery also remained in limited use. Weapons like trebuchets were not as powerful as cannon, but they still had some important advantages. With their high, arcing trajectories they could provide indirect fire in the same fashion as mortars, and their lower 'muzzle velocities' meant that they could hurl large gunpowder bombs or canisters of incendiary or caustic chemicals; projectiles too fragile to be launched from mortars or cannon. Catapults were also more affordable and portable than conventional artillery. If an adequate supply of lumber was available on site, only the metal fittings and cordage required transport with the rest of the weapon being assembled as needed. For these reasons, such 'outdated' weapons were used in sieges into the 17th century.[13]

Naval warfare in the region also appeared to be archaic. The Mughal Empire and other Indian states adopted technology and tactics very similar to those used by Mediterranean navies in the 15th and 16th centuries, fielding large fleets of war galleys long after most European powers had transitioned to 'blue water' navies of galleons, frigates and ships of the line. There were technical limitations that shaped this policy. The less advanced design of the blast furnaces in general use in India and the lower quality of Indian coal made the large scale manufacture of cast iron guns difficult. Cheaper cast iron guns were essential in the introduction of large warships with broadside batteries in Europe. A blue water navy armed with all bronze cannon would have been unpleasantly expensive, even for the Mughals.

Yet the Mughals had other, more compelling reasons to stay with the older system. Unlike rising Western states, the Empire had no need of or desire for overseas colonies. The vast wealth and diverse resources of India made the

---

12 For more detailed discussion of the construction of medieval and early modern Indian fortresses, see Sidney Toy, *The Fortified Cities of India* (London: Heinemann, 1965) and Konstantin Nossov, *Indian Castles 1206–1526: The Rise and Fall of the Delhi Sultanate* (New York: Osprey, 2006).

13 Trebuchets were still seeing limited use in sieges at least as late as the Mughal invasion of Ahmednagar in the early 17th century. See H.M. Elliot, ed., *The History of India as Told by its Own Historians,* Volume 6, Chpt. 80 (*Akbarnama* of Shaikh Illahdád).

Mughals—and many of their rivals—largely self-sufficient. They required few resources from abroad. Their most significant deficit was in precious metals; South Asia had only meager natural deposits of gold and silver. This lack was made good simply by keeping the ports open and allowing the merchants of Europe, Africa, and the Middle East to bring hard currency in exchange for Indian goods, both raw materials and manufactured items.

In this context a galley-based navy that emphasized defense of home ports and coastal waters made perfect sense. Like China, which found itself in a similar strategic and economic position, the Mughals understood a blue water fleet to be an expensive luxury, not a necessity. There was no need for raids on the Spanish Main or blockades of the Channel ports. The Empire had a more powerful weapon. It effectively used control of the Indian trade to reward or to coerce European states, and if need be, play them off against each other. Seemingly outdated weapons and technology remained in use in India, both on land and at sea, not due to a lack of will or imagination but because those devices were still suited the demands of that unique environment. In response to the specific challenges and opportunities of their setting, the Mughals and other South Asian states also used older tools in new and innovative applications.[14]

Warfare in early modern South Asia was anachronistic in many ways; not because it was trapped in the past, but because it anticipated a number of important military developments in the West and the rest of the world in the 19th and 20th centuries. These advances included changes both in how battles were fought and how armies operated behind the scenes. The unique hazards of the Indian battlefield led to an emphasis on the use of small unit tactics, the use and avoidance of antipersonnel weapons, field fortifications, and camouflage. The Mughals also invested greatly in logistics and support operations, closely integrating them into their military and government organization. They used uniformed military personnel to carry out these functions in a manner more similar to the standard practice of modern armies than to that of their early modern contemporaries.

While cavalry troopers and units often wore distinctive and color-coded uniforms, the nature of infantry combat in Mughal India placed a premium on foot soldiers being as inconspicuous as possible. It is common knowledge that the English word 'khaki' is derived from *khak*, the word for 'earth' in Persian, Urdu, and related languages. Khaki clothing originated from a change in military uniforms that started in India and other European colonies during the late

14 Mughal naval tactics and strategy are discussed in more detail in Andrew de la Garza, "Command of the Coast: The Mughal Navy and Regional Strategy," *World History Connected* 12, no. 1 (February 2015).

19th century, one that reflected a change in how soldiers fought. While brightly colored uniforms were practical in an earlier era of linear warfare and dense formations, they became a serious handicap in colonial guerrilla wars, or even in pitched battles between European armies, where the increasing firepower and lethality of modern weapons made cavalry obsolete and forced infantry to disperse into smaller and smaller units.

Newer uniforms were designed with concealment in mind, using earth tones and muted grays and greens. Both the Mughals and their rivals had come to a similar conclusion centuries earlier, as the volume of both small arms and artillery fire on Indian battlefields increased. Infantry typically wore drab civilian clothes, and many of them probably used more advanced camouflage. Camouflage clothing was widely used by hunters in early modern South Asia—not just solid greens and earth tones but complex embroidered and block printed patterns that resembled modern military and sporting apparel. Some hunters, like modern military snipers, also used textured 'ghillie suits' that mimicked clumps of vegetation. The military applications of such clothing were obvious, especially in a culture that understood hunting as not just as a pursuit of sustenance or recreation, but as a way of training for war.[15] The widespread use of camouflage, small unit tactics, and ever more complex field fortifications—the wagon laagers of Babur's day gradually gave way to networks of trenches, foxholes, and sandbagged fighting positions as gunpowder weapons became more numerous and deadly—created scenes that would have been more familiar to soldiers from the 20th century than to their early modern counterparts.

Another military technology that can be traced back to India is rocketry. Rockets, along with other gunpowder weapons, originated in China and were introduced to South Asia in the late medieval era. The Mughals and other Indian armies added a number of refinements to the basic Chinese model including: metal casings that fragmented into lethal shrapnel on impact; blades mounted on the forward end that may have balanced the rocket in flight and allowed it to 'stick' to or near a target after an accurate shot even if the fuse was still burning; whistles designed to frighten enemy cavalry horses and elephants; and, tubular hand-held launchers that were primitive versions of the modern bazooka.

Rockets were inaccurate, but they were also inexpensive, numerous, and highly portable. A single soldier could carry several at a time. A horse could

15 The use of camouflage in hunting is depicted in many Mughal-era artworks. For examples, see Stuart Welch, *India: Art and Culture, 1300–1900* (New York: Metropolitan Museum, 1985), 268 and Andrew Topsfield, *Paintings from Mughal India* (London: Oxford, 2008), 117.

carry dozens, and an elephant or wagon could carry hundreds. The destructive power of rockets combined with their superior mobility and flexibility compared to other forms of artillery was one of the factors that made Indian battlefields so dangerous, and caused tactics to evolve differently from the state of the art in Europe.

At the onset of the colonial era, rockets used by native armies were an unpleasant and deadly surprise for the British. Recognizing the utility of this weapon, the British eventually developed their own versions, starting with the famous Congreve rocket. Other Western powers eventually followed suit. Indian rockets from the Mughal era were essentially the distant ancestors of all of the rockets and missiles used by modern militaries up to the present day.[16]

The most strikingly anachronistic aspect of Mughal warfare, however, was not its weapons but its organization. Critics of the Mughal military system frequently described it as cumbersome and inefficient, emphasizing the deliberate pace of its operations and the large numbers of camp followers and supposedly surplus personnel who accompanied armies on campaign. These assumptions, however, were not well grounded in fact. The Imperial war machine was often slow to move, but it moved with a purpose. Caution and attention to detail, along with the policy of risk aversion that developed in response to the growing violence and cost of battle in the new style of warfare, not a lack of capability, dictated Mughal armies' slow and steady pace. As Stephen Peter Rosen (a historian who is otherwise skeptical of the capabilities of the Mughal military) explained, their abilities in support compared to other acknowledged masters of logistics in warfare. "The Mughal army adopted military practices that have been associated with a 'Roman' or 'American' way of war—a brute force, engineering approach to military affairs."[17] Most of the vast number of 'camp followers' who marched with the Mughal army were skilled and well-paid professionals—pioneers, porters, animal handlers, craftsmen, cooks, clerks, physicians, engineers, policemen, couriers, scouts, and spies—who were vital to the army's success. The result was an all-weather, all-season army that could work and fight in any terrain from the mountains and steppes of Central Asia to the jungle and rainforests of the southern and eastern frontiers.

---

16 A detailed discussion of Indian rocketry can be found in Khan, *Gunpowder and Firearms*, 22–30.
The use of primitive rocket launchers is described by W.G. Orr in "Armed Religious Ascetics in Northern India," *Warfare and Weaponry in South Asia*, Ed. Jos Gommans and Dirk Kolff (New Delhi: Oxford, 2001), 185–201.

17 Stephen Peter Rosen, *Societies and Military Power: India and its Armies* (Ithaca: Cornell University, 1996), 150. Mughal logistics are discussed in more detail in Chapter 6 of *The Mughal Empire at War*.

Like other early modern armies, the Mughal army did business with a wide variety of private individuals and enterprises in order to fill all of these roles. Yet compared to their European counterparts, the Mughals relied much less on private contractors to support their military. Most of the responsibility for logistics and other support functions remained with uniformed military personnel.

Logistics and non-combat operations were more tightly integrated into the organization of the Mughal military than they were in most other early modern armies. Once again, this behavior was more reminiscent of the late 19th and 20th centuries than the era of the European Military Revolution. Quartermasters and managers in other 'back office' capacities like manufacturing, engineering, medicine and personnel were typically commissioned officers. In the enlisted ranks, porters, pioneers, and other support workers were often paid on the same pay scales used for the infantry and listed on the army's muster rolls. The civilian and military administrations were closely interconnected in the Mughal Empire, with high ranking government officials holding both civil service and military ranks and titles.

The Mughal army did not have a General Staff in the modern sense, but its leaders were keenly aware that the requirements of warfare went far beyond the soldiers and weapons that were present on the battlefield. A vast network of farms, factories, stockpiles, and supply depots supported the Mughal military. Both people and objects were obsessively counted, tracked, and measured. Officers, cavalry, infantry, and sailors were all organized by detailed, numeric rank and pay scales. The army kept detailed muster rolls and rosters for both human soldiers and working animals, sorting them by name, number, rank or breed, age, physical characteristics, and identifying marks. Weapons and equipment were also counted and sorted by type and by place and date of manufacture. Some of the muskets and other firearms used by Mughal troops were marked with and tracked by serial numbers, one of the earliest known examples of this practice.

In their approach to both combat operations and support functions, the Mughal army—and those of many of its rivals in South Asia—were as advanced as any in the world. Through trial and error and by rational design and inquiry they evolved to fit the unique circumstances and challenges of their setting. In many ways, their practice differed from accepted standards in Europe and elsewhere, but this variance was the result of practical considerations and calculations, not ignorance. In other ways their tools and methods were very similar to those used in the West, but they were informed consumers, not blind imitators.

On a much larger scale, the Mughal military system is a starting point for further assessment and reassessment of not just the Empire as a whole but of how political, social and intellectual establishments functioned in early modern South Asia and elsewhere in the Muslim world. It should be readily apparent that India was in reality fully engaged with the early modern world and not left behind in an extended medieval era. Both the Mughals and their rivals in the region proved this connection by their conduct of warfare; the ability to manage complex systems, to learn, to rationalize, and to integrate both old and new technology and doctrine from diverse sources, and to reshape them to meet their own specific challenges and objectives. The states and armies that emerged from this process were not without flaws. Tradition and orthodoxy could not always be ignored, and progress was often made not by inspiration but by harsh lessons learned.

The Mughal Empire—like its contemporaries in Europe—was not a truly modern state, but it was moving towards modernity, towards the future and away from the past. The Mughals and Indians were not unique in this respect. The rise of powers like the Ottoman and Safavid Empires profoundly influenced trade, government, and warfare in the West. The Military Revolution in Europe was not driven just by the competition of Western states with each other. It was also a response to the military, economic, and political power of Asian and Muslim states and empires. Even after the fall of the Mughal Empire, the military tactics and organization of the Mughal era profoundly influenced battles fought on the frontiers between Asia and Europe.

While many post-Mughal states in India were unable to maintain existing systems, much less adapt them to changing conditions, more capable polities like the Maratha Confederacy gained superiority over their rivals with military tactics and technology of remarkable sophistication. When British invaders confronted the Marathas, the British found themselves facing an enemy in possession of weapons and equipment equal to those of European armies, and employing techniques that were in some ways advanced beyond their own. The Marathas used small unit tactics, fire and maneuver, effective field fortifications, anti-personnel weaponry, and interlocking fields of fire in applications that would seem more familiar to military observers of the 20th century than officers of an *ancien regime* or Napoleonic army. Yet, these advances were inspired by the lessons of the past as the Marathas applied the innovations of the Mughal system to the technology and circumstances of their time. Randolf Cooper, in his groundbreaking history of the Anglo-Maratha wars, notes: 'Western historians are uncomfortable with the idea of South Asian leadership in military science, but that is what Maratha doctrine represented.'[18] In an earlier era, the same could be said of the Mughals.

18 Randolf G.S. Cooper, *The Anglo-Maratha Campaigns and the Contest for India* (New York: Cambridge University, 2003), 293.

# PART 2

## *19th Century Warfare*

CHAPTER 5

# If You Can Be Seen, You Can Be Killed: The Technological Increase in Killing Zone during the American Civil War

*Edward B. McCaul, Jr.*

Today, it is a military maxim that if you can be seen on the battlefield, you can be killed. The capability to kill anyone who can be seen on the battlefield first occurred during the American Civil War. For a Soldier, armed with a rifled musket, combat range increased from 50 to 350 yards.[1] The previous infantry weapon, a smoothbore musket, was so inaccurate that Ulysses S. Grant stated in his autobiography, concerning his experiences during the Mexican American War, that "At a distance of a few hundred yards a man might fire at you all day without your finding it out".[2] Artillery had a similar increase in its combat range. For field artillery, the maximum range for case shot increased from 1680 yards (9/10ths of a mile) for an 1857 12-pounder field gun to 4180 yards (almost 2½ miles) for a 3-inch rifled gun.[3] For siege artillery, the maximum range for a 24-pounder smoothbore iron howitzer firing a 17-pound shell was 1322 yards (3/4ths of a mile) while a rifled 20-pounder Parrott could fire an 18 ¼-pound shell 4400 yards (2½ miles).[4] This exponential expansion of the killing zone, beyond what a man could see without the aid of binoculars or a telescope, which made no one safe on the battlefield, was a direct result of new technology. Technology had made the battlefield a much more deadly place where to be seen was fatal.

The advantages of rifling had been well known for years, but the military showed little interest in it. This lack of interest was due to the fact that for rifling to be effective the projectile had to be the same diameter as the groove

1 The effective ranges of a smoothbore musket and a rifled musket came from the "Flintlock Smoothbore Musket" TVT 21–82 and ".58 Rifled Musket" TVT 21–84 videos done by the Audio Visual Instructional Technology Division United States Military Academy.

2 U.S. Grant, *Personal Memoirs of U.S. Grant*, Volume I (Charles L. Webster & Company, 1885), 95.

3 Joseph Roberts, *The Hand-Book of Artillery* (New York, NY: D. Van Nostrand, 1863), Reprinted, Decatur, MI: Invictus, 1999, 73 and 166.

4 Roberts, *The Hand-Book of Artillery*, 75 and 171.

 DOI:10.1163/9789004393301_007

diameter of the weapon when it was fired, but it could not be greater than the bore diameter when the projectile was being rammed home. Limited use had been made of rifled muskets, but they were only used by a few selected units. Kentucky and Pennsylvania long rifles were famous for their accuracy and long range, with their combat range being slightly three times greater than that of a smooth bore musket. However, these rifles were not acceptable military weapons. The problems were that these rifles had a substantially smaller bore than the military smooth bore musket, had a rate of fire of about 1½ rounds per minute compared to a rate of fire of 4 rounds per minute for a military smooth bore musket, and did not have a bayonet.[5] In addition, it took longer to train a man to properly fire a rifled musket than it did to train him to fire a smooth bore musket. There were rifles that the American and British armies used such as the Hall Breech Loading Rifle and the Ferguson Breech Loading Rifle. The problems with these rifles were that they were more complicated to manufacture, required more training of the men, were never fully accepted by the generals, and were, at least initially for the Hall rifle, still flintlock. The major problem was how to manufacture an adequate number of weapons quickly enough to re-equip a large army when each weapon was handmade and the parts were not interchangeable. This problem was solved by John Hall at the Harpers Ferry Armory.

The ability to manufacture a large number of weapons with interchangeable parts was realized in 1834 when identical weapons were manufactured by machine tools at the Harpers Ferry Armory and Simeon North's factory in Connecticut. This was a milestone as it was the solution to a problem that the military had been trying to solve for years. However, it took some time for this technological advance to spread within the United States and Europe. The first real indication the Europeans had of this technological breakthrough was during the 1851 London Crystal Palace Exposition. Once the European military saw how weapons could be quickly produced with interchangeable parts by machine tools and gauges, they began to order large numbers of machine tools from the United States. As a result, the armies throughout Europe and North America replaced their smoothbore muskets with rifled muskets. In the United States this change began to take place in 1855 under Secretary of War Jefferson Davis.[6] This new ability to rapidly make interchangeable parts for rifles made

5 The information on the comparison between a Kentucky and Pennsylvania long rifle and a military smooth bore musket came from "The Kentucky Rifle in the American Revolution" TVT 21–83 done by the Audio Visual Instructional Technology Division United States Military Academy.

6 David C. Clark, ed., *Arms for the Nation* (Greensburg, PA: South Greensburg Printing Co., Inc., 1994), 24.

it possible to arm the large armies of the American Civil War. While European armies had large stocks of muskets, the United States did not, and without the industrial capability to make large quantities of rifles the American Civil War would have been fought by much smaller armies.

The problem of quickly loading a bullet in a muzzle-loading rifle was solved by French Army officers who invented a conical shaped lead bullet with a hollow base and a wedge shaped iron plug. When the rifle was fired the expanding gases forced the plug into the bullet causing it to expand to groove diameter. This type of bullet became known in the United States as the Minié Ball named after Captain Claude-Étienne Minié, one of the men responsible for the development of this bullet. Beyond a substantial difference in combat range, Minié balls could inflict more damage than the round musket ball. The British Tower Musket, better known as the Brown Bess, was .75 caliber and fired a one ounce .72 inch diameter musket ball. Americans designed their smooth bore musket after the French Charleville musket which was a .69 caliber and fired a musket ball that was slightly smaller in diameter and weighed less than an ounce. When rifled muskets were introduced it was found that a .69 caliber rifle could fire a Minié ball that had a diameter of .69 inches.[7] A .69 inch Minié ball weighs approximately 2.2 ounces, over twice the weight of the musket ball fired by the British. This allowed the diameter of rifles used by the United States to be reduced to .58 caliber. Even then, a .58 caliber Minié ball weighed approximately 1.5 ounces, which was still heavier than the old British musket ball.[8] In addition, due to the fact that a Minié ball sealed the barrel and did not allow the expanding gas from the burning black powder to go around it, Minié balls had an initial velocity of about 1,000 feet per second compared to an initial velocity of about 900 feet per second for a musket ball fired from a smoothbore musket.[9] A higher velocity, a greater effective range, a more aerodynamic projectile, and a heavier projectile made rifled muskets firing a Minié ball more deadly by substantially increasing the killing zone for the individual soldier and by additionally inflicting more lethal injuries.

One other item of technology that made Civil War era muskets more reliable, but did not increase their killing power or their rate of fire, was the percussion cap. Prior to the percussion cap the black powder charge in the musket was ignited by a piece of flint striking a metal frizzen. This method had

7 One caliber equals one inch. Thus, a .69 caliber minie ball was .69 inches in diameter.

8 The weight of a .69 caliber minie ball and a .58 caliber minie ball was determined by weighing examples of both from the author's collection.

9 The initial velocities from a smoothbore musket and a rifled musket came from the ".58 Rifled Musket" TVT 21–84 video done by the Audio Visual Instructional Technology Division United States Military Academy.

obvious drawbacks, such as not working during rainy weather. While inventors had been working on a number of replacement systems, the solution was the percussion ignition system invented in 1807 by Scottish minister Alexander Forsyth. The final step needed to complete the percussion system was when in 1814 American Joshua Shaw invented a copper thimble cap into which was placed mercury fulminate. However, the American Army did not begin producing percussion cap muskets until 1841, and continued producing flintlock muskets until 1844.[10]

While there are many famous incidents in which individuals were killed at long range by a rifled musket, two stand out.[11] During the Mexican American War, Company E, 1st Dragoons had been issued some percussion Model 1841 Hall rifles. At Buena Vista, their commander, Captain Steen, saw a Mexican officer, about 400 yards away, observing the unit. He asked that his company's best shots fire at the officer. The soldiers did and successfully killed the officer.[12] General John Sedgwick's death is another example. During the Battle of Spotsylvania, General Segwick was observing the placement of his artillery and troops when Confederate infantry started a long range fire at them. No one was being hit, but some of the Union soldiers were dodging when they heard a bullet. General Sedgwick chided these men by saying "They couldn't hit an elephant at this distance". Shortly after he made that statement, he was struck in the head by a bullet, instantly killing him.[13] While no one knows who actually killed General Sedgwick nor how far away the shooter was, it has been estimated that the shooter was at least 500 yards away, too far for anyone to be sure of who they were aiming at, especially considering the crude optics that were in use at that time. Both examples are shots that a smooth bore musket was incapable of achieving. The deadly zone for infantry weapons had increased beyond the distance in which a man could determine who he was firing at.

While the rifled musket increased the killing zone, rifled artillery increased it even more. However, it was not only an increase in range that made rifled artillery more deadly, but the adoption of new types of fuzes. However, the advent of rifled artillery did not increase the number of projectiles, which still

10 The information about the history of the percussion cap and United States production of flintlock and percussion muskets came from *Arms for the Nation* edited by David C. Clark (S Greensburg, PA: South Greensburg Printing Co., Inc., 1994), 13–19.

11 While rifling also increased the combat range for carbines, the extreme long range firing was done by the rifles with their longer barrels.

12 John D. McAulay, *Rifles of the U.S. Army* (Lincoln, RI: Andrew Mowbray Publishers, 2003) 10–11.

13 The information on the death of General John Sedgwick came from Martin T. McMahon's "The Death of General John Sedgwick," *Battles and Leaders of the Civil War: Retreat with Honor*, Volume IV (Secausus, NJ: Castle, 1983), 175.

included: canister, solid shot, case shot, and shell. Canister made an artillery piece into a large shotgun and was only used for close-in defense. Solid shot, as indicated by its name, was a solid piece of metal. Smooth bore artillery had one advantage over rifled artillery firing solid shot, in that a spherical round would bounce a number of times after it hit the ground, while the best that could be expected from a cylindrical shaped projectile was that it would skid across the ground. Case shot, or shrapnel, was invented by Lieutenant Henry Shrapnel of the British Army. It was a thin-walled projectile filled with "musket balls" and a small bursting charge. The projectile needed a time fuze to work, as to be effective it had to "explode" over and in front of the target.[14] However, due to its small bursting charge, case shot was only effective if the fuze was properly timed. If it was not, and the case shot projectile was on the ground, its killing radius was extremely limited. Shells were thick-walled projectiles filled with gunpowder. They were equally deadly in the air or on the ground.

While the principle behind the invention of the Minie ball was also used by artillery, there was another problem that had to be solved before rifled artillery could become a reality. Artillery's problem was compounded by the fact that a projectile which would expand to groove diameter placed a large amount of pressure on the artillery tube and would frequently cause the tube to explode. This problem was eventually solved by a better understanding of metallurgy, reinforcement of the tubes, a new process of cooling a tube from the inside out developed by Captain Thomas J. Rodman, and, eventually, by the development of commercially available steel.

Rifled artillery created the need for new fuzes for two reasons. First, when a rifled projectile was fired the projectile quickly sealed the tube of the gun, stopping the flames from the propelling charge from reaching the fuze. However, some flames did get through, so the failure rate of flame-ignited time fuzes in rifled artillery was greater than in smooth bore artillery. Second, when a spherical projectile was fired, it was impossible to determine what part of the projectile would hit first. In a cylindrical rifled projectile that took the rifling, the front nose of the projectile would strike first. Knowing this allowed inventors to develop percussion fuzes that were activated by the set back force that occurred when the projectile was fired, and which would transmit a flame to the charge inside the projectile when it struck the ground. Fuzes became so important that by the time of the Civil War they had become the soul of modern artillery.[15]

14 Case shot needed time fuzes until the advent of the variable time fuze in World War II.

15 The phrase "soul of modern artillery" comes from Charles Bormann, *The Shrapnel Shell in England and in Belgium* 2nd ed. (Brussels: Louis Truyts, 1862), 15.

There were two types of mechanical fuzes during the Civil War – concussion and percussion. Percussion fuzes were the most common as concussion fuzes did not work very well. Percussion fuzes made rifled artillery deadly at long ranges since the projectile would explode upon impact, assuming it hit nose first, and break into a number of flying fragments. These projectiles were effective at ranges beyond that possible for case shot due to limitations on the number of seconds a time fuze would burn.[16] Consequently, shells were the principal exploding projectile used for extreme ranges and were mainly responsible for the exponential increase in the killing zone.

Because of these changes, Civil War commanders could no longer safely gather their staff on a hill overlooking the battlefield, while remaining out of range of infantry weapons. This precaution was in contrast to the Napoleonic time period, as in the Civil War any large group would be a prime target for long range rifled artillery. General Leonidas Polk discovered this during the Battle of Atlanta. On the 14th of June 1864, he went with some of his staff to observe the Union lines while staying safely behind Confederate lines. His group was spotted by members of the 5th Independent Battery, Indiana Light Artillery commanded by Captain Peter Simonson.[17] The decision was made to fire at them as it was assumed that the party included high-ranking officers, even though they were too far away for individuals to be distinguished. When the projectiles began to explode around the group, everyone quickly dispersed except General Polk, who decided to walk away in a dignified manner. Unfortunately for General Polk, he was hit by a fragment and died from his injuries.

Beyond making the battlefield more dangerous, rifled artillery and the new mechanical fuzes were responsible for demonstrating that high masonry forts were obsolete. The ability of rifled artillery to fire a projectile which would penetrate the wall of a multi-story masonry fort at ranges well beyond that possible by smooth-bore artillery doomed this style of fort.

Long range accuracy also made the bombardment and destruction of cities a greater possibility. The Union was able to bombard Charleston, South Carolina at ranges that were previously inconceivable and created such destruction that period photographs of Charleston make it look like a bombed out city from World War II. This bombardment was accomplished by rifled Parrott guns,

16 Attempts were made during the war to increase the length of time a time fuze would burn but none of the designs were very successful.

17 Unfortunately, the Union battery commander, Captain Peter Simonson,was killed a few days later. Information on the interception of Confederate signal messages can be found in the *War of the Rebellion: The Official Records of the Union and Confederate Armies* (Washington, D.C.: U.S. Government Printing Office, 1890–1901), Series I, Volume XXXVIII, Part III, 70. Henceforth cited as *ORA*.

varying in size from a 30-pounder to a 400-pounder, firing exploding shells. One 30-pounder, with a range of 6,700 yards, fired 4,594 shells in 69 days before it burst. Out of those 4,594 shells 4,253 reached the city.[18] This firing went on day and night and, in the end, Charleston was a ruined city. The ranges these guns were capable of achieving and the destruction they produced were something no one had ever experienced before.

However, long-range fire from rifled artillery was not limited to large, area targets. During the Battle of Fredericksburg, Colonel Charles Wainwright was able to bring accurate fire, with exploding projectiles, upon a single Confederate Whitworth gun at a range that he estimated to be 2,700 yards by using a telescope to aim his 3-inch Ordnance rifle.[19] Throughout the Siege of Petersburg, General Henry Abbot, commander of the Union's siege artillery, regularly used a telescope to locate targets invisible to the naked eye. Once located, he directed his gunners to use terrain features near the target, visible to them, to aim their guns.[20]

Long-range artillery fire additionally helped to determine the course of a number of battles. During the Battle of Antietam, General "Stonewall" Jackson was ordered by General Robert E. Lee to investigate the possibility of turning the right flank of the Union Army. Jackson decided against attacking, not because of the presence of Union infantry, but rather due to the "numerous artillery so judiciously established in their front".[21] One can only imagine what General George McClellan's reaction would have been if Jackson had attacked and been successful. At the Battle of Fredericksburg, Jackson was again stopped from making an attack by Union artillery as their fire "so completely swept our front as to satisfy me that the proposed movement should be abandoned".[22] When the Confederate batteries bombarded the Union lines in preparation for "Pickett's Charge" at Gettysburg, the Union artillery batteries, with the exception of those in the Union center under General Winfield Hancock, did not engage in counter-battery fire. As a result, these batteries still had full loads and were able to subject the attacking Confederate infantry to a heavy and concentrated fire. This fire, as well as fire from other artillery batteries, was heavy and accurate enough that the units on the flanks of the attacking Confederates either fell back or moved toward the center. General Lee stated in his Gettysburg report

18 *ORA*, Series I, Volume XXVIII, Part I, 31–32.

19 Charles S. Wainwright, and Allen Nevins, ed, *A Diary of Battle* (New York: Harcourt, Brace & World, 1962), 145–146.

20 Henry L. Abbot, *Professional Papers Corps of Engineers No. 14, Siege Artillery in the Campaigns Against Richmond* (New York: D. Van Nostrand, 1986), 90.

21 *ORA*, Series I, Volume XIX, Part I, 33.

22 *ORA*, Series I, Volume XXI, 634.

that the units on the left flank were wavering from the concentrated artillery fire when they gave way after being attacked on their flank by Union infantry.[23] The impact of long range artillery fire was not one sided. Confederate artillery placed on Hazel Grove at Chancellorsville played a major part in forcing the Union Army to retreat. Another aspect that has been too often ignored is that the trenches constructed by both sides, in the latter part of the war, were made strong enough to protect the men from artillery fire, in addition to rifle fire.

Although there were numerous problems with artillery, its projectiles and fuzes during the war, the mechanical fuzes developed during this time period did not substantially change until the development of the variable time fuze in World War II. The major restrictions on Civil War-era artillery were the lack of a recoil system and the limitations of black powder as an exploding force within a projectile. Even with these constraints, the killing range for artillery had increased well beyond a man's ability to see the damage he was causing. Further, there were issues with the muzzle-loading rifles used by the infantry. The two major issues were that the men had to stand up to be able to quickly load the weapon and that their rate of fire was still four to five rounds per minute.

ILLUSTRATION 5.1 Charleston, South Carolina, Library of Congress

23 *ORA*, Series I, Volume XXVII, Part II, 321.

However, both of these issues would disappear with the advent of breech loading rifles.

The combination of aimed, long-range rifle fire and accurate, long-range artillery fire with the new mechanical fuzes made the battlefield a deadlier place and changed warfare forever. A general and his staff could no longer safely view a battlefield from a nearby hilltop nor could units maneuver in the open without taking casualties – men were no longer willing to needlessly expose themselves, and they began to dig without orders for safety. Technology had increased the killing zone beyond anyone's previous experience and, now, if a man could be seen, he could be killed. Modern warfare, with a battlefield in which no one could be seen and on which long-range destruction was the norm, had been created (Illustration 5.1).

CHAPTER 6

# Without Experience or Precedent: Transformational Technology and the Light Draft Monitors

*William Roberts*

Within a year after the Battle of Hampton Roads in March 1862, the US Navy was building dozens of ironclads based on John Ericsson's *Monitor* concept, including light draft vessels able to attack the Confederacy along its rivers. The light draft monitor program began in October 1862, when Ericsson sent the Navy a design for simple, cheap ships, and offered to build some in three months. Instead of accepting Ericsson's simple design, the Navy chose to "improve" the ships by repeatedly adding capability and complexity. The "improvements" delayed the urgently needed ships by sixteen months, and added so much weight that when completed they would not float. A promising program had become the "light draft fiasco."

This examination of the light draft program found the causes of its failure in three key areas: transformational thinking, technological ambition, and concurrent development and production. Transformational thinking stressed a new technology's disruptive effects and minimized its drawbacks, devaluing experience and seeing a future in which all that was past would be swept away. The second factor, technological ambition, was the human urge to maximize capability through incremental improvement. Although each improvement individually has little impact on completion, their cumulative effect was profound. The third element was concurrent development and production, in which the urgent desire to obtain an advanced capability prompted developers to start building the product before the design was fully developed. These three elements, which brought the light draft monitor program to grief, also illuminate current Navy programs such as the Littoral Combat Ship (LCS) and the Gerald R. Ford (CVN 78) class aircraft carriers.

## 1 Light-draft Origins

Gustavus Vasa Fox, the Assistant Secretary of the Navy, was the most influential proponent of transformational thinking. Fox witnessed the Battle of Hampton

 | DOI:10.1163/9789004393301_008

Roads in March 1862 and came away a true-believer in the new *Monitor* ironclad and in its inventor, Ericsson. Based on the original *Monitor*'s extremely limited combat experience, Fox endorsed a grandiose program that included monitors of "20 feet [draft] for [operations against] foreign nations; 10 feet for coast defence and harbor work; 6 feet for rivers."[1] Fox's over-enthusiasm for monitors supported Andrew Gordon's analysis of how theory took over when experience was lacking. New technology, Gordon found, helped to discredit empirical doctrine, and the adherents of the new technology would be the most evangelizing rationalists. The monitors' allure was enhanced by the attractions of an alternative force structure: By harnessing new technology, the monitors offered the opportunity for the Navy to prevail in the specific strategic context it faced, and to save money as well.[2]

The *Monitor* bandwagon shifted into high gear after Hampton Roads, and, by mid-1862, Ericsson was fully occupied designing and building the *Passaic*-class coastal monitors as well as two large ocean-going monitors.[3] Fox, who thought of Ericsson as a genius, pressed him to square the circle: that is, to design a heavily armed, strongly armored monitor with a draft of only four feet. That request was not technically possible, and in mid-September 1862, Ericsson quit trying. When he learned of this issue, Fox wrote, "Tell him to take six feet. He can have it easy. Our series are not complete without them."[4] The additional two feet of draft made the difference, and, by early October, Ericsson was able to send his design to the Navy.

Ericsson knew shipbuilding much better than Fox did, and his light draft design (Illustration 6.1) showed it. His ship was "a plain, oval tank with a flat bottom and upright sides", surrounded by a timber "raft" to give "stability and impregnability." This austere vessel minimized the need for iron and machinery, and, thus, minimized the stress it placed on shipbuilding and marine

1 Gustavus V. Fox to John Ericsson, September 27, 1862, John Ericsson Papers, American Swedish Historical Foundation Microfilm Edition (Philadelphia: Historic Publications, 1970), Reel 4. Fox to Ericsson, August 5, 1862, Ericsson Papers, Reel 4. Fox to Ericsson, August 8, 1862, Naval History Society Collection—Gustavus Fox Papers, MS 439, courtesy of the New-York Historical Society (hereafter "Fox Papers"), Box 3.

2 Andrew Gordon, *The Rules of the Game: Jutland and British Naval Command* (Annapolis: Naval Institute Press, 1996), 579. Alternative force structure: Steven Wills, *The Perils of Alternative Force Structure*, http://cimsec.org/perils-alternative-force-structure/28259 accessed October 6, 2016.

3 William H. Roberts, "'The Name of Ericsson:' Political Engineering in the Union Ironclad Program, 1861–1863," *Journal of Military History* 63 (October 1999): 823–844.

4 Alban C. Stimers to Fox, September 17, 1862, Fox Papers, Box 4; Fox to Stimers, September 20, 1862, Unofficial, ibid., Box 5.

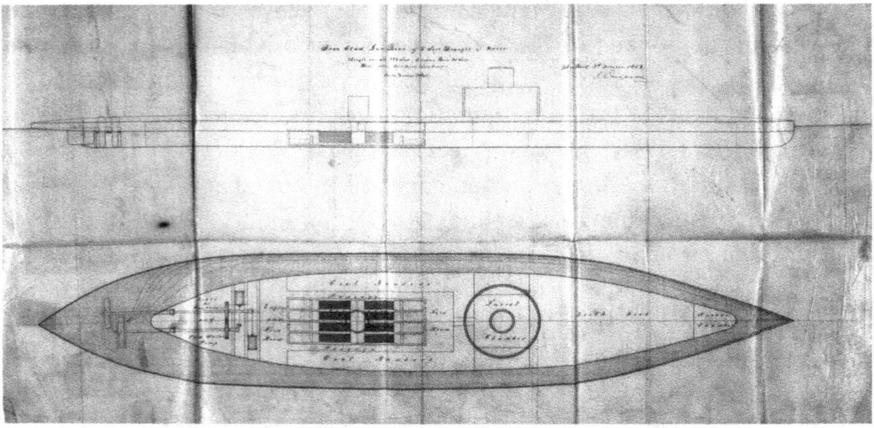

ILLUSTRATION 6.1 Plan and longitudinal section of John Ericsson's design for an "Iron Clad Gun Boat of 6 Feet Draught of Water."
CREDIT: ENCLOSURE TO LETTER FROM JOHN ERICSSON TO GIDEON WELLES, FEBRUARY 24, 1863, NATIONAL ARCHIVES RECORD GROUP 45, ENTRY M124, ROLL 433: 67.

engineering resources, already stretched thin by the Navy's other demands. Ericsson later testified that he could have built ships to his design in ninety days.[5]

Fox sent Ericsson's sketches to the Office of the General Superintendent of Ironclads, the Navy's "project office" for ironclads, to be translated into contract specifications. There the designs came to Chief Engineer Alban Crocker Stimers, who had supervised the building of the original *Monitor* for the Navy, and caught Fox's eye while serving aboard her in the Battle of Hampton Roads. Stimers had more ironclad building experience than anyone else in the Navy and, at the time, a good relationship with the irascible Ericsson.

In May 1862, Stimers reported for duty as General Inspector of Ironclads, a position nominally responsible only for quality assurance. Thanks to Fox's sponsorship, however, he was the technical leader and *de facto* head of the so-called "Ironclad Bureau."[6] He brought with him a reputation as a man who got things done and an ambition to become Engineer-in-Chief of the Navy.

5 Ericsson testimony, United States Congress, *Report of the Joint Committee on the Conduct of the War*, "Light-Draught Monitors," 38th Cong, 2d Sess (Washington: Government Printing Office (GPO), 1865 (hereafter "Light-Draught Monitors"): 69. The plans, dated 9th October, 1863, are enclosed in Ericsson to Gideon Welles, February 24, 1863, National Archives Record Group (NARG) 45, Entry M124, Roll 435: 67.

6 Francis H. Gregory testimony, "Light-Draught Monitors," 74. Deposition of Alban C. Stimers, August 11, 18, and 19, 1873, NARG 123, Records of the Court of Claims, Entry 1, General Jurisdiction Case Files 6326/6327. Case 6326 is "Alexander Swift et al. v. U.S.;" case 6327 is "Alexander

The *Passaic*-class monitors, then being built, were needed to take the offensive against the Confederacy and to deter Great Britain. They were behind schedule, though, and the Navy Department's official letters exhorted Stimers and his superior, Rear Admiral Francis H. Gregory, to hurry the ships to completion. Stimers' unofficial correspondence from Fox was studded with even more forceful letters. Even before Stimers became General Inspector, Fox had written, "We must have no delay, time is the great object....*time, time, time*." The Assistant Secretary never relaxed this pressure.[7] There could be no doubt in anyone's mind of the urgency of the Navy's need.

That urgency made the light draft program an outstanding career opportunity for Stimers, who could gain great professional credit simply by building light drafts as fast as possible. Even better, Stimers would be insulated from technical failures because the basic design was Ericsson's. Moreover, Ericsson and Fox intended that the shipbuilders would receive a basic plan and written specifications and would do most of their own design work. That buffer would both further protect Stimers from technical failure and give enough design diversity to increase the probability that at least some of the ships would succeed.

Stimers, though, viewed the program and the opportunity in a different light. He saw himself as a designer of ironclads, and for reasons discussed elsewhere, the light drafts were his last real chance to make his mark as such.[8] Fox's transformational thinking and technological ambition reinforced Stimers' personal ambition for recognition as a ship designer. The combination first delayed the light draft program and then utterly derailed it.

In the autumn of 1862, technological ambition showed in the Assistant Secretary's enthusiastic push for more capable ships, and Stimers readily implemented Fox's ideas. During October and November 1862, Stimers added capability (and weight) incrementally, but, in early December, he realized that more buoyancy would be needed. To get it, he had Second Assistant Engineer Theodore Allen completely redesign the ships to make the iron hull twelve feet longer and three feet wider and to expand the wooden anti-ramming raft as well.[9]

---

Swift and the Niles Works v. U.S." Principles, arguments, evidence and records were commingled, so items are hereafter cited as "Swift v. U.S."

7 Fox to Stimers, April 23, 1862 (emphasis in original), Fox Papers, Box 5. Emphasis in original.

8 Stimers testimony, "Light-Draught Monitors," 104–105. William H. Roberts, *Civil War Ironclads: The U.S. Navy and Industrial Mobilization* (Baltimore: Johns Hopkins University Press, 2002), 109–112.

9 Stimers testimony, "Light-Draught Monitors," 97. Allen described the modifications in Deposition of Theodore Allen, Swift v. U.S., August 11, 1873, 7–10.

Stimers took his new light draft plans to Washington in mid-January 1863, where he proposed, and Fox approved, a system of "water compartments," or ballast tanks. They would be filled to increase the ship's draft when navigating unfamiliar areas so that if the ship went aground, the water could be pumped out to refloat the ship. Stimers later asserted that Rear Admiral Joseph Smith of the Bureau of Yards and Docks suggested the tanks in January, but Stimers had corresponded with Delaware shipbuilders Harlan & Hollingsworth about them in December 1862, and Allen was likely working on them even before Stimers met with Fox.[10]

The water compartments, sandwiched between the iron hull and the wooden raft (Illustration. 6.2) added complexity, weight and cost; filling and emptying them added still more complexity, weight and cost in the form of piping, valves and pumps. The impact of technological ambition and undisciplined design is clear: Stimers and Fox spent three valuable months "improving" the light drafts to be larger, more complex, more expensive, and more time-consuming to build.

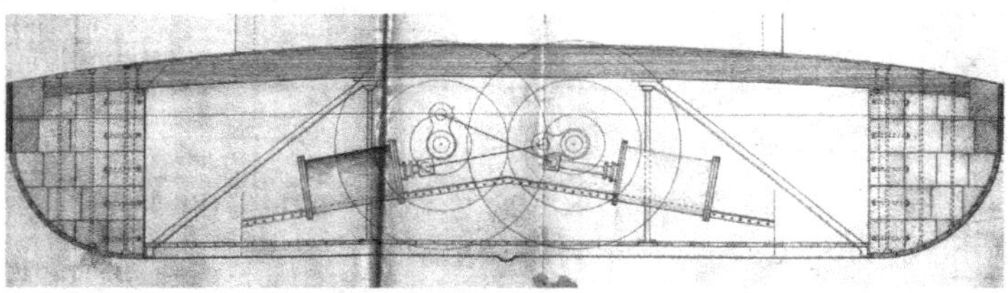

Ericsson's original design

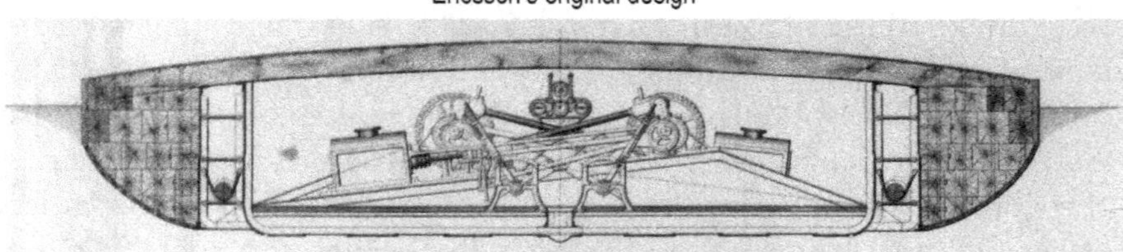

Stimers' elaborated design

ILLUSTRATION 6.2 Ericsson's "tank and raft" raft design (top) and Stimers' design with "water compartments." (bottom).

CREDIT: ERICSSON ORIGINAL FROM ERICSSON TO WELLES, FEBRUARY 24, 1863, NARG 45, ENTRY M124, ROLL 433: 67. STIMERS' DESIGN FROM NARG 19; DASH FLAT: PLAN NUMBER 139-15-1K.

10 Allen took 22 days to redesign the ships between December and mid-January. He could not have completed another major redesign between January 17 and February 1, when Stimers sent the list of weights to Fox.

## 2 *"The bids are enormous:"* Contracting for the Light-Drafts

On February 1, 1863, Stimers sent Fox a detailed estimate, showing that the light drafts would cost about $260,000 apiece.[11] Stimers then provided a plan, general specifications, and a detailed weight list to the Bureau of Construction and Repair, which advertised for the ships on February 10, 1863. Contractors were invited to visit Stimers' office or the Navy Department to examine the plans and specifications. The bids would be opened on February 25, 1863.[12]

On February 13, however, Stimers was sent to Port Royal, South Carolina. In his absence, Fox wrote to Ericsson on February 21 to confirm Ericsson's approval of the light draft design.[13] The reply, dated February 24, was a shock. Ericsson had not seen the plans until after he received Fox's letter, and, he wrote, "The moment I had Stimers' plans before me this morning, my course was clear, viz. That of repudiating all responsibility in regard to the Light Draughts." He recommended that Fox obtain bids on the original October 1862 plan, but acknowledged that Fox probably could not do so. "Possibly also you are so well pleased with the water compartment idea that you may say, go ahead. If so I say God speed you! Only do not hold me responsible."[14] On the same day, February 24, Ericsson telegraphed Secretary of the Navy Gideon Welles that his principles had been "frittered away by changes," and also wrote to Welles to describe his concerns in more detail.[15]

The next day, Fox received another shock—the light draft bids ranged from $380,000 to $395,000, half again what Stimers had estimated. Fox once more turned to Ericsson, writing, "The bids are enormous in price and the time seven months." He opined that the Navy would have to advertise again, and asked Ericsson if he could furnish plans. On February 27, Ericsson replied with an offer to build four vessels to his own design for $375,000 apiece with delivery in four months.

Fox replied that Ericsson's offer of $375,000 was only $11,000 less than the other bids "which you consider too expensive." Later the same day, the Assistant Secretary asked, "What price is your offer for cheap ones and what is the time? You better send the offer by mail for half a dozen—cheap and quick."[16]

---

11 Stimers to Fox, February 1, 1863, Fox Papers, Box 7. "Light-Draught Monitors," 107–108.

12 NARG 19, Entry 405 Proposals and Advertisements of Sales.

13 Fox to Ericsson, February 21, 1863, Unofficial, Fox Papers, Box 5A.

14 Ericsson to Fox, February 24, 1863.

15 Ericsson to Welles, February 24 1863, Telegram, Fox Papers, Box 7; Ericsson to Welles, February 24, 1863, NARG 45, Entry M124, Roll 435: 67.

16 Fox to Ericsson, February 25, 1863, NARG 45, Entry M209, Roll 25, 70: 401. Ericsson to Fox, February 27, 1863, NARG 45, Entry M124, Roll 435: 14. Fox to Ericsson, February 28, 1863, Entry M209, Roll 25, 70: 414; Fox to Ericsson, February 28, 1863, ibid. 70: 418.

On March 3, Welles received Ericsson's formal proposal to build six light draft monitors to the plans he had sent in October 1862. Ericsson confirmed that the vessels would cost $375,000 each, and that he would deliver them in four months.[17]

The Navy rejected Ericsson's offer, which Fox likely saw as regressive and expensive—regressive, because it abandoned cherished improvements, and expensive, because Ericsson wanted $375,000 for a bare-bones design. In addition, Fox saw Stimers' design as more mature than the one shown in Ericsson's three sketches, and reverting to Ericsson's design would mean admitting that Fox and Stimers had wasted precious months.[18] In hindsight, it was another unfortunate decision driven by technical ambition.

By April 2, 1863, twelve light drafts had been ordered to Stimers' design, for completion in four to six months, at an average price of $393,500. Fox gave the early ships to experienced builders, with the first two contracts going to Harlan & Hollingsworth of Wilmington, Delaware, and Merrick & Sons of Philadelphia, Pennsylvania. Besides assisting Stimers with the light draft hull design, Harlan & Hollingsworth had built a *Passaic* class monitor and were building one of the *Tippecanoe* class, while Merrick had teamed with Charles Cramp to build the broadside ironclad *New Ironsides*. The next two ships went to Reaney, Son & Archbold, who had built two *Passaic* class monitors in Chester, Pennsylvania, and to Snowdon & Mason of Pittsburgh, then building a *Tippecanoe* class. Thus, three of the first four shipbuilders had already completed at least one ironclad.

To fulfill their contracted delivery dates, the successful bidders had to begin construction immediately. The first step, described elsewhere, was for them to order the iron and machinery they would need, and they did so based on the plans and specifications they had seen.[19] Clearly the shipbuilders were satisfied with the information they received—if a contractor had thought that the specifications were inadequate, he would not have bid for a ship.[20] In this

17 Ericsson to Welles, March 3, 1863, NARG 45, Entry M124, Roll 436: 39. Given the conditions, four months compared favorably with Ericsson's October 1862 estimate of 90 days.

18 Fox testimony, "Light-Draught Monitors," 82.

19 The process is described in Roberts, *Civil War Ironclads*, 59, 64–66. The specifications were already changing, however; when Stimers went to Charleston, he left instructions "to re-write the specifications so changing them as to conform to the plan finally decided and making them more minute in their provisions." Norman W. Wheeler, "Chief Draughtsman, Lt. Drft Monitors," to Gregory, March 7, 1863, NARG 19, Entry 186, Box 33.

20 Charles Cramp stated that he bid based on crude drawings and sketches. Charles H. Cramp, "Certain Incidents in the Evolution of the Modern Warship." *Proceedings of the Numismatic and Antiquarian Society of Philadelphia*, 1907: 118. If true, Cramp freely took a serious business risk, but his recollections are often unreliable.

decentralized system, the shipbuilders expected to do most of their own design work.

Stimers was in Port Royal when the first contracts were let in early March 1863, so First Assistant Engineer Isaac Newton implemented the decentralization plan in Stimers' absence. He told the contractors on March 20, 1863, that the government was making drawings, but, if lack of drawings would slow the work, they could make their own drawings and send copies to the General Inspectorate for approval.[21] Decentralization appeared to be off to a good start.

## 3 *"To Go Into Fights and Come Out Whole"*: Better Is Bigger

Soon after, however, the monitor program received a major shakeup. Twelve light draft contracts had been awarded when, on April 7, 1863, Rear Admiral Samuel F. Du Pont's force of ironclads attacked Charleston Harbor; the squadron briefly engaged the Confederate defenses, and then withdrew, with four monitors and the hybrid *Keokuk* "more or less disabled."[22] Stimers and his workmen repaired much of the damage within a day, but Du Pont decided not to renew his attack.

That decision, and disagreement, over the extent of the monitors' injuries, led to a virulent controversy.[23] Of more immediate importance was the monitors' poor performance under fire, which revealed "faults of design which only such experience could point out." Stimers wanted to correct those faults, both in existing ships and in the monitors under construction.[24]

The first priority was altering the ships already in service, for which Stimers organized work gangs.[25] Second priority went to the *Tippecanoe*s, but the most advanced ships of that class were so close to completion that the extensive rework was extremely disruptive. Stimers later cited combat effectiveness—to "enable people to go into fights… and come out whole"—as the reason for the

---

21 Newton to Aquila Adams, March 20, 1863, NARG 19. Entry 1251, Mar. 1863–Sept. 1866, Volume I (March 20, 1863, to January 26, 1864): 1; Newton to Merrick & Sons, March 24, 1863, ibid. 1: 10.

22 First report of Rear Admiral Du Pont, April 8, 1863, *Official Records of the Union and Confederate Navies in the War of the Rebellion* (28 vols) (Washington, DC: GPO, 1894–1922) (hereafter *ORN*), 14: 3.

23 Roberts, *Civil War Ironclads*, 94–100.

24 Stimers to Welles, April 14, 1863, in *Report of the Secretary of the Navy in Relation to Armored Vessels*, House Executive Document 69, 38th Cong, 1st Sess, 1864, 81.

25 Roberts, *Civil War Ironclads*, 103, 104–105.

changes, but neither Stimers nor Fox recognized how much the improvements would cost, or how much they would delay the vessels.[26]

The changes effectively halted light draft construction, although the Navy continued to award contracts. Fifteen had been given out by the time Stimers issued revised general plans to the builders on May 26, 1863.[27] Before the plans were issued, Fox told Stimers that the builders would have to take the revised specifications at the original price, with the understanding that Stimers would "make all these changes now instead of as the work progresses."[28] Forcing the contractors to take the new specifications at the old price was manifestly unfair, but the fifteen builders agreed to the new terms, and so did five more whose contracts were awarded in June 1863.

Unfortunately, the light draft design was not nearly as mature as Fox thought it was. As Allen noted, "the working out of the details" showed the vessels to be too shallow for their length, so the hulls were stiffened and other weaknesses remedied.[29] In addition to being under intense pressure to modify three classes of monitors as rapidly as possible, and himself the subject of a time-consuming Court of Inquiry, Stimers was also designing a fast ironclad steamer and a coast defense monitor.[30] Overburdened, Stimers appears to have ordered the changes to the light drafts without calculating their effect.[31]

Some of the changes did indeed show lessons learned from the battle—Stimers added a ring of armor to protect the turret base, thickened the pilot house, added below-decks backup steering gear, and strengthened the rudder protection. Both technological and personal ambition were evident, however, in the ways in which the General Inspector took advantage of the opportunity to further elaborate the light draft design. The post-Charleston plans included circulating pumps, air pumps, thrust bearings and spring bearings for the propeller shafts, and other items that did not appear in the pre-Charleston plans. Each improvement added complexity, weight and cost, but none came from combat lessons learned: As Stimers later admitted, "some of these changes

26 Deposition of Alban C. Stimers, Swift v. U.S., August 11, 18, 19, 1873; Roberts, *Civil War Ironclads*, 118–121.

27 Stimers to Gregory, May 18, 1863, NARG 19, Entry 64, 2: 317; Fox to Ericsson, May 25, 1863, Fox Papers, Box 5A.

28 Fox to Stimers, May 11, 1863, Unofficial, Fox Papers, Box 5A (emphasis in original).

29 Allen deposition, Swift v. U.S., August 11, 1873.

30 Stimers to Fox, April 30, 1863, Fox Papers, Box 7.

31 Negative evidence cannot be conclusive, but if calculations were made, neither contemporary investigators, later litigants, nor modern researchers have found them.

were made without special reference to the effect of shot upon the vessel, but were rather the result of studying the subject."[32]

Stimers appears to have equated "good design" with "elaborate, heavy, and expensive," and his hunger for recognition as a designer of ironclads dovetailed with Fox's technological ambition.[33] As a result, he later testified, "... the very first thing we did in preparing plans for this class of vessels, was to make the plans very different from what was contemplated by the contract." No-one questioned Stimers making changes "as the work progress[ed]"—he routinely gave his instructions directly to the contractors, and thanks to Fox, the Navy's contracting bureaus rubber-stamped whatever Stimers wanted.[34] A court filing decades later summed up the case: The shipyards "contracted to build a vessel according to the plans and specifications they had seen." Stimers intended "that they should build a vessel which should in outward general appearance be the same," but which would be "substantially different and far more difficult and expensive...."[35]

The Navy tried to compensate for the delay by tacitly extending the existing contracts, restarting the period of performance when the new general plan was issued on May 26, 1863. Even so, the builders could make little progress, because Stimers had already unilaterally reversed the decision to decentralize light draft design. Stimers' move was based in part on his experience as inspector of the *Monitor*, where he found that a lack of clear specifications made his job very difficult. Centralized design would also enable the Monitor Bureau to control better the ships' weight, to improve their logistics through parts commonality, and to prevent contractors from cheapening their product.[36] By late April 1863, Stimers had established a pattern of rejecting contractors' engine and boiler designs. The contractors quickly determined that if Stimers would not approve their designs, they could save effort by not making

32 Comparison of Plans of the Light Draft Monitors Casco, Squando, Waxsaw, Modoc, Suncook and Yazoo, NARG 19, Entry 186, Box 33, s.v. *Squando*. The plan of *Casco* was sent to her builder on March 20, 1863; the plan of *Squando* was sent on May 26, 1863. Stimers deposition, Swift v. U.S., August 11, 18, 19, 1873, 266. After the Civil War, Stimers became a consulting engineer and testified for the plaintiffs in several lawsuits against the Government.

33 Roberts, *Civil War Ironclads*, 109–112.

34 "First thing" Stimers deposition, Swift v. U.S., August 11, 18, 19, 1873, NARG 19, Entry 186, Box 33, s.v. *Squando*, 254. "Some of the parties I notice suggest to Stimers to get such authority [for changes] from the Dept, which will be given as soon as we know what they [the changes] are." Fox to Ericsson, January 12, 1863, Fox Papers, Box 5A.

35 NARG 123, Entry 1, Case 16816 (McKay & Aldus), Supplemental Brief and Argument for Claimant, printed booklet, n.d., 18. A major issue, discussed elsewhere, was the use of fixed price contracts for poorly defined products. Roberts, *Civil War Ironclads*, 199–200.

36 Stimers to Nelson Curtis, August 10, 1863, NARG 19, Entry 1251, 1:376–378.

any, and using Government plans would insulate them from risk if there were design problems.

Stimers' insistence on using the Navy's plans might have worked if the drawings had been ready, but they were not. The post-Charleston need to modify several classes of monitors placed insupportable stress on the Inspectorate's already-overloaded draftsmen, and the light-draft plans fell farther and farther behind. The backlog meant that many of the plans the contractors needed were not even issued until long after the projected completion dates of their original contracts. As shown in Table 6.1, less than two-thirds of the drawings had been issued by the time the first ships should have been finished. It is evident that the General Inspector overreached both the state of the art and the capabilities of his staff. In doing so, he cost the Navy the speedy construction that Fox craved.

Some changes were made explicitly, but, as noted, many were simply incorporated into the Inspectorate's drawings—examples include the massive turret base ring; the vents and flood valves for the ballast tanks; the thicker pilot house; the secondary steering gear; the gun carriages; and many extra bulkheads, ventilation pipes, and braces. Other drawings changed items ranging from the main condenser shell to the way in which the pilot house was attached, and some things were changed more than once. As Stimers testified in 1873, the plans were "very different from what was contemplated by the contract," and "nearly all such changes would add to the cost, in addition to making delays."[37]

Stimers' assistant, Allen, later testified, "the machinery could have been made to work equally well with much less expense," and others ranging from Benjamin Franklin Isherwood, Chief of the Bureau of Steam Engineering, to shipbuilder Nathaniel McKay spoke in a similar vein.[38] Besides being more expensive, the elaborate designs were heavier. Examples of added weight included a complex custom-designed "vacuum engine" instead of a simple commercial air pump; extra thick piping; a larger-than-necessary condenser; and more and heavier braces and bulkheads. James Maynard, who supervised building the *Wassuc* for George Lawrence, was asked, "Was the boat that you contracted to build a very complicated one?" He answered, "It was before it was finished."[39]

37 Stimers deposition, Swift v. U.S., August 11, 18, 19, 1873; NARG 19, Entry 186, Box 33, s.v. *Squando*, 273. Ibid. 254, 258.

38 Allen deposition, Swift v. U.S., July 25, 1877, 115, 126–127, 131.

39 Deposition of James A. Maynard, November 24, 1891, NARG 123, Entry 1, Case 16835, Lawrence v. U.S., 7–8.

TABLE 6.1 Light Draft Monitor Drawings[a]

| Month | | Sheets of Drawings | Cumulative % issued |
|---|---|---|---|
| March 1863 | ("As bid" specification and general plan issued) | 16 | 1.1% |
| April 1863 | | 36 | 3.4% |
| May 1863 | ("As contracted" specification and general plan issued; delivery clock restarted) | 100 | 10.1% |
| June 1863 | | 222 | 24.8% |
| July 1863 | (Finish first ships—original schedule) | 188 | 37.3% |
| August 1863 | (Revised specification issued) | 275 | 55.5% |
| September 1863 | (Finish first ships—revised schedule) | 133 | 64.3% |
| October 1863 | (Revised specification and general plan issued) | 155 | 74.6% |
| November 1863 | | 112 | 82.0% |
| December 1863 | | 157 | 92.4% |
| January 1864 | | 0 | 92.4% |
| February 1864 | | 76 | 97.5% |
| March 1864 | | 38 | 100.0% |
| July 1864 | First monitor (Tunxis) commissioned | | |
| September 1864 | Tunxis decommissioned for rebuilding | | |

| Month | | Sheets of Drawings | Cumulative % issued |
|---|---|---|---|
| December 1864 | First torpedo boat (Casco) commissioned | | |
| March 1865 | First operational torpedo boat (Casco) | | |
| June 1865 | Second monitor (Squando) commissioned | | |
| July 1865 | First operational monitor (Squando) | | |

a NARG 19, Entry 1259, Record of Monitor Drawings Sent to Contractors. United States Navy Department, Naval History Division. *Monitors of the U.S. Navy 1861–1937*, Washington: GPO, 1969. *ORN*, Ser. 2, Vol. 1, "Statistical Data of Ships" *s.v.* ships' names.

However they were communicated, the changes were almost always implemented without negotiation. This placed significant liability on the government in the long term, but it imposed an immediate financial burden on the contractors, who had to do much more work than they had estimated with no increase in their progress payments from the Navy. The contractors made their pain clear to Fox and Welles, and in July 1863, Welles cracked down, telling Rear Admiral Gregory that all direction to the contractors was to go through Gregory. To evade this restriction, Stimers sent orders to the Local Inspectors of Ironclads and told them to direct the contractors. Changes continued unabated until Gregory discovered the dodge in October 1863 and forced Stimers to stop.[40]

Thanks to Welles and Gregory, the flood of changes began to ebb and the builders could try to catch up. Shipbuilding conditions had worsened markedly, though, even in the few months since the contracts were issued. Inflation was biting deeply and materials were expensive and hard to get. During the spring and early summer of 1863, labor had been readily available, but by autumn 1863, it was scarce and expensive. Work on the ships slowed.

40 Welles to Gregory, July 11, 1863, Welles Papers, Letterbook May 23–September 30, 1863, 233. Gregory to Stimers, October 28, 1863, NARG 19, Entry 1235, July 1861–December 1864, 13: 25.

## 4 The Light Draft Crisis: *"It will be impossible for her to float"*

In late February 1864, Stimers again bypassed Admiral Gregory to propose to Fox that the Navy establish a Bureau of Iron Clad Steamers, with Stimers at its head. Fox refused to expend political capital on a campaign that was unlikely to succeed, and his reply to Stimers' proposal was revealing. The urgently needed light drafts were not finished because of the "additions, alterations and improvements," which he himself could have and should have prevented. The lack of completed ships, Fox wrote, was "fatal," both to the monitor program and to a separate ironclad bureau.[41]

As Stimers was promoting his aspirations to become a Bureau chief, the Confederates were finishing an ironclad in a cornfield. CSS *Albemarle* was weak and slow, and a light draft would completely overmatch her. However, no light draft had yet been launched when *Albemarle* attacked Plymouth, North Carolina, on April 19, 1864. The ironclad sank one Union gunboat and forced others to withdraw, enabling the Confederates to capture the town. Word of the *Albemarle*'s role in the fall of Plymouth quickly reached the North, increasing the urgency of completing a light draft.[42]

Under intense pressure from Fox, Stimers and his assistant Allen went to Boston at the beginning of May 1864 to complete the *Chimo* "with the greatest possible dispatch."[43] Stimers showed theatrical urgency and a deficient understanding of the balance between risk and reward by launching the ship at night instead of waiting twelve hours for the next favorable tide in daylight.[44] After the May 5 launching, Stimers commuted from New York, while Allen stayed in Boston to supervise *Chimo*'s completion directly.

On May 5, the same day *Chimo* was launched, wooden gunboats repulsed *Albemarle*'s attack on New Bern, North Carolina. The Confederate ironclad remained a serious threat, however, and Stimers and Allen continued to force the work on *Chimo*. The first official indication of trouble was a May 12 letter from Gregory, reporting that *Chimo*'s draft was excessive, and that if the report

41 Fox to Stimers, February 25, 1864, Private, Fox Papers, Box 8.

42 Fox to Stimers, April 23, 1864, "Private," Fox Papers, Box 3.

43 Telegram, Stimers to Aquila Adams, April 30, 1864, NARG 19, Entry 1251, 2: 227; Deposition of Theodore Allen, 4, 5, 6, 7, 12 August 1873, NARG 123, Entry 1, Case 16816, Folder GJ16816 #5, 36. "Greatest possible dispatch" NARG 19, Entry 1254, 51:107. Roberts, *Civil War Ironclads*, Chapter 6 states "early March 1864;" this is my transcription error "Mar" for "May."

44 "Launching of the Steam-battery 'Chimo,'" *The New York Times*, May 8, 1864, http://www.nytimes.com/1864/05/08/news/launching-of-the-steam-battery-chimo.html.

were correct, "it will be impossible for her to float" with her turret and guns.[45] Three weeks after the launching, on May 31, Gregory wrote directly to Welles, saying that *Chimo*'s draft made it "a matter of doubt" whether she could perform efficiently.[46]

The admiral understated the case. With no ammunition, crew, or stores, and only part of her coal supply aboard, *Chimo* averaged 2" of freeboard. Fully loaded, only the turret and the arched deck along the ship's fore-and-aft centerline would have been above water, as shown graphically in Illustration 6.3. Naval Constructor W.L. Hanscom called it, "Rather a small margin for a man to go to sea with."[47]

Stimers tried hard to put the best face on the situation. On May 31, 1864, Stimers told Fox, "If they were six inches more out of water…they would be decidedly the best vessels we have for this war."[48] The ships, however, were *not* "six inches more out of water," and Fox had finally had enough. On June 3, Fox

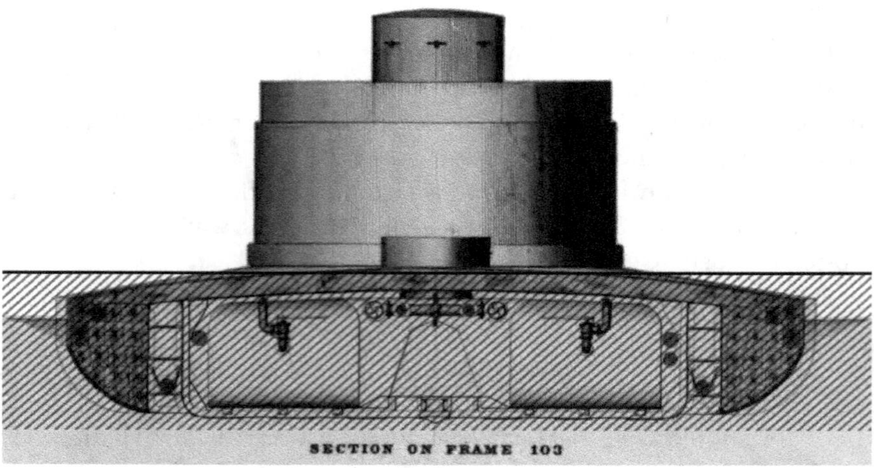


ILLUSTRATION 6.3 "Rather a small margin:" a light draft monitor ready for sea.
CREDIT: NARG 19; DASH FLAT: PLAN NUMBER 139-15-1K.

45 Gregory to Lenthall, May 12, 1864, NARG 19, Entry 1235, 15: 9.

46 Gregory to Welles, May 31, 1864, NARG 19, Entry 1235, 24: 35–36. *Chimo* "cannot be sent to sea without extreme hazard." Gregory to Welles, June 8, 1864, ibid. 24: 36.

47 Fox to Benjamin F. Wade, December 15, 1864, in "Light-Draught Monitors," 3–5; Aquila Adams testimony, ibid., 12–13; Hanscom testimony, ibid., 5–6.

48 Stimers to Fox, May 31, 1864, Fox Papers, Box 9.

asked Ericsson to develop a remedy, and on June 4 Ericsson agreed.[49] Ericsson's return to center stage precipitated Stimers' eclipse.

Stimers and Gregory visited *Chimo* together, likely on June 5. On Monday, June 6, they wrote to Welles with a plan, undoubtedly developed by Stimers, to modify the vessels by adding a wooden deck.[50] Stimers then returned to New York and a chilly meeting with Ericsson, with whom he had quarreled. After several days of hesitation, Welles abolished Stimers' position of General Inspector and turned to Ericsson and to Chief Engineer William W.W. Wood, the General Inspector of Steam Machinery, to retrieve the situation.

By this time, in mid-June 1864, *Chimo*, *Casco*, and *Tunxis* had been launched, but work on the other seventeen ships was suspended while the Navy considered its technical options. *Tunxis*, the second light draft to be completed and the first to be commissioned, was much closer to the design draft but was almost lost at sea in mid-July 1864, because her lighter hull was too weak.[51] The *Albemarle* was foremost in Fox's mind, and he pressed Ericsson to get *Tunxis* "into the Sounds of North Carolina, where there never is any sea and her height out of water is no object."[52] Stimers and an assistant were sent to investigate, but there was no quick way to correct *Tunxis*'s longitudinal weakness.[53] An anecdote recounted by Ericsson's partisan biographer likely reflected Stimers' mood—Stimers, he wrote, cut his name out of a plate that stated that *Tunxis* was built from Stimers' designs.[54]

Stimers' imprint on the light draft program was not so easily eradicated. Two intertwined issues, one technical and one political, faced the service. The technical challenge was how to modify twenty ships, in various stages of completion, so that they could safely be sent to sea. The political and administrative

49 Fox to Ericsson, June 3, 1864; Ericsson to Fox, June 4, 1864, Fox Papers, Box 8.

50 Gregory and Stimers to Welles, June 6, 1864, NARG 19, Entry 186, Box 33. The wooden false deck gave the "six inches more out of water" but would not have survived combat. *Chimo* carried such a false deck for her sea trials but it was discarded afterward. "The New Light-Draft Monitors," from the Boston *Journal* of June 23, 1864, reprinted by *The New York Times*, June 26, 1864. http://www.nytimes.com/1864/06/26/news/the-new-light-draft-monitors.html.

51 A month before, Fox had asked Ericsson, "Why is the Tunxis about right though deep and the Chimo all wrong[?]" Fox to Ericsson, June 27, 1864, Telegram, NARG 45, Entry M209, Roll 27: Vol. 74: 300. *Tunxis*' short river voyage from Chester to Philadelphia was successful but four days later, she returned from sea "very near sinking." Commodore Cornelius K. Stribling to Gregory, July 15, 1864, NARG 19, Entry 1241, 26:15.

52 Fox to Ericsson, July 28, 1864, "Unofficial and Private," Fox Papers, Box 3.

53 Gregory to Fox, July 28, 1864, Fox Papers, Box 8.

54 William Conant Church, *The Life of John Ericsson* (2 vols.) (New York: Charles Scribner's Sons, 1891), 2: 30.

challenge was dealing with the consequences of spending millions of dollars with nothing to show for the expense.

## 5 Technical Recovery

To make the light drafts safe for sea, the Navy had to increase their buoyancy by removing weight, increasing freeboard, or doing both. The technical problems were promptly addressed, and in mid-June, Welles approved removing the turrets from three ships to convert them into torpedo boats. While perhaps uncertain about whether he wanted monitors or torpedo boats, Welles was quite clear about *who* he wanted, and that was Ericsson. Starting in late June 1864, the key technical players in the redesign were Ericsson and Wood, who conversed daily, and Newton, who assisted Wood.[55] By June 17, Wood's proposal to raise the monitors' sides had gained traction.[56] By the end of June, the technical recovery split into two threads—one for the five vessels that lost their turrets and were converted to torpedo boats, and one for the fifteen vessels that were completed as monitors.

The torpedo boat conversions required relatively little design and shipyard work. The major elements were to remove the turret and its machinery; reduce the deck armor aft; remove the ballasting system and pumps; replace the engine auxiliaries with simpler, lighter ones; and install a gun platform and a spar torpedo.[57] Some torpedo boat conversions actually saw commissioned service, although none saw significant action.

The monitor alterations were much more extensive than the torpedo boat conversions. In broad, the deck was cut free and the hull built up under it—the change added the weight of 22" worth of hull and raft, but since the original armor belt was raised bodily to maintain its position relative to the deck, the weight of armor did not increase.[58] Weight-saving changes included removing the valves and piping from the water compartments, but the ballast pumps were left in place to dewater the engine room in the event of flooding. As modified, the light drafts drew from about 7'8" to 8'3" and were ready for sea. Their deeper hulls gave them acceptable freeboard.

55 Ericsson to Newton, June 27, 1864, Newton Papers, Box 2, Folder 42.

56 Fox to Gregory, June 17, 1864, Fox Papers, Box 8.

57 Wood to Harlan & Hollingsworth, August 30, 1864, NARG 19, Entry 1251, 2:380.

58 Wood to Gregory, with enclosure, July 8, 1864, NARG 19, Entry 186, Box 33, s.v. *Squando*.

## 6 Political Consequences

Within the Navy, Welles conducted his own informal investigation. On June 10, 1864, he recorded receiving Gregory's report on *Chimo*, telling his diary, "I am not satisfied with Stimers's management, yet Fox has in this matter urged what has been done."[59] It promptly became public knowledge that the light draft program had been suspended, and on June 29, 1864, the Senate directed the powerful Joint Committee on the Conduct of the War to investigate. The Committee's inquiry did not begin in earnest until mid-December 1864, and it was January before the Committee chair, Senator Benjamin F. Wade, asked Welles for an explanation. Welles replied, "The Navy Department, in this emergency, was compelled to feel its way, without experience or precedent in any quarter to guide it." He forcefully defended his department, describing the urgent need for light draft ironclads and the lack of examples to follow.[60]

By the end of January 1865, the Joint Committee had heard sworn testimony from John Lenthall, Chief of the Bureau of Construction and Repair, Fox, Ericsson, Gregory, Wood, and many less-prominent figures. After a brief respite, testimony resumed on February 11, 1865, when Stimers himself appeared. Testifying after most of the other principals, Stimers knew what questions might be asked and how the others had answered them. He tried hard to deflect blame by asserting that he had consulted fully with Ericsson and submitted his plans to Lenthall and to Chief of the Bureau of Steam Engineering Benjamin Franklin Isherwood.[61] Lenthall and Isherwood, who asked to testify to dispute Stimers' statements, were the final witnesses.

The Committee decided that the ships failed because of "errors in the calculations… attributable to Mr. Stimers and those in his employ." The ships, however, were 23 inches too deep, and miscalculations could have accounted for no more than 5 inches of that. The fault was not with the calculations, and Fox knew it. As early as February 1864, the Assistant Secretary had lamented that the light drafts that "would have given us the vitals of the South" were still unfinished, thanks to "additions, alterations and improvements." Fox rightly blamed himself, and he spoke for many later program sponsors when he wrote that he "repent[ed] in bitterness of disappointment."[62]

59 Gideon Welles, *Diary of Gideon Welles Secretary of the Navy under Lincoln and Johnson*, Howard K. Beale, ed. (New York: W.W. Norton, 1960), entry for June 10, 1864 (II: 52–53).

60 Welles to B.F. Wade, January 16, 1865. NARG 45, Entry 10, Letters to Congress, Volume 12; labeled on spine "Congress No. 14 May 19, 1862 to Mar. 3, 1869," 213.

61 Stimers testimony, "Light Draught Monitors," 95.

62 Fox to Stimers, February 25, 1864, Private, Fox Papers, Box 8.

By the time the Committee's report was issued in mid-1865, though, the war at sea was long over—light draft ironclads were no longer needed, and the Navy's political leadership got off lightly. The Committee did not seriously examine how Fox's obsession with the monitors and his technological ambition affected the program, preferring to say simply that "it was unwise to order the construction of so many vessels upon precisely the same plan, without first testing the questions involved, by the construction of one or two…"[63] Few of the light drafts saw service and all were laid up in reserve by 1867. By 1875, all twenty had been scrapped.

The three elements that doomed the light drafts—transformational thinking, technological ambition, and concurrent development and production—were similarly present in two major U.S. Navy shipbuilding programs that were ongoing as this work went to press: the Littoral Combat Ship (LCS) and the Gerald R. Ford class of aircraft carriers.

The LCS was described by a supporter as a "transformational network system," with "battle modularity," the ability to interchange single-mission modules for antisubmarine, anti-surface and mine warfare. "The combination of high sustained battle force speeds and battle modularity could potentially transform the role of small combatants…."[64] According to the Navy, the LCS program was a "bold departure from traditional Navy shipbuilding programs based on its use of innovative acquisition, construction, manning, training and operational concepts."[65] Transformational thinking, and a blinkered concentration on the supposed benefits of an alternative force structure, has been prominent throughout.

The LCS program involved two radically different vessels—the odd numbered ships had conventional hulls, while even numbered ships were trimarans. The Navy originally intended to build two of each type, compare them, and purchase 48 more ships to a single design. However, less than a year after the second LCS was commissioned in 2010, the Navy awarded contracts for ten ships of each design. After accepting six LCS, the Navy was still making changes, and it recently announced that the first two ships of each type would be relegated to testing and training.[66] A layman's summary of the Fiscal Year 2015

63 "Light-Draught Monitors," 2.

64 Robert O. Work, *Naval Transformation and the Littoral Combat Ship* (www.csbaonline.org/wp-content/…/2004.02.18-Littoral-Combat-Ship.pdf, accessed July 7, 2016), iii.

65 United States Navy Fact File: Littoral Combat Ships—Fleet Introduction and Sustainment—LCS. http://www.navy.mil/navydata/fact_display.asp?cid=4200&tid=1650&ct=4, accessed September 20, 2016.

66 Testing and training: "Navy Adjusts LCS Class Crewing, Readiness and Employment." http://www.navy.mil/submit/display.asp?story_id=96574, accessed September 21, 2016.

report of the Defense Department's Director of Operational Test and Evaluation would be that neither the ships nor the "plug-and-fight" mission modules worked.[67]

There are many parallels between the LCS program and the light draft monitors. Both programs were, for their time, transformational, and were promoted as such. The two also embraced untested alternatives to existing force structures. In addition, they attempted to reduce the time to get ships into service by concurrent design and construction. Finally, both contracted for block buys of ships before the lead ship had been tested.

The LCS and the light drafts differ, however, in two significant ways: First, the Navy of the 1860s was making its first foray into such a numerically large and technically risky program. Second, the light drafts were urgently needed by a Navy at war. As Fox wrote, "The want of a light draft iron-clad has been so imperative that the Department was justified in taking great risks to obtain one."[68] The Navy of the 2000s had no such excuses. One analyst even wrote, "the risks of concurrent design and construction are well known, [so] the LCS provided a reminder rather than a lesson."[69] The Joint Committee verdict from 1865 applies to the LCS over 150 years later: *"It was unwise to order the construction of so many vessels upon precisely the same plan, without first testing the questions involved, by the construction of one or two…"*

Similarly, the aircraft carrier Gerald R. Ford (CVN 78) was designed with an Electromagnetic Aircraft Launch System (EMALS) to take the place of steam-powered catapults, and an Advanced Arresting Gear (AAG) for aircraft recovery. The combination of AAG and EMALS was supposed to reduce manning, increase the aircraft sortie rate by 33 percent, and enable aircraft not designed for carrier operation to land on and launch from the carrier. Technological ambition and transformational thinking were again evident, but so was concurrent development and production—the production contract for AAG was awarded in 2009, but the first test with an aircraft did not take place until April 2016.

In May 2016, the AAG system was two years behind schedule and its cost had risen 186 percent. The AAG failed once every 20 aircraft recoveries, against a

---

Design changes: *Defense Acquisitions: Assessments of Selected Weapon Programs*. GAO Report to Congressional Committees GAO-16-329SP, March 2016, 106.

67 http://www.dote.osd.mil/pub/reports/FY2015/pdf/navy/2015lcs.pdf, accessed May 25, 2016.

68 Fox to Ericsson, 28 July 1864, Ericsson Papers, Reel 4.

69 Ronald O'Rourke, *Navy Littoral Combat Ship (LCS)/Frigate Program: Background and Issues for Congress*, Congressional Research Service Report for Congress RL33741, May 20, 2016, at https://www.hsdl.org/?view&did=793504, 56–57.

requirement of once every 16,500 recoveries.[70] Gerald R. Ford was not commissioned in 2016 as scheduled, in part due to her inability to operate aircraft. Defense Undersecretary for Acquisition Frank Kendall wrote, "With the benefit of hindsight, it was clearly premature to include so many unproven technologies" in the ships.[71]

After his experience with the light drafts, Gustavus Fox would have agreed.

---

70 National Defense Authorization Act for Fiscal Year 2017, Senate Report 114–255, 114th Cong, 2d Sess, 11–13. https://www.gpo.gov/fdsys/pkg/CRPT-114srpt255/pdf/CRPT-114srpt255.pdf, accessed May 25, 2016. Ronald O'Rourke, *Navy Ford* (CVN-78) *Class Aircraft Carrier Program: Background and Issues for Congress*. Congressional Research Service Report for Congress RS20643, May 27, 2016. https://www.fas.org/sgp/crs/weapons/RS20643.pdf.

71 Pentagon Orders Review of Aircraft Carrier. By Reuters 2016-08-29 22 http://maritime-executive.com/article/pentagon-orders-review-of-aircraft-carrier.

CHAPTER 7

# The Uganda Railway and the Fabrication of Kenya

*Robert H. Clemm*

The territory of what would become the British colony of Kenya was little regarded by Europeans during the mid-to-late 19th century. At that time, it served as little more than a barrier to cross to places more renowned and important. For explorers who wished to verify if the reports of a snow-capped mountain in Africa were true, it was simply a land to traverse on the way to Mount Kilimanjaro. For British officials in the aftermath of the Berlin Conference of 1884–1885, it was a land necessary to possess not for its own intrinsic worth, but only as a means to stabilize "effective occupation" and to preserve British dominance. For officers of the Imperial British East Africa Company, it was an obstacle to surmount to gain access to the much richer lands of Uganda. The construction of the Uganda Railway, however, radically changed the perception, and, by extension, the nature and history of Kenya.

Dr. Guilmartin, to whom this volume is dedicated, urged historians to recognize the importance of technology to our discipline generally and specifically in the realm of military history. His definition of technology as "any physical or mental tool created by a person or society for the purpose of problem solving" fits as far as the history of the Uganda Railway seems to suggest. The Uganda Railway was a piece of technology crafted to solve the joint political and economic concerns of the British Government in eastern Africa. In linking the coast firmly to Uganda it would solidify British control over a region contested by German colonial enthusiasts, and would ensure the prosperity of the region through the expected transport of cash crops to the coast.

As much as the Uganda Railway seems to present yet another example of the importance of technology generally, and the railway specifically, to the process of imperialism, it goes well beyond that. The "Lunatic Express," as the Uganda Railway was nicknamed, illustrates the power of technology to create and transform well in excess of our own intentions. While its creators simply wished to solve the technical question of linking important regions via a stable transportation network, the railway fundamentally transformed the land over which it crossed. The transformation went beyond that of the physical landscape, which would be leveled and etched with rails and ties, and extended to the very understanding of what Kenya was. To return to Dr. Guilmartin's definition, the Uganda Railway was designed to answer a particular problem, but

 | DOI:10.1163/9789004393301_009

ended up answering a problem none of the British imperialists had considered – what to do with Kenya?

The Uganda Railway, therefore, is not simply the prototypical story of technology in the service of empire, but evidence of how technology can powerfully reshape our understanding of the world. As much as we believe technology is a tool to be utilized for our benefit, the Uganda Railway underscores how technology could just as equally reshape us in the process of being used. In the case of Kenya, as was recounted by later governor of the East African Protectorate Charles Elliot, "this line [the Uganda Railway] has literally created a country."[1]

In an imperial context the role of the railway was self-evident in both its ideological and practical effects. The railway was not only a product of an industrial Europe capable of projecting military force overseas, it also heralded the arrival of civilization in the lands which Europeans occupied. It is not surprising, therefore, to find many of the most rapturous colonial visions to be bound up intimately with a railway project. From French visions of a trans-Saharan railway to Cecil Rhodes' belief that Britain should establish a "Cape-to-Cairo" line of colonies bound by a single railway, this piece of technology appeared integral to the colonial project itself.[2] Beyond the fevered dreams of colonial enthusiasts, it seemed the practical import of the railway was equally self-evident.

In colonial warfare during this period, where conflicts were as much a struggle with climate, disease, and logistics as they were against hostile indigenous peoples, the railway seemed the lynchpin of colonial conquest. As C. E. Callwell noted in his *Small Wars,* a seminal work on imperial combat during the period, wars in the colonial context were not:

> the question of pushing forward the man, or the horse, or the gun, that has to be taken into account, so much as that of the provision of the necessaries of life for the troops when they have been pushed forward.[3]

1 Sir Charles Eliot, *The East Africa Protectorate* (New York: Barnes & Noble, 1966), 208.

2 The Frenchman Lionel Wiener declared "it is especially railroads that bring civilization behind them." Donald Headrick, *Tools of Empire* (New York: Oxford University Press, 1981), 194. For a general overview of the use of railways in Africa, including discussion of the Trans-Saharan railway, see Chapter 14 of the aforementioned work.

3 Colonel C.E. Callwell, *Small Wars: Their Principles and Practice* (London: Harrison and Sons, 1906), 57. His assessment would form the basis of an entire chapter of his work entitled "The influence of the question of supply upon small wars and the extent to which it must govern the plan of operation."

It was the preeminent concern with logistics, as Callwell would note, that "constitutes one of the most distinctive characteristics of this class of warfare."[4]

One of the more famous conquests of the colonial period, the British expedition led by Herbert Kitchener to defeat the Mahdist forces, uniquely reflects the railway as the solution to this "distinctive characteristic" in the struggle across the Sudan. His methodical advance, which took two years to reach the Mahdist capital, was slowed by his construction of multiple railways along his line of advance. For Kitchener, these railways were of absolute military necessity in order to ensure proper logistical supply.[5] The greatest effort was expended in building a 225-mile railway to connect the Sudanese towns of Wadi Halfa and Abu Hamad. Instead of bringing supplies laboriously across the Sudanese desert via camel and steamer over the course of 18 days, the railway cut travel time to a mere 24 hours. It also freed Kitchener's forces from the vagaries of the Nile River as he could now move and supply his forces at will.[6]

The criticality of the Sudan Military Railway to Kitchener's operation would lead to it being dubbed "the deadliest weapon ever used against Mahdism" with one of its steam engines now residing outside a railway museum in Atbara, Sudan.[7] That Kitchener's forces, nearly 25,000 strong by the time of their arrival at the gates of the Mahdist capital, arrived in such strength and supply was a testament to the importance of the railway in his conquest and as a "Tool of Empire" generally. What is lost in these stories of self-evidence, however, is the manner in which the railway began to transform perceptions of Africa as much as it did the landscape and metrics of transportation.

The origins of the Uganda Railway lie not in military success or grand colonial visions, but in the over-ambitious appetite of a failed company. British engagement on the east African coast began through their long relationship with Oman and their presence on the Omani expansion to the coastline of eastern Africa under Said bin Sultan. Through the use of diplomats and "advisors" the British were able to achieve a dominant political position within Zanzibar to

---

4 Callwell, 57.

5 It is impossible to prove, but Kitchener may have been influenced by Callwell's work given that it was published a decade prior to his campaign and the end of his chapter on supply concerns the disastrous Khartoum relief expedition which, "was essentially a campaign against nature, a struggle against the difficulties arising from supplying the wants of troops traversing great stretches of desert country." Callwell, 68–70.

6 Edward M. Spiers, "Introduction: The Reconquest Reappraised," in *Sudan: The Reconquest Reappraised* (London: Frank Cass, 1998; 1–10), 3.

7 Spiers, 3. For a tale of traveling, in the present, along Kitcheners route see: Minty Clinch, "Along Kitchener's route from Sudan" in *Financial Times* May 16, 2014 https://www.ft.com/content/b3a50556-d9f0-11e3-9b6a-00144feabdc0.

such an extent that they forced the Sultan to outlaw the slave trade.[8] Such an informal arrangement with the Sultan of Zanzibar suited British interests. The British government wanted to solidify a hold over as wide a coastal area as possible, while having to carry the most minimal of political and military costs.

With the beginning of the "Scramble for Africa," largely brought about by the effort of King Leopold II of Belgium to acquire a colony of his own, however, the strategic calculus for the British interests within Africa changed. At the Berlin Conference of 1884–1885, called largely to settle issues relating to European claims to African territory, a new standard was set for European engagement in the interior. Under the banner of "effective occupation," the British position within Africa was not only over-extended, but fundamentally weakened. The agreements which came out of the Berlin Conference meant that the "informal empire" the British had acquired under the auspices of the Sultan of Zanzibar would be insufficient to hold off the covetous dreams of German colonialists.[9]

In order to maintain their position on the East Africa coast, the British turned to William Mackinnon. Mackinnon was already a successful businessman who had established a prosperous shipping company in India.[10] He also possessed an active interest in further development in eastern Africa having already once approached, at that time unsuccessfully, the British government for permission to establish a company to administer the region. While the British government demurred initially, the new political situation presented a melding of interests between those of the British government and William Mackinnon. The British, in keeping with many a colonial adventure, sought to use Mackinnon's newly created Imperial British East Africa Company to solidify their position and hold off their German competitors. As for Mackinnon, he

8 For a good overview of the early British involvement in Oman and Zanzibar, see M. Reda Bhacker, *Trade and Empire in Muscat and Zanzibar* (London: Routledge, 1992) and for later British dominance see L.W. Hollingsworth, *Zanzibar Under the Foreign Office 1890–1913* (London: MacMillan & Company, 1953).

9 Despite the Conference establishing the British had control over the territory of Uganda, for example, that position was threatened by German expansion northwards along the coast of Lake Victoria-Nyanza. What was settled in a conference could be undone by such treaties signed with African leaders on the ground; which had British and German officials racing one another to have such treaties signed.

10 As the owner of the British India Steam Navigation Company, he already had a regular line of trade between Zanzibar and India. For a good overview of his company's work in the region, see P.I. McDermott, *British East Africa or IBEA: A History of the Formation and Work of the Imperial British East Africa Company* (London: Chapman and Hall, 1893).

responded out of a mixture of patriotism and of profit-motive, and officially formed the company in 1888.[11]

The Imperial British East Africa Company was born into a maelstrom of competing claims to eastern Africa and found it necessary to walk in the direction of Uganda well before it could crawl. The main reason for the about-face by the British government for Mackinnon's efforts was the growing threat posed by Carl Peters' German East Africa Company. Peters, a colonial enthusiast and founder of the German East African Society, embarked on a determined effort "to forestall foreign rivals by an aggressive policy of rapid land-grabbing in East Africa."[12] In 1883–1884, Peters travelled with three colleagues to the coast of Africa disguised as mechanics in order to stake a German claim to the region. Part of the reason for their disguise was that the German Foreign Office was unsupportive of their efforts as the government was still committed to leaving Africa alone.[13] Leaving Zanzibar on November 4, they "travelled through Uzingua along the Wami River into Uluguru, making bogus treaties as they moved along."[14]

After returning to Germany and presenting Chancellor Otto von Bismarck, who had little interest in colonies, with a *fait accompli,* Peters was granted an Imperial charter in 1885. Upon receipt of this charter, Peters proceeded to take two steps: to form the German East Africa Company with him as its managing partner; and immediately to send exploratory caravans to acquire more territory.[15] At its very outset, therefore, Mackinnon's company found itself in desperate need to expand quickly both to stabilize their fragile position on the coast and to prevent being boxed in by Peters' competing claims.

By far, the greatest prize for both of these companies was to establish a predominant position in the rich and populated region of Uganda. Located around the northern coast of Lake Victoria-Nyanza, the region of Uganda was highly regarded. As Henry Morton Stanley, the African explorer made famous by his expedition to rescue Dr. Livingstone, once alleged: "Since my return from

11 As the company later declared, in attempting to garner subsidies and support from the British government, the Company "were prosecuting national purposes in deference to the declared policy of the State." FO 2/58, 75.

12 W. Henderson, *Studies in German Colonial History* (London: F. Cass, 1962), 17.

13 Anne Perras, *Carl Peters and German Imperialism* (Oxford: Clarendon Press, 2004), 51.

14 G.C.K. Gawassa, "The German intervention and African resistance in Tanzania," in *A History of Tanzania*, ed. I.N. Kimambo and A.J. Temu (Nairobi: East African Publishing House, 1969), 98.

15 Kimambo and Temu, *A History of Tanzania*, 101, and Albert F. Calvert, *German East Africa* (New York: Negro Universities Press, 1970), 2. One of the first pushes was towards the north in an attempt to acquire not only Mount Kilimanjaro but also Mount Kenya.

Africa, I have persistently maintained that so long as there are no facilities of transport between Lake Victoria and the sea, nothing in the way of progress can be hoped for from East Africa…It is the 'Pearl of Africa' [Uganda] that is our object."[16] The Imperial British East Africa Company recognized that the coastal lands, while valuable, "could be little more than a barren acquisition" without control of the valuable interior.[17] Uganda was reported to be "populous, productive, and highly cultivated…and by far the most civilized."[18] More importantly, from the point of view of the British government, a strong British position in Uganda would protect the Nile basin assuring that "the immediate interests of the Company were identical with those of the nation."[19]

Given the competing interests and the shared estimate as to the value of Uganda, what followed could best be described as a race. The Imperial British East Africa Company and German East Africa Company pushed their administrators and caravans to maximum effort in an attempt to expand their holdings and lay claim to this pearl.[20] Mackinnon, utilizing the services of Frederick Lugard, who would later become Lord Lugard and Governor General of Nigeria, eventually beat out his rivals. Lugard was able to reach Uganda and established sufficient political connections to lay claim to the region and box out his German competitors. This success was seen as the crowning achievement of the Company's expansionary efforts, and one, it was hoped, that would assure the stability and profitability of the company for years to come.

Snatching the pearl of Uganda out of the clutches of Germany, however, proved to be a pyrrhic victory. Solving the problem of political control only created a graver and more intractable problem; namely, how the company would be able to efficiently reach and develop their newly claimed prize. Company caravans had proven useful for reaching and for demarcating territory for the Company to acquire, but had little utility beyond that. No caravan was guaranteed to succeed whether because of climate, such as the inhospitable Taru desert, or because of problems with the caravan itself.[21]

---

16 Henry Morton Stanley, "The Uganda Railway," *Saturday Review*, 1 June 1895, 719.

17 McDermott, 104.

18 Ibid., 104.

19 Ibid., 118.

20 One of the first actions the Company took upon arrival in Mombassa in 1888 was to equip a 700-man caravan to advance into the interior and established a station 250 miles from the coast. McDermott, 208.

21 The Taru Desert starts 44 miles inland from Mombasa at Samburu and ends at the town of Voi 104 miles inland. A waterless area, caravans were forced to push forward and, barring success at water holes near Mongu, would have to make the entire trip based on what water they could carry. The desert was so formidable that even after the construction of the railway the trains were limited to twelve cars and each train was forced to haul its own water supply. See E.H. Sadler, "Notes on the Geography of British East Africa" in *Journal*

The problems were only magnified once the caravans reached the interior. Food and supplies were never assured, and the populations of the central highlands were feared. All of these obstacles in time and in risk would be borne in the higher costs to transport anything along the caravan routes.[22] The caravan was not the proper vehicle for a claim of "effective occupation" and hindered any future plans for developing the colony.[23] Given that most of the economic hopes for Uganda rested on visions of large plantations producing cash-crops, a more regular and efficient means of travel was an economic necessity.

Thus, the Uganda Railway represented a clear tool by which the Imperial British East Africa Company intended to solve their problem. Unfortunately, while the answer to their problem seemed obvious, the means to achieve it proved elusive. Already wracked with the costs of claiming and administering such an expansive territory, the company did not have the means to construct a railway sufficient for their needs. The company thus found itself stuck in a trap of their own making. The only means by which they could pay for a railway, let alone make their company a profitable venture, would be via the mass export of goods from Uganda which necessitated a railway. Faced with this intractable problem, and unable to garner sufficient subsides from the British government, the company would end its days protesting that it had been hung out to dry after sacrificing itself for the public interest.[24] Facing a deficit of £193,757, and unable to make the territory profitable, the Imperial British East Africa Company relinquished their rights to the government in April, 1893, and, finally, collapsed as an entity by 1895.[25]

---

*of the Royal African Society* 11, no. 2 (January, 1912; 173–186), 179 and Robert W. Foran, *A Cuckoo in Kenya: The Reminiscences of a Pioneer Police Officer in British East Africa* (London: Hutchinson & Company, 1936), 60–61. Caravans themselves were no guarantee as a railway survey party once came across abandoned goods for an company station since the porters had simply deserted before reaching their destination. See M.F. Hill, *Permanent Way: The Story of the Kenya and Uganda Railway* (London: Hazell Watson and Viney Ltd, 1949), 71.

22 A caravan from the coast to Nairobi would cover the 330 miles in 28–30 days while, after the construction of the railway, would take only 24 hours. Additionally, the costs of transporting anything by porter meant an increase in 3 pounds per porter and thus "the Uganda Railway materially reduced the cost of living up-country." Foran, 73.

23 Based on his travels to Uganda, Sir Gerald Portal believed that Mombasa could never be the outlet for commerce if the porter remained the basic form of transportation. See: Sir Gerald Portal, *The British Mission to Uganda in 1893* (Westport, Connecticut: Negro Universities Press, 1970), 47.

24 McDermott, 246.

25 The consensus of many contemporaries echoed Hill who felt that the Imperial British East Africa Company's work "was widely recognized as one of the greatest philanthropic achievements of the latter part of the nineteenth century." Hill, 127.

Crafting a railway line from the coast to Lake Victoria-Nyanza proved a difficult undertaking which, although begun inauspiciously, eventually was achieved. In an effort to pressure the British government to provide funds for such a railway, the Imperial British East Africa Company created the ignoble "Central African Railway." As William Mackinnon noted in a letter to Lord Salisbury of 17th December 1890, the Company,

> Already anticipated the engagement of the Brussels Conference to a certain extent by erecting a land telegraph-line to connect the coast towns; by providing for the construction of 60 miles of narrow-gauge railway into the interior.[26]

While the directors laid the first seven miles of this railway to much public acclaim, the end result was not equal to the initial fanfare. Three years after the beginning of this railway, Sir Gerald Portal was sent on a mission to Uganda, and commented that the "little 24-inch tramway" was never used "except for occasional picnic parties from Mombasa."[27] Despite the assurances of *The Times* that, "it is not, after all, a very serious matter to build four or five hundred miles of railway over land that costs nothing," building such a railway was a massive undertaking by the British government.[28]

To begin with, very little was known of the territory the railway would cross given how little care it had received during the "Scramble." Henry Morton Stanley noted, in the same article on Uganda as the "Pearl of Africa":

> I take but little interest in the region through which the railway must run [to connect the coast to Lake Victoria], because of itself it is scarcely worth a thought. By itself it is not worth the luxury of a railway.[29]

As Sir Guildford Molesworth, the consulting engineer for Railways to the government of India, admited in a speech, "So little is known of the country that

26 Hill, 47. The mention of the Brussels Conference refers to the agreed upon conventions by the European powers to help put an end to the slave trade in Africa.

27 Portal, 28–30. His account of his experience on this tramway is nothing if not entertaining. The railway consisted of open-bed trollies which ensured that passenger luggage had an unfortunate habit of falling off in transit. The "end of the line" was not so much a station as a pile of loose luggage and boxes of unopened materials intended for the extension of the line.

28 Quoted in Hill, 60.

29 Henry Morton Stanley, "The Uganda Railway," *Saturday Review*, 1 (June 1895), 719.

many of the views I express must be mere guess-work."[30] An initial survey expedition was established in 1891 that sketched out several routes of advance for the railway, completing its work and sending a report to the government by 1893.

By January 1896, with the engineering team in place, the construction began.[31] Despite the railway traversing nearly the entirety of Kenya, it was telling how little care was paid to the land itself. As one author described the process:

> The early designers of the railways paid no heed to the economic development of the native tribes clustered around the skirts of Mount Kenya; no heed to the development of the Highlands of Kenya by white settlement, for such a possibility was then but casually envisaged; and only incidental to the development of the populated native areas through which the railway passed and reached. Their task was to design a cheap railway, along the easiest and shortest route, from Mombasa to the shore of Victoria Nyanza.[32]

And such a railway being "cheap" was indeed a relative term. Although early estimates as to the cost of the railway were calculated to be roughly £2,000,000, the eventual costs for building a railway would balloon to £5,500,000. The debate over expending even the estimated sum caused a huge argument within Parliament further illustrating how little attention was paid, whether by engineer or politician, to Kenya proper.

The largest defense for spending such public monies concerned the importance of Uganda as a political object. Lord Salisbury argued that if the British government would not act they would face a grave strategic threat as "there are four if not five Powers that are advancing towards the upper waters of the Nile." G. Curzon was more blunt in stating simply, "if we do not construct a railway to the lake the Germans will."[33] Amid all these arguments, however, it is telling

30 "Sir Guilford Molesworth, "The Uganda Railway," *International Engineering Congress* (Glasgow, 1901), 1–3.

31 For correspondence and government reports on the construction of the Uganda Railway see: FO 403/224, FO 403/258, FO 403/277, FO 403/291, FO 403/307, FO 403/317, FO 403/329, and FO 403/339.

32 Hill, 90.

33 Hill, 130–131. In October, 1897 Lord Salisbury went further in indicating to the chief engineer of the project, George Whitehouse, that "The line should be pushed on as soon as possible. It's a political necessity that we should get to the lake [Victoria-Nyanza] as soon as possible as the French were doing something on the Nile." Quoted in Henry Lubega, "History of Uganda Railway," *Daily Monitor* April 19, 2015 http://www.monitor.co.ug/Magazines/PeoplePower/History--Uganda--Railway/689844-2689814-ojyaeoz/index.html.

how little Kenya was ever mentioned when the construction bill was passed by a vote of 255 to 75.

If the survey and parliamentary debate proved difficult, no less taxing was the actual construction of the Uganda Railway. As one author related, "it seemed as if the very spirit of Africa resented the intrusion of the white man's railway."[34] One of the initial problems was the lack of ready labor. The British territory was wracked by a revolt from 1895–1896 which hindered not only the early construction of the railway, but also highlighted the potential danger of relying on African labor to complete the project.[35]

From the outset the British government chose to rely on Indian colonial expertise in railway development. Indeed, throughout the entire project, one can see the "stamp" of British India on the colony as much as anything else. The members of the initial survey party which demarcated the route from Mombasa to Port Kilindi were all drawn from India. The senior staff of the railway committee, including the chief engineer, served as part of the government of India or had experience in the sub-continent. Even more critically, it would be from India that the Uganda Railway project would draw the technical and lower-level staff necessary to bring the project to a satisfactory conclusion.

Such was the dependence on India for men with these skill sets that it was a source of ready stress and concern that railway projects within India might "steal" men away from east Africa.[36] Even at a practical level, Indian experience suffused the project as the Indian railway gauge was adopted, and Indian railway operating procedures as well. During construction, they also relied on the provision, by the Indian government, of a series of locomotives to aid in transportation.[37] As Sir Harry Johnson described it, the Uganda Railway meant, "the

---

34 Hill 158. Not only was this a matter of the difficult terrain and climate, but also dangers caused by malaria and the tste fly. And these were the dangers forseen by the engineers, not including the famed incident involving the "Man Eaters of Tsavo"made famous by Bruce Patterson's book that was turned into the 1996 film *The Ghost and the Darkness*. See Bruce D. Patterson, *The Lions of Tsavo: Exploring the Legacy of Africa's Notorious Man-Eaters* (New York: McGraw-Hill, 2003).

35 As Sir Guilford Molesworth would relate on an early report to the Imperial British East Africa Company, "the habits of the country render it doubtful whether it would be possible to hold out inducements to the aborignes to labour systematically at railway work." Thomas R. Metcalf, *Imperial Connections: India and the Indian Ocean Arena, 1860–1920* (Berkeley: University of California Press, 2007), 197.

36 As George Whitehouse would note to London, "The men I am now mostly in want of are clerks, signalers, and that class of subordinate, and I know of no place but India where these can be obtained," and worryingly noted, "it appears to be very difficult to get good men, as there seems to be a great demand for them in India." Metcalf, 193.

37 Although, as George Whitehouse would relate in his diary, they were not the modern locomotives he had requested but cast-offs from the Indian government. "I managed to

driving of a wedge of India two miles broad right across East Africa from Mombassa to the Victoria Nyanza."[38]

Yet it was in the pursuit of labor that one can see the glimmers of the transformative power of the Uganda Railway. As Indian laborers were imported into Kenya to help in construction, British officials began to conceive of a new vision for the land they were developing. As one official related in a report to the Foreign Office, "East Africa is, and should be, from every point of view, the America of the Hindu."[39] While such connections between India had been formed in the past and indeed sparked William Mackinnon's interest in the region in the first place, it is telling how important the railway was in fostering hopes for settlement and development. Kenya had been, as was made clear in the parliamentary debates, an afterthought, but now was being reshaped, by the construction of the railway, into a hoped for bastion of Indian settlement. Moreover, the hoped for settlement, it was believed, would consequently transform the entire region into a productive and prosperous colony. With the final railway spike driven in at Port Kilindi, in 1901, the railway was completed, but the fabrication of the colony of Kenya had just begun.

In the context of Kenya, the Uganda Railway reshaped attitudes towards the territory due to its practical effects, and, even more importantly, conceptual power. Practically, the railway served to regulate and to regularize traffic across the colony, to facilitate the development of inter-colonial as well as international trade, and to increase the level of governmental control across the colony of Kenya. The railway not only linked the designated points of the coast and Uganda, but also drew all of central Kenya into a web of imperial connections. In this regard, the railway served as a visible reminder of the dominance and power of the colonizer as well as an indication of the intended permanence of the imperial venture.

As Daniel Headrick has noted, it was the usefulness of technology in terms of transportation, communication, and logistics that proved crucial to imperial success as much as the application of military technology. In this case, the railway greatly facilitated the expansion of imperial oversight over the territory while ensuring that military force could now be efficiently brought to bear on any troublesome areas of the colony. The railway went hand-in-hand with this greater application of military force with pacification efforts by the King's

get 22 outdated locomotives from India, many of which were unable to cope with the demanding conditions of the Uganda Railway Project." Quoted in Henry Lubega, "History of Uganda Railway."

38 Metcalf, 188.

39 Report by Sir Harry Johnson quoted in Hill, 220.

African Rifles, which were transported and supplied by the Uganda Railway.[40] Thus, the railway was not simply a practical measure of dominance over the landscape, but an indication that Great Britain could summon the wealth and power of an industrialized nation-state to stabilize and to ensure their colonial rule. But, the railway was not only an imposition of physical power but also illustrated the power of technology to reshape human horizons.

The Uganda Railway served, to adapt a term from Wiebe E. Bijker, as a technological frame which would reshape Kenya. Bijker approaches the study of technology from the belief that all technology must be understood as being a part of a system, and "one must study how technologies are shaped and acquire their meanings in the heterogeneity of social interactions."[41] His insights are valuable as he seeks to move beyond positivist models of scientific and technological development, and instead move towards a more contingent view of these processes.[42] In the case of the Uganda Railway this acknowledgement of contingency, or the unintended, is particularly apropos given how little attention was paid to Kenya prior to the construction of the railway.

But, it is the concept of a technological frame that seems most useful in explaining the conceptual power of the Uganda Railway. According to Bijker a technological frame is built up via interactions around an artifact, or piece of technology, beginning with the concepts surrounding its use in problem solving.[43] These interactions could help explain how society could shape a technological design, and, more relevant to the Uganda Railway, how "technology structures the social environment."[44] Even more critical is how social actors

40 In tracing the British engagement with the Kikuyu in the Central Highlands of Kenya, Peter Rodgers notes that the effective suppression of the Kikuyu was tied to the movement of troops to Nairobi in 1902. Beyond that Nairobi, as a central stop along the Uganda Railway, ensured these forces "were capable of relatively rapid deployment to most parts of the Protectorate, a factor which broadened the horizons of possible military activity by virtue of proximity alone." Peter Rogers, "The British and the Kikuyu 1890–1905: A Reassessment" in *The Journal of African History*, 20, no. 2 (1979; 255–269), 266.

41 Wiebe E. Bijker, *Of Bicycles, Bakelites, and Bulbs: Toward a Theory of Sociotechnical Change* (Cambridge, Massachusetts: MIT Press, 1995), 6.

42 Ibid., 14.

43 Here the intersections are quite clear between Bijker's theories and those of Dr. Guilmartin. In his own description, Bijker indicates a similarity between his concept of a technological frame to that of Thomas Kuhn's "paradigms." In both regards the nature of the capability of the railway to solve the problem of connectivity also served to connect, and thus transform, thoughts about Kenya. Wiebe E. Bijker, "The Social Construction of Bakelite: Toward a Theory of Invention," in *The Social Construction of Technological Systems: New Directions in the Sociology and History of Technology*, Ed. Wiebe E. Bijker, Thomas P. Hughes, and Trevor Pinch (Cambridge, Massachusetts: MIT Press, 2012; 155–182), 168.

44 Bijker, "The Social Construction of Bakelite," 173.

will attribute "various meanings to an artifact."[45] The Uganda Railway, by the very act of its construction, enabled British officials and settlers to ascribe new meanings to the territory of Kenya.

These new meanings served to proscribe certain ways of thinking about Kenya. The technological frame not only "liberated" Europeans to think about the territory in a new way, it also served to "constrain" such thinking along a particular developmental path. As Sir Charles Eliot would later relate, after the completion of the Uganda Railway, "it changed the centre of gravity; it became at once the natural and only road up country."[46] Once the railway was built it was impossible to conceive of development along any other avenue than the one the railway created.

These multitudinous effects of railways have been well documented in other contexts as well.[47] When first appearing in Europe, for example, the railway changed European attitudes towards space, time and conceptions of what constituted "home."[48] The railway, thus, created an instrumental logic to the colony of Kenya where previously there had only been indifference.

The Uganda Railway not only facilitated transport across Kenya, but also "without it there would have been no social nor economic nor political progress."[49] The railway served as the chief impetus for development in the first manner through improved transportation. One of the difficulties that the Imperial British East Africa Company had floundered upon was the exceedingly high cost, in time and money, of transporting goods by porters.

---

45 Bijker, 159.

46 Eliot, 51.

47 In a colonial context, there is a whole sub-set of historiography of Indian Railway Studies with a recent compendium of new approaches in Ian J. Kerr, *27 Down: New Departures in Indian Railway Studies* (New Delhi: Orient Longman, 2007) as well as more specific cultural studies such as that by Laura Bear, *Lines of the National: Indian Railway Workers, Bureaucracy, and the Intimate Historical Self* (New York: Columbia University Press, 2007). But, it is not just in a colonial context that such works have been produced. For an examination of railways serving to construct national identity in South America, see A. Kim Clark, *The Redemptive Work: Railway and Nation in Ecuador, 1895–1930* (Wilmington, Delaware: SR Books, 1998). Even living in an age in which we see trains as anachronistic, scholars are exploring such questions as to how graffiti can shape the experience of passengers. For a compendium of such more recent research, see Benjamin Fraser and Steven D. Spalding, *Trains, Culture, and Mobility: Riding the Rails* (Lanham, Maryland: Lexington Books, 2012).

48 See Wolfgang Schivelbusch, *The Railway Journey: The Industrialization of Time and Space in the 19th Century* (Berkley: University of California Press, 1986). For another cultural look on the effect of railways, see Michael J. Freeman, *Railways and the Victorian Imagination* (New Haven: Yale University Press, 1999).

49 Hill, v.

For example, a caravan from the coast to Nairobi would cover the roughly 330 miles in 28–30 days; while, after the railway was constructed, the trip would take only 24 hours. Beyond speed, the "Uganda Railway materially reduced the cost of living up-country."[50]

It also made agricultural export a viable practice. Rates for shipping along the railway from the interior to Mombasa compared, "most favorably with similar rates in any other part of the world."[51] Functionally, without the regularity and ease of the railway system there would be no way of conceiving of, let alone achieving, economic productivity within Kenya. As Winston Churchill would put it rather bluntly, "civilization must be armed with machinery if she is going to subdue these wild regions to her authority, iron roads, not jogging porters; tireless engines, not weary men."[52] But, even more important than transportation costs was how the railway transformed what was believed possible in the colony.

A good example was the region around Kilindi as related in A.G. Anderson's *Our Newest Colony*. He related how the land was first-class, but suffered from lack of settlement due to the infrequency of rain. Anderson's contention was that this problem could be solved, and thus transforming the region into one of settlement, through the ability of the railway to deliver "over 2,000 gallons of water…for a few shillings."[53] But, by far the greatest example of the transformative power of the railway on Kenya was in the so-called "White Highlands" around Nairobi. As Sir Charles Eliot would relate, "The importance and, one may say, the political existence of East Africa are due to this line without which the Highlands could not be utilized."[54] The regions around Kikuyu territory had long been recognized as an area of fertile soil and good climate, but, absent a railway, those remarks had simply remained curious observations.

The arrival of the Uganda Railway meant the region could now become the primary area of agricultural export and settlement. As early as 1901 the East African Agricultural and Horticultural Society established itself and began

50 Foran, 73. Foran surmised that costs were £3 per porter (with a shilling per porter per pound of weight they had to carry) which would drastically increase the transportation costs.

51 Anderson, 34. For example, Anderson notes that to ship Potatoes in bulk – ten tons – one could ship them from Nairobi to Mombasa for about 7 shillings and six pence per ton. Comparing such rates to an equivalent amount carried by porters helps underscores the necessity of the railway for making any large-scale agriculture possible in Kenya.

52 Quoted in Hill, 297.

53 Anderson, 109.

54 Hill, 263.

publishing a farming quarterly. By 1904, experimental farms had been established at Nairobi and three agricultural shows were held.[55]

Even as the railway helped produce such developments, it served as the main artery from which future plans were made. Branch lines were planned to help facilitate settler access to timber at the Uasin Gishu Plateau in the hopes that doing so would make it even easier for settlers to plan and to develop the land around the White Highlands.[56]

By way of contrast, the lack of a railway proved detrimental to any efforts to develop the Northern Frontier District. To be fair, at the outset, the Northern Frontier District (NFD) was a "harsh and arid region" which suggested many obstacles to the development of this region of the colony.[57] In addition to the difficulties of climate, there were no easily useful routes of transportation. Caravan routes passed far south of this region and there were no rivers beyond the limited navigability along the Juba River.

If climate and transportation were problems to be overcome, they were made all the worse by the lack of defined borders.[58] While accords were signed between the British and their European neighbors, the boundary between Abyssinia and Kenya was imprecise and open to the interpretation of traveling herders.[59] A survey expedition along the boundary was undertaken in 1902–1903 under Philip Maud, but it was not until 1907 before a draft treaty was prepared.[60] To indicate how far "behind" the NFD was from the rest of the colony, the Uganda Railway had reached the Victoria Nyanza in 1901.

The lack of a clear border only underscored the region's "chronic security problems."[61] Abyssinian raiders were a constant worry as the Northern Frontier District abutted the unstable southern border of the Ethiopian kingdom.

55 Hill, 286.

56 Eliot, 179.

57 Barber, James. "The Moving Frontier of British Imperialism in the Lake Rudolf Region 1890–1919." *Ethnohistory* Vol 53:1 (Winter, 2006, 143–172), 143.

58 Governor Eliot described the "line drawn on most maps" as "quite imaginary," and believed "the delimitation of the Abyssinian frontier" could "precipitate" the settlement and development of the district. Eliot, 121 and 183.

59 Barber, 145.

60 In fact, the administration of this territory was so porous that the initial survey was completed only when Archibald Butter, leading a private expedition to southern Ethiopia, allowed Philip Maud to join him. Even this treaty would be unsatisfactory leading to a further delineation mission sent under Maj. Charles W. Gwynn. See Charles W. Gwynn, Major, "A Journey in Souther Abyssinia," *Geographical Journal* 38 (1911), 113–39, and Charles W. Gwynn, Major, "The Frontiers of Abyssinia: A Retrospect," *Journal of the Royal African Society*, 38, (1937), 150–61.

61 Barber, 144. In 1963 the Earl of Lytton in an address to the House of Lords noted that "It is virtually a desert…This desert is so poor and so unimportant that it is called a District

Lower Ethiopia, largely settled by nomadic herding populations, presented a fluid boundary that was often violated. Even more threatening, these raids appeared to be moving further southward with each successive year, with a real danger that the "White Highlands" might be threatened.[62] Even at the southern-most edges of the NFD, which was within range of the Uganda Railway, the British were unable to adequately control much beyond the railway line.[63] These problems only escalated prior to the First World War. With the illness of Emperor Menelik II the Abyssinian control over the southern regions deteriorated even further leading to "a spate of bandits and ivory hunters" crossing into the NFD.[64] Col. Thesiger, commanding a KAR detachment in the NFD, warned action had to be taken "unless we are going to be content to sit on a powder magazine and smoke pipes until something happens."[65]

While diplomatic efforts continued, the violence in this region led to the deaths of several European officers. The conflict was grave enough that one officer, William Lloyd-Jones, received the last Distinguished Service Order given before the Great War.[66] These losses pushed the British government to consider radical commitments to try to correct the perceived breakdown across the entire border region.[67]

---

only. It is not even a Province. The Province of which it is a part is so unimportant that it is not given a separate map in the Kenya atlas."

62 These continued as Governor Eliot's memoirs, published in 1905, that the "previous six years" had seen a steady rate of Abyssinian raiding. He feared that if they continued at their present pace they would, by 1911, "be on the Uasin Gishu plateau and the slopes of Mount Kenya." Eliot, 184.

63 These were not seen as mere economic violations, though the officials did want the revenue, but as affronts to the authority of the British government. See R.W. Beachey, "The East African Ivory Trade in the Nineteenth Century," *Journal of African History*, 8, no 2 (1967), 269–290; Peter T. Dalleo, "The Somali Role in Organized Poaching in Northeastern Kenya, 1909–1939," *The International Journal of African Historical Studies*, 12, no 3. (1979), 472–482; Michael L. Stone, "Organized Poaching in Itui District: A Failure of District Authority, 1900 to 1960," *International Journal of African Historical Studies*, 5, no 3 (1972), 436–452; and George L. Simpson, "Frontier Banditry and the Colonial Making Process: The East Africa Protectorate's Northern Borderland prior to the First World War," *International Journal of African Historical Studies*, 29, no 2 (1996), 279–308.

64 Simpson, 279.

65 Simpson, 290.

66 He was wounded in action and recounted his exploits in a post war memoir. See Willian Lloyd-Jones, *kar: Being an Unofficial Account of the Origin and Activities of the King's African Rifles* (London: Arrowsmith, 1926).

67 While the war put a stop to these plans the problem was so grave that a meeting in Nairobi in 1914 had decided on tripling the strength of troops in the Northern Frontier District as well as creating two flying columns to better handle the mobile raiders. They also

Beyond these problems, however, it was the lack of a railway, and the technological frame it would have provided, which circumscribed the vision of what was possible there and limited the development of the region as a whole. While regions abutting the Uganda Railway were developing, with consequential administrative growth, the Northern Frontier District garnered no such sophistication. Absent any technological frame the NFD remained an inchoate territory to British officials. In official documents and maps of the period, the contrast could not be plainer with the NFD being depicted as wild and uncontrolled. Roads, more paths than anything else, were depicted in marches rather than miles. These differences, while appearing minor, signified a completely different conception of this area of the colony.

In public documents, the territory along the Uganda Railway was depicted in miles. Indicating distances in this manner suggested orderly transport at a steady rate of speed, and the assumption that there would be no disruption in transit. Every announcement in the *Kenya Gazette* on ticket prices and train schedules further solidified that the southern region was settled, orderly, and efficient. In contrast, wells were still depicted along the routes of the Northern Frontier District, in much the same way as they had on the sketch maps of explorers, suggesting that traveling in this region was a step back in time as much as it was in development.

Absent the organizing logic of a railway, the Northern Frontier District appeared out of step with the rest of the colony with control only being assured by men with guns, rather than the strong metal ties of the railway. The continuing effect of the lack of a railway can be seen in the delayed development of the Northern Frontier District. As late as the 1960s, there were still debates within Britain as to whether the government should even bother trying to develop the region.[68]

The role of the railway in development was best illustrated by how the Uganda Railway transformed the relatively unimportant stop of Nairobi into the center of commerce and governance. As indicated earlier, the survey for the Uganda Railway sought to find the easiest and most cost-effective route through Kenya in order to reach Lake Victoria-Nyanza. As Sir Charles Eliot would relate in his work on the protectorate and his governorship there:

---

suggested purchasing vessels to patrol the Juba river and improving roads in the NFD at an initial cost of £64,017 and annual expenditure of £45,261. Simpson, 305.

68 In 1963 the Earl of Lytton in an address to the House of Lords noted that "It is virtually a desert…This desert is so poor and so unimportant that it is called a District only. It is not even a Province. The Province of which it is a part is so unimportant that it is not given a separate map in the Kenya atlas." Barber, 144.

> [The Uganda Railway] was not built, as is generally the case, in order to connect a series of known points, but was driven through an entirely unknown country. Hence it does not so much traverse the best districts, as supply a series of points from which these districts may be easily reached.[69]

There is perhaps no greater example of this fixation on cost and on ease than the station of Nairobi. For ease, the surveyors of the railway decided to follow closely the caravan routes of ivory and slave traders within eastern Africa. Nairobi had been a caravan stop located halfway between the coast and the Victoria-Nyanza. Its name, "Enkare Nairobi" meaning "Cold Water" in Maasai, underscored its importance as a stop for caravans dependent on water along their travels.[70] That the placement of the station was motivated more by cost-concerns than a predetermined plan for development is that its location was horrid. As late as 1906, there were debates over whether to move the entire town of Nairobi 10 miles up-country to overcome the problems of flooding.[71] If poor placement did not suggest much for the growth of a town, neither did the fact that it was placed "327 miles from the nearest place where even a nail could be purchased."[72] Like the Northern Frontier District, there was little to suggest this spot would become the locus of all political, military, and economic power in Kenya. What Nairobi had, however, was the Uganda Railway.

"Nairobi was not even a native village, only a barren camping site" when the railway was first built, but the railway proved crucial in the transformation of the town and the colony.[73] Even the construction of the railway and Nairobi station necessitated residences for Indian laborers and European officials, along with the beginnings of a town to service their needs. A town plan from 1902, a mere one year after the completion of the railway itself, indicated a European Bazaar as well as buildings set aside for detachments of the King's African Rifles.[74] Such planning showcased how the logic of the railway was beginning to drive forward development given that at that time Nairobi only possessed 30 settlers.[75] Despite such simple beginnings, the town of Nairobi

69 Sir Charles Eliot, *The East Africa Protectorate* (New York: Barnes & Noble, 1966), 18.

70 C. Blevin and J. Bouczo, "Nairobi: A Century of History 1898–1997," in *Les Cahiers de l'IFRA* (Nairobi: Institut français de recherché en Afrique, 1997), 30.

71 Foran, 122 and Blevin, 33, 35.

72 Blevin, 34–35.

73 Foran, 73.

74 See *Plan of Nairobi Town Showing Proposed New Sites of Military Lines and Other Government Buildings*, Mombassa, Public Works Department, 1902. MPK1/74.

75 Sorrenson, 25.

only continued to grow. Each train that arrived in Nairobi signified the stabilization of the colony and normalized the land. What had previously been a wilderness, best exemplified by the famed "man-eaters of Tsavo" which had hindered construction of the Uganda Railway, became a jumping off point for big game hunters in governmentally established "game reserves."[76] Even the very act of riding the railway, with all of the regularity and symbols of a similar act in Europe, made Kenya appear to be a place of regular order and worthy of settlement and development.[77]

> It was a telling sign of how critical the railway was believed for development that the capital was shifted from Machakos, which had been the administrative center since 1887, to Nairobi. The former station, and center of governance for the territory, had been by-passed by the route of the Uganda Railway and was left to wither away.[78]

Nairobi developed to such an extent that it became socially stratified with prosperous Europeans residing on "The Hill" while lesser Europeans, let alone Indians and Africans, were confined to "The Plain." The social stratification was probably best observed by Robert Foran who originally arrived in Nairobi as a big-game hunter and was welcomed on "The Hill." When he returned to the colony as a police officer, however, he found himself unwelcome as a mere public functionary. As he noted,

---

76 Early on the selling of licenses was seen as a useful form of governmental revenue as "the thousands of wild animals in this region could be turned to profitable account through the issue of game licensese to sportsmen and local residents." Foran, 268. For various coverage of the development, and problems, of organizing the game reserve system see: Edward I. Steinhart, "Elephant Hunting in 19th Century Kenya: Kamba Society and Ecology in Transformation," in *Journal of African Historical Studies*, 33, no. 2 (2000), 335–349 and Michael L. Stone, "Organized Poaching in Itui District: A Failure of District Authority, 1900 to 1960," in *InternationalJournal of African Historical Studies*, 5, no. 3 (1972), 436–452.

77 Even recent travelers on the railway have indicated elements that, while anachronistic to modern travelers, are suggestive of how riding the railway was a process of suggesting British dominance over the territory as much as political or military conquest. "For a traveler, these days, a ride on the railway seems an odd blend of colonial heritage and a continent's modernity. Just after the express leaves Nairobi or Mombasa at 7 P.M., for instance, an attendant in a white uniform will perambulate through the corridors of the first-class section ringing out a tune on a small xylophone. That is the signal that dinner is to be served, on tables laid with white linen and gleaming silverware." Alan Conwell, "Kenya's little track of iron that almost wasn't," *New York Times* April 7, 1983 (http://www.nytimes.com/1983/04/07/world/kenya-s-little-track-of-iron-that-almost-wasn-t.html).

78 Bleven, 34–35.

> A man might belong to the leading Clubs in London, be a member of the British aristocracy, have been presented at Court, hold His Majesty's commission, be University or Public School, but if serving as a subordinate official in British East Africa the social bars were rigidly raised against him.[79]

This social stratification was only furthered by the development of the town itself. In 1901, the Nairobi Club was founded which included extensive facilities including a racecourse laid out by the East Africa Turf Club.[80] As early as July, 1904 there was a "Race Week" in Nairobi with cricket matches, dinner parties, polo games, concerts, and other events.[81] Nairobi even became a boom town for Indian settlement with the National Bank of India establishing a small branch in the Treasury Office.[82] As Robert Foran would note, by the end of his tour in Kenya in 1908, Nairobi was "beginning to look like a real town."[83]

If Nairobi was becoming a real town, so was Kenya becoming a true colony. As A.G. Anderson would write in his propagandistic *Our Newest Colony*, Kenya was "a land where the out-crowded and heavily burdened tax-payer of the older civilization may seek a more simple and yet brighter life."[84] That these were the descriptions applied to a colony that had been such an afterthought was only due to the Uganda Railway. The "lunatic line," which had been passed amid rancor in order to achieve a basic political object, served as the life-line that helped transform the land of Kenya into one of settlement. The railway, therefore, was a creation which solved a particular problem, but also served to create a solution for Kenya that went beyond the visions of their own creators.

To return to Dr. Guilmartin's formulation, the Uganda Railway perfectly encapsulates his belief that technology was "any physical or mental tool created by a person or society for the purpose of problem solving." As documented, however, the solution of linking Uganda to the coast also created a new problem to solve; what to do with Kenya itself. But, the Uganda Railway also serves to reiterate why Dr. Guilmartin felt that studying technology was crucial to military history. Studying technology in all its details, rather than leaving it as a

79 Foran, 118.

80 Hill, 287.

81 Foran, 151. What occurred in Nairobi was also emulated across several of the other towns along the route of the Railway with larger towns establishing their own cricket, hockey, polo, and golf clubs. Anderson, 20–21.

82 Foran, 129.

83 Foran, 307. He even noted in passing that new shops were being set up in Nairobi including a safari outfitting company and a photography studio.

84 Anderson, 1.

"black box" as some historians do, often serves as a means of freeing historical study from the dominance of assumed narratives. Dr. Guilmartin proved this with *Gunpowder and Galleys* in which an analysis of the technology of triremes pushed back against the Mahanian narrative that Mediterranean warfare was "backward."[85] Studying the Uganda Railway provides a similar service against colonial narratives. In the case of the "Scramble for Africa" historians generally view the application of military force as something which ends with the conquest of the colonies themselves. Setbacks, such as the British defeat at Isandlwana, are noted but generally the focus is on the inexorable success of European arms. Historians then leave Africa, almost as G.F.W. Hegel argued, only to return for large-scale rebellions or the world wars.

By focusing attention on the Uganda Railway, however, military historians could instead come to see force as an ever-present constant in colonial Africa. The forces of the King's African Rifles were not stationed in Nairobi simply to solidify that settlement but served as the "tip of the spear" for further extensions of colonial control.[86] The "technological frame" of the Uganda Railway not only transformed attitudes towards Kenya as a whole, but also the territory along the railway line as well. Military force proved crucial in pacifying the regions around Nairobi to transform it from a waypoint along a caravan route to the central area of settlement as the "White Highlands." Military force pacified the region along with forcing tribes into the colonial reservation system designed to support European plantation agriculture.[87] Thus, technology provides a window by which historians can view a more comprehensive picture of imperialism. Probably the greatest inducement for historians to study technology is that contemporaries often are as aware of the transformative power of technology as our discipline should be. In the case of the Uganda Railway a fitting summation of its importance was provided by Sir Edward Grigg, later governor of Kenya, in a speech at the African Society:

85 See John F. Guilmartin, *Gunpowder & Galleys: Changing Technology and Mediterranean Warfare at Sea in the 16th Century* (Annapolis, MD: Naval Institute Press), 2003.

86 See G.H. Mungeam, "Masai and Kikuyu Responses to the Establishment of British Administration in the East Africa Protectorate" in *Journal of African History* 11, 1 (1970; 127–143) and Peter Rogers, "The British and the Kikuyu 1890–1905: A Reassessment." In *Journal of African History* 20, 2 (1979; 255–269).

87 For an examination of how this process played out with the Massai, see Richard Waller, "The Maasai and the British, 1895–1905 and the Origins of an Alliance" in *Journal of African History* 17, 4 (1976; 529–553) and for the Giriama see Cynthia Brantley, *The Giriama and Colonial Resistance in Kenya, 1800–1920* (Berkeley: University of California Press, 1981).

"Into this empty paradise, only at the beginning of this century, there enters a British Government and British colonists, all on the back of the Uganda Railway – which alone made it possible – for the railway is the beginning of all history in Kenya. Without it, there would be no history of Kenya. This country was really nothing but a corridor... [it was] not conquered by force of arms. It was conquered by one of the greatest forces of our modern civilization: it was conquered by a railway...The railway brought settlers and our Government in its track, and it was the railway which created Kenya as a Colony of the Crown."[88]

88 Hill, v.

# PART 3

## *The World Wars*

CHAPTER 8

# German Technology and the Origins of World War II in East Asia

*Robyn Rodriguez*

Throughout the 1930s, over one hundred German military advisors, many of whom had been some of Germany's most innovative officers in the First World War, went to China to train and to modernize Chiang Kai-shek's army.[1] Chiang had long been infatuated with the grandeur of the German military, the country's rapid unification, and its rise to great power status. He wanted to create a modern China that could compete—militarily and industrially—with Japan and the western powers. Chiang sought out German military officers, many of whom found themselves unemployed after the down sizing of the German military, due to the restrictions of the Versailles Treaty, to assist him in the quest to modernize his military. The German advisors believed that by aiding Chiang's military modernization, they were building a future ally and trading partner for Germany.

Beginning in 1928, the German advisors began training Chiang's army. They imported not only German machines and tactics, but also ideas about the relationship between nationalism, technology, and power. The 1932 battle between the Chinese and Japanese forces in the cosmopolitan center of Shanghai was a pivotal moment for the German military advisory mission. Despite their ultimate withdrawal, the performance of Chiang's divisions was a testament to the progress they were making in the conduct of modern warfare under the German advisors.

With the embrace of German technology, Chiang's military became a source of national pride during and after the 1932 battle. Following the battle of Shanghai, the German advisors continued to train Chiang's troops and began planning for the next, inevitable Sino-Japanese war. The importation of German technology gave Chiang, and China, the tools and, more importantly, the confidence needed to fight Japanese aggression as the 1930s progressed. The confidence in his German-trained troops and German-developed war plans

1 Note: Pinyin will be used to transliterate Chinese names except in cases of commonly known names and places, like Chiang Kai-shek, Yangtze, and Manchuria. Citations have also retained the original spellings used in the primary sources.

 | DOI:10.1163/9789004393301_010

has long been an overlooked factor in Chiang's decision to go to war against Japan after the Marco Polo Bridge Incident in 1937, rather than quickly negotiate a settlement as in previous incidents.

## 1 Importing Western Technology in the 19th Century

In the nineteenth century, Japan and China both sought western technology as a means of defense against the aggression of foreigners, notably from the European powers. As historian David Ralston points out, 'the introduction of European-style military techniques and institutions into… China and Japan was a significant first step, probably an irreversible one, in the modernization of these countries.'[2] Western technology could not be imported in a military vacuum. Machines and ideas aimed at military modernization needed to be situated in the broader social context and would have a great impact outside the military sphere.

Japan was the most successful of the non-western countries at transforming its military and society in the second half of the nineteenth century so that it could compete with the Great Powers militarily and economically in the twentieth century. Japan's leaders who initiated military change recognized that the might of European militaries rested on a strong economy and an educated, energetic population, and thus, anticipated the fundamental changes needed in Japanese society to support modernization. The Japanese relied heavily on the French and Germans to transform their military in the second half of the nineteenth century. By the turn of the twentieth century, Japan had a European-style military supported by a modern social, political, and economic infrastructure.[3]

China encountered more friction as it sought to emulate the European style militaries in the nineteenth century. Despite being far advanced compared to the barbarian and backward Europeans upon first contact in the 1500s, by the nineteenth century, China suffered one military defeat after another to the western powers. In response, Chinese leaders sought to import European hardware and to build western arsenals, but the Chinese ruling class was unwilling to make the deeper social changes to accommodate new technology. Defeat

2 David B. Ralston, *Importing the European Army: The Introduction of Military Techniques and Institutions into the Extra-European World, 1600–1914* (Chicago: University of Chicago Press, 1990), 173.

3 Ralston, *Importing the European Army*, Chapter 6.

in the Sino-Japanese War and in the Boxer Rebellion at the end of nineteenth century demonstrated China's desperate plight, and the ruling elites began working to reform the military after 1900.

While they began to introduce European style military training and organization, the Manchu ruling elites were largely unwilling to accept, let alone promote, the socio-economic changes required to cultivate a professional military. Ralston eloquently explained that military reform in China 'amounted to little more than exchanging Confucian roles for German epaulets.'[4] Ultimately, the socio-political currents unleashed by the importation of European training, education, and organization resulted in revolution in 1911 that toppled the Qing dynasty, and initiated a civil war that lasted nearly four decades.[5]

In the wake of the Qing dynasty, warlords, whose power rested on the military might of their armed forces, became the principal political actors in China. Historian Edward Dreyer has highlighted the divergence between the superficial appearance of the warlord armies as modern in terms of terminology, uniforms, and occasionally their weaponry, and the deeply rooted pre-modern social structure of the warlords' armed forces, which were essentially feudal in organization. In the 1920s, Chiang Kai-shek, with his army, became essentially the most powerful warlord and the first serious contender to unify China.[6]

Chiang rose from the ashes of a China torn by revolution, seeking to modernize his army and the Chinese nation-state. He looked to Germany as a country in which the military had played a central role and sought out German military and economic advisors to aid him in transforming China. Chiang was attempting to modernize China through the militarization of society, to create a new social order and a modern industrialized nation-state with a world-respected military, all while maintaining traditional Chinese culture and identity.[7] He had come of age during a time of outside abuse of Chinese sovereignty and territory, when Chinese armies suffered numerous defeats at the hands of western or westernized military forces. Having studied in Japan, he witnessed first-hand the military prowess of a recently modernized Asian country. He sought to import German technology to modernize his military to defend against Japanese encroachment and to turn China into a great power.

4 Ralston, *Importing the European Army*, 133.

5 Ralston, *Importing the European Army*, Chapter 5.

6 Edward L. Dreyer, *China at War, 1901–1945* (New York: Longman, 1995), 5, 56, 76–78; Ralson, *Importing the European Army*, 140–141.

7 William C. Kirby, *Germany and Republican China* (Stanford: Stanford University Press, 1984), 47.

## 2 The 1932 Battle of Shanghai

In January 1932, after the Mukden Incident, Chiang found himself again facing Japanese aggression. This time it was much closer to the capital in Nanjing, as the Japanese military threatened the metropolitan center of Shanghai. Chiang's troops had been working with the German advisors for barely four years when they were tasked with defending Shanghai. Only two years earlier, they could barely hold their own against the ragtag warlord armies in the Chinese Civil War; yet, by 1932, they had advanced considerably under the tutelage of the German advisors. For over a month in Shanghai, Chiang's forces waged a strong defense against the Japanese onslaught. The tenacity of the defense surprised the Japanese aggressors and international observers alike. The 1932 battle marked an important developmental moment for Chiang's forces. Though Chiang's soldiers ultimately had to withdraw from the city, their defense of Shanghai for over a month against the more modern, materially superior Japanese forces became a mark of Chinese pride.

The strong Chinese defenses in Shanghai in 1932 displayed the influence of their German advisors. The Chinese defenses included an elaborate trench system, barbed wire, machine gun nests, and obstacles to hinder the Japanese advance. One historian, Donald Jordan, described the Chinese defensive tactics as 'flexible and resilient, the Chinese seemed to follow the *tao* of falling back when necessary to survive, then returning to fight again.'[8] The Chinese defenses, designed by Chiang's German advisors, embodied elastic defense-in-depth developed by the Germans in the First World War, and it proved to be effective against the Japanese attackers in Shanghai.

In the middle of the battle on 13 February, one Chinese general reported back to leadership in Nanjing that the Japanese had attempted to land reinforcements at Wusong. The Japanese pre-invasion artillery bombardment had little effect upon the Chinese defenders, protected in their deep defensive positions. When the Japanese landed, the Chinese counterattack pushed them back and inflicted heavy casualties on the invader.[9] Despite the heavy attack by superior Japanese artillery, the German-designed Chinese defenses were able to hold.

8 Donald A. Jordan, *China's Trial by Fire: The Shanghai War of 1932* (Ann Arbor: University of Michigan, 2001), 115. For a first-hand account of the Chinese defenses, see Hauptmann Krummacher's report on the Battle of Shanghai, 1932, *Bundesarchiv-Militärarchiv*, Freiburg, Germany, (BA-MA) MSG 160/2.

9 '*Die Ereignisse des jap./chin. Konflikten ab* 27.1.32' BA-MA MSG 160/2.

The German advisors trained and armed Chiang's 87th and 88th divisions, which manned the defensive lines north of Shanghai. An American army officer who observed the battle of Shanghai credited the divisions 'for the determined stand at Chiangwan,' and 'the German training which it is learned was based, in the preparation of defense positions, on the supposition that Chinese, poorly equipped, must prepare to hold lines against an enemy greatly superior in mechanical [arms].'[10]

The 87th Division utilized German defense-in-depth tactics to successfully defend Liuhe for much of the battle, preventing the Japanese forces from outflanking the larger Chinese defensive network. This stubborn defense forced the Japanese to continually bring in reinforcements. In his report on the battle, an Italian observer reported that the Chinese defense 'proved more tenacious than the Japanese and foreigners had expected.'[11] He noted the fact that time and again the Japanese were surprised by the ability of the Chinese divisions—a markedly inferior force in terms of training and armament, yet still able to give a good account of themselves. The German advisors recognized the material inferiority of the Chinese troops, but planned the defenses accordingly in the urban environment in order to minimize the striking power of the more modern and mechanized Japanese forces. The Japanese 9th Division war records reported that 'the attack revealed the presence of a defense system prepared with great care. This shows how the enemy had, with considerable foresight, foreseen the future action.'[12]

The defense of Shanghai may have lasted longer, had the Chinese commanders of the 88th Division, defending Jiangwan, followed the advice of their German advisors more closely. Early on in the battle, one Chinese general expressed his desire to attack the Japanese forces. The lead German advisor, retired General Georg Wetzell, the good Clausewitzian that he was, reminded his Chinese colleague that defense is the stronger form of war. He recognized that the Japanese benefitted from air superiority and were protected by naval gunfire as well. In the Great War and in the Chinese Civil War, he witnessed units attacking day after day without making any forward progress.[13]

---

10 Captain Mayer's report of 2 March 1932 MID report 2657-I-276, National Archives II, College Park, MD, (NARA II). It is quoted in Jordan, *China's Trial by Fire*, 155.

11 'The Shanghai Incident: January 20-March 5, 1932. Causes of the Conflict and The Military Action,' MID report 2657-H-377-6, NARA II RG 165, Entry No. 65 Microfilm M 1216, roll 7, 52.

12 'The Shanghai Incident: January 20-March 5, 1932. Causes of the Conflict and The Military Action,' MID report 2657-H-377-6, NARA II RG 165, Entry No. 65 Microfilm M 1216, roll 7, 1.

13 '*Die Ereignisse des jap./chin. Konflikten ab* 27.1.32,' BA-MA MSG 160/2.

Nonetheless, the Chinese commanders of the 88th Division launched a number of attacks against the Japanese forces. These aggressive thrusts managed to push the invaders back periodically, but at great cost.[14] After nearly a month of this offensive-defense at Jiangwan, the 88th Division required reinforcements to replace its losses. The 87th Division, which was successfully defending Liuhe utilizing German tactics, was ordered to withdraw in order to reinforce the 88th Division. Only then were the Japanese able to breakthrough at Liuhe.[15]

Despite the withdrawal, the month-long defense of Shanghai was a Chinese victory of sorts. Much of its success was due to the foresight of Chiang's German advisors. They had planned the Chinese defenses, trained the 87th and 88th Divisions, and continually advised the Chinese leadership through the duration of the battle. During the battle, the German advisors were in constant contact with Chiang Kai-shek, the Chinese High Command, and the war minister. The German advisors working directly with the 87th and 88th Divisions went to the front lines. The head of the advisory mission, General Wetzell, even traveled to Shanghai to observe the battle firsthand in order to better advise Chiang on future action. The German advisors made recommendations to the Chinese High Command as well as lower level leaders on troop movements and actions, strengthening the defenses at decisive points throughout the region of Shanghai, and peace negotiations with Japan and the international community.[16]

## 3 Nationalism and Technology:

Despite the breakthrough of the Japanese forces through the Chinese defensive line after the withdrawal of the 87th to reinforce the 88th division, the fierce resistance of the Chinese troops became a mark of national pride. It was a testament to progress. The incursion of Japanese troops bred unity among the Chinese and gave them a common cause around which to rally. Jordan wrote that 'for many Chinese, to merely survive in positional warfare such pounding from Japanese naval and army tanks and cannon and bombing for over

14 'The Shanghai Incident: January 20-March 5, 1932. Causes of the Conflict and The Military Action' MID report 2657-H-377-6, NARA II RG 165, Entry No. 65 Microfilm M 1216, roll 7, 3–4, 32, 38.

15 'The Shanghai Incident: January 20-March 5, 1932. Causes of the Conflict and The Military Action' MID report 2657-H-377-6, NARA II RG 165, Entry No. 65 Microfilm M 1216, roll 7, 3–4, 32, 38.

16 '*Die Ereignisse des jap./chin. Konflikten ab* 27.1.32' BA-MA MSG 160/2.

a month seemed to be China's first modern victory.'[17] For the Chinese, their performance in the battle of Shanghai was a moral victory. The battle proved that the Chinese forces, aided by their German advisors and utilizing German arms and tactics, were indeed capable of fighting in modern warfare. This perceived military success, largely the result of the importation of German technology, began to accomplish Chiang's goal—a fusion of modern technology and Chinese tradition.

Dr. John F. Guilmartin, Jr., defined technology broadly to include 'any physical or mental tool created by a person or society for the purpose of problem solving.' Technology is vital to the study of military history. In warfare, competitors rely on technology, whether it is in the form of a machine or an idea, to achieve superiority. Dr. Guilmartin considered ideology a technology, as it was a form of social engineering. The shared worldview was a mental tool that could be used to organize societies and define power relationships.

In the west, ideologies defined political goals and motivated fighting forces. Nationalism, one of the defining ideologies of the modern western world, became a driving force in the wars of the French Revolution and afterwards. In the Great War, millions of men sacrificed themselves in the name of nationalism. Chiang, impressed with the passionate German nationalism that took root long before the country was unified, wanted to cultivate a sense of nationalism in China that was rooted in its long history, yet forward looking. He hoped that developing a sense of nationalism would aid him in his quest to politically unify the country.[18]

Though China may have been considered a nation long before 1900, it was only in the last decades of the nineteenth century that the Chinese people became aware of themselves as such, in part due to a growing awareness of the country's susceptibility to outside aggression. The roots of this western-style nationalism in China extended back to the turn of the twentieth century when Chiang came of age as a military officer. Under the Manchu reform movement in the early 1900s, a number of military schools were established in China with western instructors and a western-style education for military officers. As David Ralston pointed out, with this western instruction, young Chinese officers were exposed to the political and philosophical ideas of the west including nationalism.[19]

17 Jordan, *China's Trial by Fire*, 235–236.

18 Jay Taylor, *The Generalissimo: Chiang Kai-shek and the Struggle for Modern China* (Cambridge MA: Belknap Press of Harvard University Press, 2009), 102–103, 108. Kirby, *Germany and Republican China*, 45.

19 Ralston, *Importing the European Army*, 135–136.

Chiang sought to import not only German armaments and tactics for his military, but also a Prussian-German style nationalism.[20] Chiang had long been intrigued by German nationalism. He hoped to emulate Bismarck's 'blood and iron' policy, in which Prussia forged a unified Germany through a series of wars.[21] Military might was at the heart of Prusso-German power. A common question among Prusso-Germans was whether they were soldiers or citizens first as the two identities were inextricably linked. Chiang recognized the central role that the German military played in unification and national identity and wanted to emulate that in China.

Chiang's China was fractured among the domains of warlords. He saw his military as a tool for creating a new nation-state.[22] Military service was one means of indoctrinating the population with a sense of national identity. Military victories against the warlords and the communists provided a means of conquering and of forcing national unity. Military successes against a foreign power, like Japan, bred national pride and support of the military, which led to the nation becoming invested in the military.

While Ralston examined the exposure of Chinese military officers in the late nineteenth century to western nationalism among other ideologies, it was also important to consider the role of technology in western political and philosophical ideas. Throughout the modern era, Europeans measured themselves based on the superiority of their scientific thought and technology compared to non-western peoples. Throughout the industrial era, as historian Michael Adas points out, 'European thinkers concluded that the unprecedented control over nature made possible by Western science and technology proved that European modes of thought and social organization corresponded much more closely to the underlying realities of the universe than did those of any other people or society.'[23]

Europeans had an unquestioning faith in the power of technology, which Chiang and many of his military officers were exposed to during their military education by western instructors. It was only the calamity of the Great War

20 Kirby, *Germany and Republican China*, 47.

21 Kirby, 47.

22 Taylor, *The Generalissimo*, 101; Kirby, *Germany and Republican China*, 47; Historian Bernd Martin acquired a private English translation of the GMD's Office of Military History 'Summary of the Work of the German Military Advisory Group in China' (*Te-kuo Chu Hua Chün-shih Ku-wen-t'uan Kung tso Chi-yao*) (Taipei, Taiwan: Bureau of Military History of the Ministry of National Defense, Republic of China, 1971), BA-MA MSG 160/12 2.

23 Michael Adas, *Machines as the Measure of Man: Science, Technology, and Ideologies of Western Dominance* (Ithaca: Cornell University Press, 1989), 7.

that forced the Europeans to begin questioning this faith in technology.[24] In the Great War, technological innovation and nationalist competition collided in a continental calamity in which millions of men sacrificed themselves on the altar of nationalism.

Many of the German advisors in China had been trained and indoctrinated with the western faith in machines, but witnessed their destructive nature in the Great War as battles devolved into industrialized slaughterhouses. The German advisors recognized the power of technology, such as the railroad, communication network, or highway system, to physically tie China together for both military and civil purposes as they had done in Germany but they also intimately understood the destructive potential of the new weapons systems that the Europeans were rapidly developing and that were being adopted by the Japanese, to include aerial bombardment and poison gas. Many of the German advisors in China had been some of the most innovative thinkers in the First World War and were involved in developing defense-in-depth and storm troop tactics as a means of mitigating the deadly effectiveness of the new machines on the battlefield.

The European fascists, who grew out of the destruction of the First World War, believed in embracing technology's destructive capability in war as a means of national regeneration. The German advisors in China were dispersed across the right of the political spectrum. Most were conservative nationalists, but a handful were ultra-nationalists and Nazis. No matter their precise location on the right, historian William Kirby points out that for Chiang, the employment of German advisors was 'compatible with what may be called his effort at "conservative modernization."'[25] The right-leaning tendencies of the German advisors were conducive to Chiang's beliefs about development of the nation-state.

Historians have long debated the fascist nature of Chiang's nationalist China. Some scholars have argued that Chiang was a Confucian fascist while others, like his biographer Jay Taylor, believed he was not a fascist in ends or means.[26] Whether or not Chiang should be assigned to the category of fascist was less important for this chapter than a consideration of the components of fascism that he did want to import to China. Historian Philip Morgan defines fascist movements as

24 Adas, *Machines as the Measure of Man*, Chapter 6.

25 Kirby's, *Germany and Republican China*, 46–47.

26 Taylor, *Generalissimo*, 101–103. For a summary of the historical debate on Chiang as a fascist, see Hans J. van de Ven, War and Nationalism in China (New York: Routledge, 2012), 164–166, and Fredrick Wakeman, 'A Revistion view of the Nanjing Decade: Confucian Fascism,' The China Quarterly, no. 150, (June 1997), 395–432.

> radical hyper nationalist cross-class movements with a distinctive militarist organization and activist political style… [that] sought the regeneration of their nations through the violent destruction of all political forms and forces which they held to be responsible for national disunity and divisiveness, and the creation of a new national order based on the moral 'spiritual' reformation of their peoples, a 'cultural revolution' achievable only through the 'total' control of society.[27]

Chiang used military might and political finagling to secure his position as the most powerful warlord in China. The communists became the greatest threat to Chiang's consolidation of power and China's internal unity. Chiang sought to stamp out the communists using his military to encircle and annihilate them on the battlefield. While he came up short in violently destroying the enemy, they did retreat temporarily in the 1930s to the remote outskirts of China.

Chiang lacked the totalitarian control of China that Mussolini and Hitler had over Italy and Germany. While he may not have cultivated the blind following of and obedience to the leader, he did appreciate the idea of total control over society. The Blue Shirts and the New Life Movement certainly appeared fascist in form and function.[28] The Blue Shirts, the secretive paramilitary group within the Guomindang, was similar to the Italian Black Shirts and the Nazi Brown Shirts. It promoted Chiang's desired new Chinese society and sought to shore up support for Chiang but it failed to create the mass political movement reminiscent of the European fascists.

The New Life Movement, which Chiang launched in 1934, was an effort to reshape Chinese society to accommodate modernization. According to historian Hans J. van de Ven, the movement advanced 'a secularised, essentialised, and militarised understanding of Confucianism to promote a new imagining of China, transcending regional, cultural, religious, and personal bonds. In the army, it aimed to make China's divided military forces cohere around a single *esprit de corps*, instill a new pride, and strengthen identification with the National Government.'[29] The New Life Movement was Chiang's program for creating a new national order grounded in a cultural revolution and a spiritual reformation of the nation, which Morgan described as an essential component of fascism though Chiang lacked the vital element of total control.[30]

---

27 Philip Morgan, *Fascism in Europe, 1919–1945* (New York: Routledge, 2003), 13–14.

28 Frederic Wakeman, 'A Revistion view of the Nanjing Decade: Confucian Fascism,' *The China Quarterly*, no. 150, (June 1997), 395–432.

29 Van de Ven, *War and Nationalism in China*, 1925–1945, 165.

30 Morgan, *Fascism in Europe*, 13–14.

The European fascists were infatuated with the aesthetics of technology. Machines, such as artillery or aircraft, were symbols of power. The Wehrmacht, an industrial war machine, was on a pedestal in Nazi Germany. The Nazis lauded its weapons, like the tank or aircraft, as "manifestations of the German 'spirit'... [infused] with the German soul."[31]

In a similar merger of man and machine, Chiang wanted to utilize technology to move Chinese society from its traditional form to a modern state, melding tradition with new machines, processes, and ideas. While he wanted the military to embrace new technology, the Chinese soldier was at the center of the new army. The soldier was the vessel for carrying Chinese culture and thought as well as loyalty to Chiang, while manipulating western machines. Chiang's army became a vehicle for building a new Chinese nationalism—embracing new technology while upholding China's traditions and cultures.

The German advisors guided the development of Chiang's military forces encouraging the Chinese soldiers to maintain their unique identity but to embrace western technology. The surge of nationalism that followed the defense of Shanghai in 1932 was exactly what Chiang desired: national pride grounded in military prowess. This newfound military capability was inextricably linked to the embrace of western technology. As in the west, nationalism in China was becoming tied to technology as a new measure of success.

## 4 Planning for the Next War

After the 1932 battle, the German advisors continued to train Chiang's army. They encouraged Chiang to pursue better relations with Japan in the mid-1930s. In order to build a new military, Chiang's forces needed a respite from fighting the Communists and the Japanese. By the fall of 1934, Chiang had concluded his encirclement campaigns against the Communists, who fled in the Long March, and he began quietly pursuing friendlier relations with Japan. While improved relations with Japan were unpopular among almost all the Chinese factions, it gave the German advisors time and space to build Chiang a modern army and prepare for the next, inevitable Sino-Japanese War.

In 1934, the German advisors created a large-scale defensive exercise around Chiang's capital city of Nanjing. Building on the experience of Shanghai, the German advisors created a fully developed defense in depth battle for the exercise. While the basic elements of defense-in-depth were present in Shanghai in 1932, they had been hastily, if not haphazardly, put together. The 1934

31 Morgan, *Fascism in Europe*, 192.

exercise around the capital city tested an elaborate, well-prepared defensive system.[32] During the 1932 battle of Shanghai, there had been anxiety among the Chinese leadership about the hostilities spreading up the Yangtze River to Nanjing, which was largely defenseless. The German and Chinese leadership even considered evacuating the capital city during the fighting in Shanghai.[33] The 1934 exercise was designed to reassure leadership about the military's ability to defend the capital and execute a successful defense-in-depth.

The five-day military exercise around Nanjing proved to be quite the spectacle of military power. Representatives from thirteen of China's provinces came to observe the exercise and witness the strength of Chiang's modern divisions. The exercises tested not only the skill of Chiang's troops but also the performance of the German advisors, who trained them. The German-trained divisions performed admirably during the exercise. Chiang was so pleased after observing the 1934 defensive exercise around Nanjing that he ordered two more massive exercises for the spring and fall of 1935 to test three divisions in air defense and maneuver warfare.[34]

The exercises in 1935 were increasingly complex. These exercises involved tanks, aircraft, and flak batteries. It showcased the German technology that was imported as well as the strides the Chinese troops were making as they embraced German tactics and experimented with maneuver warfare.[35] The new weapons symbolized Chiang's growing power over the land and air of the battle space.

As the Chinese troops were displaying their proficiency with German arms and tactics, the German advisors began planning for the next Sino-Japanese war. War planning was its own sort of German craft industry. The advisors, led by retired General Alexander von Falkenhausen, were optimistic that Chiang's troops could successfully defend against a Japanese attack. In the 1930s, von Falkenhausen may have been the most experienced German military officer in Far Eastern affairs. He had taken part in the suppression of the Boxer Rebellion at the turn of the century and he also served as the military attaché in Japan before World War I. He was intimately familiar with the Japanese military and offered a bleak, albeit realistic, assessment for China regarding Japan's strength, but noted that Japan had not completely modernized its military.

32 Letter from von Falkenhausen to Brinckmann, Nanking, 25 December 1934, BA-MA MSG 160/13, 1–2.

33 *Die Ereignisse des jap./chin. Konflikten ab* 27.1.32," BA-MA MSG 160/2.

34 Letter from von Falkenhausen to Brinckmann, Nanking, 25 December 1934, BA-MA MSG 160/13, 1–2.

35 Letter from von Falkenhausen to Brinckmann, Nanking, 6 January 1936, BA-MA MSG 160/13, 4.

Japan was still in the process of mechanizing its forces.[36] Despite Japan's military superiority, the German advisors suggested to Chiang that in the event of a Sino-Japanese war, China might be able to draw support from another great power, which would require Japan to reconsider its attack.[37] The German advisors recognized that while direct support from the United States and Britain would be difficult to secure, a Japanese attack on China could draw them into the struggle—economically, if not outright militarily—given their interests in Chinese markets.[38]

The German advisors highlighted Japan's fragile economic situation to Chiang. It was one of Japan's great vulnerabilities. Von Falkenhausen's experience in Japan gave him credibility and Chiang trusted him. He told Chiang that Japan could not financially sustain a long war against China. An attack on China would likely mean that Japan could no longer rely on monetary support, or weapons deliveries from the United States and England. Thus, it was in China's best interest to hold out against Japan and make the war last as long as possible. The prospect of a long war would require Japan to come to the negotiating table.[39]

## 5 A Dangerous Gamble

The German advisors speculated that the Japanese leadership, upon encountering real resistance in China, would be willing to negotiate a settlement rather than continue to pour military resources into China to fight a long war. The German advisors also recognized that China would have to demonstrate that they were capable and willing to fight Japan in order to draw support from the western powers. Chiang's troops had begun to demonstrate that they could defend against a Japanese attack at the battle of Shanghai in 1932, which drew international attention and surprised many observers, westerners and Japanese alike. Von Falkenhausen made it clear to Chiang that his forces would have to fight like that again in order to bring Japan to the negotiating table.[40]

36 Von Falkenhausen, *Grundsätze der Landesverteidigung Chinas*, Nanking, 26 January 1935, BA-MA N 246/7, 1.

37 Report from von Falkenhausen to Chiang, Nanking, 12 August 1935, BA-MA N 246/7, 7–9.

38 Von Falkenhausen, *Grundsätze der Landesverteidigung Chinas*, Nanking, 26 January 1935, BA-MA N 246/7, 2–4.

39 Von Falkenhausen, *Grundsätze der Landesverteidigung Chinas*, Nanking, 26 January 1935, BA-MA N 246/7, 3–4.

40 Report from von Falkenhausen to Chiang, Nanking, 12 August 1935, BA-MA N 246/7, 13.

The German advisors anticipated that a Japanese attack on China would descend from the north. The plains of northern China gave the advantage to the Japanese mechanized forces. They could move quickly and bring the full weight of their firepower to bear against the Chinese defenders. The German advisors planned to use the Yellow River as the main line of defense. Some troops would be stationed north of the Yellow River. Their purpose was not to defeat a Japanese attack in that territory, but rather slow it down as they fell back to the main line of defense on the Yellow River.[41] The bulk of the Chinese forces would be stationed between the Yellow and Yangtze rivers—most of which were positioned as reserves along the Yangtze to quickly deploy forward to the Yellow River, or to move to defend the coastline against a Japanese landing.[42]

The German advisors recognized the decisive role of the Yangtze River. If Japan attacked up the Yangtze River, took Nanking, and advanced to Wuhan, it would effectively cut China in two. China would lose some of its strongest industrial areas along the river, and if Wuhan fell, it would lose access to foreign supplies coming in from Canton. Despite the prevailing view of the Yangtze River as a vulnerable target in the Chinese defense, the German advisors believed it could be defended. While it was broad, its difficult fairway made it challenging for naval ships to maneuver. In the lower straits near Shanghai, it could be blockaded by mobile artillery and air power.[43]

In their war planning, the German advisors repeatedly stressed the need for Chiang's troops to fight like they had in 1932. Von Falkenhausen pointed out that the Chinese must defend forward and not cede any ground without a fight. The German advisors feared that China's resistance meant that it would suffer heavy casualties as Japan unleashed the brunt of its superior firepower and material resources; yet the potential of China losing its industrial areas, if they gave up too much ground, was of far greater concern. The German advisors believed that with a sufficiently prepared defense, 'the assailant will not be able to break the resistance at first assault, as we are comparatively strong in infantry and machine-arms.'[44]

41 Von Falkenhausen, *Grundsätze der Landesverteidigung Chinas*, Nanking, 26 January 1935, BA-MA N 246/7, 7.

42 Von Falkenhausen, *Grundsätze der Landesverteidigung Chinas*, Nanking, 26 January 1935, BA-MA N 246/7, 7–10. Von Falkenhausen, *Gesichtspunkte für den Nach—und Abschub*, Nanking, 27 May 1936, BA-MA N 246/7.

43 General von Falkenhausen to Chiang Kai-shek, Nanking 12. August 1935, 10–11 BA-MA N 246/7.

44 Von Falkenhausen, '*Denkschrift über die Verteidigung Chinas gegen Japan*,' August 1935, BA-MA MSG 160/2, 11. See also Report from von Falkenhausen to Chiang, Nanking, 12 August 1935, BA-MA N 246/7.

In 1932, Japan had not expected such a ferocious defense of Shanghai and had to bring reinforcements into the city. The German advisors suspected that once Japan encountered 'real resistance' again, it would reconsider its efforts at expansion on the Asian mainland.[45] They hoped a valiant Chinese defense would draw international support and bring Japan to the negotiating table.

The German advisors knew that Chiang would have to mobilize all of China's military forces, economic resources, and people to resist a Japanese onslaught.[46] Most of the German advisors in China had fought in the Great War and understood total war to mean the complete mobilization of society to support the war effort. They believed that the brief surge of nationalism following the 1932 battle of Shanghai showed that a Japanese incursion into China could unify China's people against an outside threat and that Chiang would be able to mobilize all of China's resources to engage in a total war.

## 6 The Outbreak of World War II

The next Sino-Japanese War, and ultimately World War II, began on 7 July 1937, with the Marco Polo Bridge Incident in northern China. That night, the Japanese were conducting military training maneuvers outside Beijing—a frequent occurrence. Several treaty agreements entitled them to conduct such maneuvers and it was common for them to do so in northern China; however, they were supposed to secure permission from local Chinese officials. On the night of 7 July, no Chinese official had been informed that the Japanese were conducting maneuvers. Hearing gunfire from the blanks discharged by Japanese rifles to simulate combat conditions, Chinese forces in the area fired upon the Japanese. There were no casualties, but the next morning a Japanese soldier was missing. The Japanese commander, assuming the missing soldier (who was later found unharmed) had been captured by the Chinese, ordered his forces to attack the town of Wanping. This attack led to several skirmishes between Japanese and Chinese troops in the area.[47] Following the shooting at

45 Report from von Falkenhausen to Chiang, Nanking, 12 August 1935, BA-MA N 246/7, 2–3.

46 Letter from von Falkenhausen to Madame H.H. Kung, dated 6 September 1935, BA-MA MSG 160/2, 11. See also Report from von Falkenhausen to Chiang, Nanking, 12 August 1935, BA-MA N 246/7.

47 Jonathon D. Spence, *The Search for Modern China* (New York: W.W. Norton), 444–446. Peter Harmsen, *Shanghai 1937: Stalingrad on the Yangtze* (Havertown, PA: Casemate Publishers, 2013), 22–25.

the Marco Polo Bridge, the local Chinese and Japanese commanders worked to negotiate a settlement; however, the ceasefire was thwarted by both the Konoe government in Japan and Chiang.

Like Chiang, the Konoe government, according to Dreyer, in line with the high army and navy leadership in Tokyo, hoped to prevent war in China in 1937 in order to focus on military modernization. While the Japanese armed forces were more modern than China's, they still could not compete with the Great Powers. The Japanese leadership hoped to avoid war with China so that they could continue modernizing their forces in order to be able to survive in a war with Russia or Britain.[48]

While the Konoe government worked to prevent further hostilities, according to diplomatic historian Akira Iriye, it succumbed to the pressure of public opinion that demanded punitive action against China for the shooting. Prime Minister Konoe supported the army's plans to send three divisions from Japan to China as an insurance against future hostilities. Ironically, the cabinet approved this measure on 11 July, the same day local commanders agreed to a cease-fire.[49]

Chiang refused to endorse the local cease-fire and called for a settlement negotiated by the central governments. He took a hard line against Japanese aggression and sent four divisions to northern China, appealing to the signatories of the Nine Powers Treaty for assistance, and declaring in a speech that no more Chinese territory would be surrendered to Japanese control.[50] Chiang decided to openly resist Japan after the Marco Polo Bridge Incident rather than settle on Japanese terms. On 11 August, he sent his best, German-trained troops to Shanghai, thus escalating the skirmishes occurring in northern China into a Sino-Japanese War.[51]

Without the German defensive plans and the confidence the advisors gave Chiang that his troops could indeed fight a defensive campaign, the Marco Polo Bridge Incident may well have remained a limited incident, like the Mukden Incident in 1931. The importation of German technology, including armaments, tactics, and war-planning played no small part in Chiang's decision to engage the Japanese in Shanghai. The German advisors were confident in the ability of Chiang's forces to defend the country, and he trusted his advisors.

48 Dreyer, *China at War*, 211.

49 Akira Iriye, *The Origins of the Second World War in Asia and the Pacific* (New York: Longman, 1987), 42.

50 Dreyer, *China at War*, 211.

51 Harmsen, *Shanghai 1937*, 31, 39–40.

Chiang's forces and their German advisors had demonstrated their ability in the battle of Shanghai in 1932 and in the subsequent military exercises.

Chiang wrote to his son, who was studying at one of the German military academies, 'Do not be distracted by the Japanese invasion. I have the means to counter them.'[52] The means were his German-trained divisions. On the brink of war, Chinese financier and former Minister of Finance, T.V. Soong, who was also Chiang's brother-in-law, wrote that the Chinese army "has been trained by the Germans and we have spent much to modernize. It knows that it is stronger and has the confidence that it will not be beaten this time."[53] The Chinese leadership and troops had great confidence in their German training and armaments.

Despite the estimations of the German advisors that Chiang's army would not be modernized until 1938, they believed that Chiang could successfully defend against a Japanese attack, especially by engaging the Japanese in terrain that was favorable for defense and negated the advantages of the more modern Japanese forces. The German advisors encouraged him to fight in Shanghai. Unlike northern China, the urban battlefield was less conducive to the mechanized Japanese forces. While the Japanese had the advantage of naval gunfire, artillery, and air power, the Chinese defensive positions surrounding Shanghai were well prepared. Japanese bombardment of the city would be sure to provoke international opinion as well.

The German advisors were instrumental in the defense of the city. Since the 1932 battle, the Chinese had set up numerous defensive measures, based on German advice, on the outskirts of Shanghai, especially near the coast. Much of the defense had been based on the plans drawn up by von Falkenhausen between 1935 and the start of the war. One advisor describing the Chinese defensive positions wrote that the obsolete connected lines of defense were gone, and single nests stood in their place. The advisors were impressed with the Chinese soldiers, who stuck bravely to their positions during Japanese attacks.[54] Despite the ability of the Chinese troops to hold their defensive positions in the face of superior firepower and the advice of the Germans, the local Chinese commanders insisted upon attacking the enemy as the 88th Division had done in 1932 in suicidal frontal attacks.[55] The German advisors believed

52 Quoted and translated by Taylor, *Generalissimo*, 146.

53 Quoted and translated by Kirby, *Germany and Republican China*, 222.

54 Untitled Report, presumably written in September 1937, in which Otto Heinrichs discusses his experiences. BA-MA MSG 160/8.

55 Harmsen, *Shanghai 1937*, 53, 71–75.

that while the tenacious Chinese defense was valiant and courageous, it was also ill advised because of the massive casualties that resulted.[56]

In what is perhaps the ultimate expression of the embrace of western technology by both the Chinese and Japanese, the 1937 battle of Shanghai became an industrialized killing ground. Chiang's troops suffered massive casualties in the defense of Shanghai, while inflicting heavy casualties on the Japanese invaders, who continually had to pour reinforcements into the city. An estimated 40,000 Japanese troops were killed in the three-month battle.[57] Casualty estimates for the Chinese forces during the battle range between 150,000 and 300,000.[58]

The German advisors were horrified at the deliberate sacrifice of life in the 1937 battle of Shanghai. One Chinese military historian who had worked with the German advisors wrote, "The German advisers were at complete variance with the Chinese concept of 'live and die in defense.'"[59] The German advisors were witnessing the Chinese struggle to survive against an opponent fueled by a western-style hyper-nationalism. Ironically, the 1937 battle in Shanghai foreshadowed the horrific battles that the Germans would soon be fighting from Stalingrad to Berlin, which included orders to stand and die on the defense.

## 7 Conclusion

In November 1937, Chiang's forces withdrew from Shanghai. The German advisors planned for China to fight a long war of attrition. While they might not have anticipated the irreparable losses of Chiang's best divisions in Shanghai, the advisors were still confident in China's ability to defend the country against Japan and to continue the war after the battle. Despite their miscalculations about Japan's willingness to negotiate once it experienced real bloodshed, Chiang still trusted his German advisors. The three-month battle in Shanghai gave the Nanjing government time to relocate to Chongqing and move a small, but significant, portion of China's industrial base to the interior. Chiang was also able to transfer his headquarters to Wuhan. During the Battle of Shanghai, the

---

56 Liu, *A Military History of Modern China*, 163.

57 Edward Drea, *Japan's Imperial Army: Its Rise and Fall, 1853–1945* (Lawrence: University Press of Kansas, 2009), 196.

58 Taylor places Chinese casualties around 187,000, *The Generalissimo*, 150, while Dreyer and Hsi-sheng Ch'i put it at 300,000. See *China at War*, 218 and *Nationalist China at War, Military Defeats and Political Collapse, 1937–45* (Ann Arbor: University of Michigan Press, 1982), 43.

59 Liu, *A Military History of Modern China*, 163.

Chinese, as a nation, rallied to fight against the brutal invaders and the subsequent atrocities in Nanjing only hardened their will to resist.

The importation of German technology, both machines and ideas, throughout the 1930s gave Chiang and his troops the confidence needed to fight Japan in 1937. Despite not having a completely modernized military or a country unified around the Guomindang, Chiang went to war against Japan. He saw no point in negotiating with the enemy, a position fueled by conceptions of empire, radical nationalism, and racial superiority that had been imported from the west.

When Chiang, encouraged by anti-Japanese public sentiment and confidence from the German advisors, decided to escalate the fighting in northern China by opening a new front in Shanghai in August 1937, he created a war for survival. The China Theater in the late 1930s augured the savagery of the war that was about to spread to the European continent. Barbarism and atrocities did not require western machines, but they were fueled by western ideas of racial and national superiority in both Europe and Asia. Western machines only amplified the scale of the slaughter.

CHAPTER 9

# Freeman's Folly: The Debate over the Development of the "Unarmed Bomber" and the Genesis of the de Havilland Mosquito, 1935–1940

*Sebastian Cox*

The de Havilland Mosquito is, justifiably, considered one of the most famous and effective military aircraft of the Second World War. The Mosquito's development is usually portrayed as being a story of a determined and independent aircraft company producing a revolutionary design with very little input coming from the official Royal Air Force design and development process, which normally involved extensive consultation between the Air Ministry's technical staff and the aircraft's manufacturer, culminating in an official specification being issued and a prototype built. Instead, so the story goes, de Havilland's design team thought up the concept of the "unarmed speed bomber" all by themselves and, despite facing determined opposition from the Air Ministry and the RAF, got it adopted by persuading one important and influential senior officer, Wilfrid Freeman, to put it into production (Illustration 9.1).[1]

Thus, before it proved itself in actuality a world-beating design, it was known in the Ministry as "Freeman's Folly". Significantly, even the UK Official History on the "*Design and Development of Weapons*", published in 1964, perpetuated this explanation, stating that:

> When … [de Havilland] found itself at the beginning of the war short of orders and anxious to contribute to the war effort they proceeded to design an aeroplane without any official prompting from the Air Ministry. They had to think out for themselves the whole tactical and strategic purpose of the aircraft, and thus made a number of strategic and tactical assumptions which were not those of the Air Staff. While the Air Ministry was still wholly devoted to the doctrine of night bombing by heavy bombers, Sir Geoffrey de Havilland conceived the idea of day bombing

1 Amongst the earliest books to put forward this hypothesis was Edward Bishop, *The Wooden Wonder* (London, Max Parrish, 1959), especially Chapters 2 & 3. See also C Martin Sharp and Martin Bowyer, *Mosquito* (London, Faber and Faber, 1967), Chapter 1.

 | DOI:10.1163/9789004393301_011

ILLUSTRATION 9.1
Air Chief Marshal Sir Wilfrid Freeman.
CREDIT: AIR HISTORICAL BRANCH (RAF)

by fast unarmed aircraft…It was …very fortunate that almost from the outset of the negotiations with the Air Ministry, the Air Member for Development and Production, Sir Wilfrid Freeman, was able to back the de Havilland proposal. He regarded it as a gamble, but thought the risks worth taking.[2]

2 M M Postan, D Hay and J D Scott, *Design and Development of Weapons* (London, HMSO, 1964) 85–86. Wilfrid Freeman's nephew and biographer, Anthony Furse, includes some detail of the discussions which took place within the Air Ministry on the question of the unarmed bomber, but sticks firmly to the line that only Freeman recognised the worth of the concept and does not accurately record the fluctuating debate within the Air Staff or the influence (except negatively) of Air Marshal Ludlow-Hewitt at Bomber Command set out below. Anthony Furse, *Wilfrid Freeman: the genius behind Allied survival and air supremacy 1939–1945* (Staplehurst, Spellmount, 2000), 70–72 and 118–126. The title of Furse's biography of Freeman accurately reflects the approach taken. Malcolm Smith in his work, *British Air Strategy between the Wars* (Oxford, Clarendon, 1984) recognises and briefly discusses the debates within the Air Ministry over the relative merits of speed and defensive armament in bombers and the Ideal Bomber paper referred to below. Smith also recognises the part played by Ludlow-Hewitt, but does not address the early papers circulated prior to 1938, nor the vital role played by Freeman. Smith, 240–247.

So, one independent-minded firm and one perspicacious officer, more or less between them, ensured that the aircraft entered service with the RAF. This is certainly a romantic view, but it is not entirely accurate.

The truth, whilst still reflecting great credit on Freeman's foresight and judgement, and placing him firmly front and centre in the development of the final Mosquito design, is very much more complicated. Rather than simply a case of a pioneering firm and a buccaneering air marshal the reality of the Mosquito's entry into service is inextricably entwined with the wider aspects of British re-armament in the 1930s, and illustrates well the very many factors involved in adopting or adapting a particular technological design as a weapon system. The story of the Mosquito's acceptance by the Air Ministry in fact illustrates clearly the many conflicting pressures which play a part in the adoption of a particular design for production, and shows how the merit of the actual design is just one, and not necessarily the most important, of many factors at play. In the case of the Mosquito, the factors were technical, tactical, strategic, political, financial, doctrinal, and industrial, and thus a microcosm of those which affected the RAF as a whole.

The genesis of the Mosquito as a design does indeed lie with the de Havilland design team, but not in 1938 or 1939. Its lineage was the de Havilland Comet Racer which won the 1934 air race from Mildenhall in the UK to Melbourne, Australia, achieving this feat whilst combining the inimical features of speed and range. The Comet was a sleek, twin-engined racer, with a stressed wooden skin, which incorporated the then-novel features of a retractable undercarriage, wing flaps, and variable-pitch propellers. There followed the de Havilland Albatross, a streamlined four-engined airliner, with the range to fly to Berlin and back.[3]

The significance of the Comet's performance and its possible military implications, at a time when the international situation showed the first ominous signs of a dangerous deterioration, had not gone unnoticed in the Air Ministry. Those historians who have seen fit to castigate the Air Staff for their supposed failure to appreciate the arguments in favour of the Mosquito appear to be unaware that the Air Ministry first investigated the question in the summer of 1935, three years before the de Havilland design team produced any serious suggestions. In June 1935, an anonymous Air Staff paper considered "The Possibility of Dispensing with the Armament in the Light Bomber Class", and quite specifically addressed the question of whether "a bomber, by stripping it of its defence, [can] be made so much faster as to obviate the need for

3 On de Havilland aircraft see A J Jackson, *de Havilland Aircraft Since 1909* (London: Putnam, 1962), *passim.*

self-defence?"[4] The paper admitted that the question had been brought into prominence by the appearance of the Comet, but it concluded that the Comet's startling performance, in terms of speed and load-carrying capacity, was a result of its very short gestation period, and that the real need was to reduce the time lag between design and production for RAF aircraft. This point led the author to a related question which was to become a constant refrain in the discussion on unarmed, or "Speed", bombers throughout the next four years. Specifically, the concern with this aircraft type was that its duration of technical superiority over enemy fighters—most especially with regard to speed—was likely to be so short-lived that its service life would be brief and its development, therefore, uneconomic.

An accompanying paper addressed the question of the influence of tactics on the design of bombers. Whilst this paper did not mention the unarmed bomber specifically, it did appear less hostile to the concept of performance. It pointed out that every design was a compromise between the inimical factors of range, load, and armament, and those of speed, rate of climb, and production cost. The author stated that, since any practical experience of war, aircraft had undergone a technical revolution, and that it was therefore impossible to predict "whether high speed and light defensive power, or lower speed and heavier defensive power will prove the better formula. To concentrate on one design would, therefore, be most dangerous".[5] The Air Staff must then be given credit for considering the problems of armament versus speed at an early stage, even if on balance they tended to favour the provision of adequate defensive firepower.

It is also important to appreciate that the term light bomber meant, at that time, single-engined biplanes such as the Hawker Hind. The point is important because of the influence of factors other than the largely tactical and technical ones considered in these early papers.

On the political front, faced with the continuing deterioration of the international situation and the increasing domestic political pressure for rearmament, particularly air rearmament, the Government had voted increased amounts of money to the RAF. The political imperative, however, was to build up the RAF's frontline strength as rapidly as possible, in order to deter potentially hostile European powers, particularly Germany, and to reduce the level

4 "The possibility of dispensing with the armament in the light bomber class", note by the Air Staff, 22.6.1935. UK National Archives [hereafter NA] AIR2/2715.

5 "The influence of tactics upon design of bombers for the air defence of Great Britain", Air Staff paper prepared for Lord Weir and submitted on 10.7.1935, NA AIR2/2715. Lord Weir was at the time industrial adviser to the Air Ministry.

of domestic criticism over the state of the nation's defences, without undertaking wholesale rearmament. The RAF was placed at the forefront of rearmament for two reasons, and both were political rather than strategic. Firstly, the main public perception of a threat to Britain was from the air, and RAF rearmament was therefore relatively popular, whereas Army rearmament was not, as it brought back uncomfortable images of trench warfare and associated continental "entanglements". Secondly, RAF rearmament was considered by the Government and the Treasury to be the cheapest and most effective form of deterring war, both in the short and the long term, at a time when concern over the economy was still as much of a preoccupation as concern over resurgent foreign nationalism.

These political factors had a profound effect on RAF rearmament, down to and including the design of aircraft. There was a conflict over the pace of RAF rearmament between the Government and the Treasury on the one hand, and the Air Staff on the other. The former needed rapid progress for political, domestic, and international reasons;[6] whereas, the latter wished to delay large-scale rearmament for sound military reasons. Foremost amongst these was the fact that, in 1935, the designs available for immediate production were predominantly light bombers of restricted range and striking power. The Air Staff harboured grave doubts about the utility of such aircraft in a European war with a major power. They, not without reason, were wedded to a concept which wished to assess the "parity", of which politicians made such play, in terms of total striking power, in essence weight of bombs on target.

The Government, anxious to still domestic critics, who tended to highlight the supposed numerical imbalance between the RAF and the Luftwaffe, and, conscious that German expansion could not simply be ignored *pro temp,* were equally firmly interested in overall numbers. Light bombers were not only cheaper and quicker as well as easier to build, but the designs were available, and the aircraft firms practised in their production. The Air Staff, however, were acutely aware that, in the event of war, large numbers of biplanes without the range to reach Germany would be useless, particularly where the Government's extant strategic policy indicated a reluctance to commit troops to continental Europe, and the forward basing of aircraft on French soil, therefore, was not likely to be politically acceptable.

6 This tension was stated explicitly in a report of 7 June 1935 by the Sub-Committee of the Ministerial Committee on Defence Requirements which referred to the Government's "duty… to make a large increase in the Royal Air Force, partly as a deterrent to Germany and partly in order to assure a more rational state of public opinion". DCM (32) 145 quoted in NA AIR 41/8, Air Historical Branch(RAF) narrative, *The Expansion of the Royal Air Force 1934–1939*, 37.

The Chief of the Air Staff (CAS), Sir Edward Ellington, initiating a review of the equipment of the RAF's bomber squadrons in the autumn of 1935, himself stated: "It seems to me doubtful whether in reality the Light Bomber, with its short range and small load, has any value in a European war against such a country as Germany".[7] The review itself came in the form of two Air Staff papers. The first, written by the then-Group Captain A. T. Harris, Deputy Director of Plans in the Air Ministry, was directed at establishing the proportion of light bombers in the RAF. Harris not only concluded that the medium bomber had great advantages over the light in most areas, but also that the medium bomber itself "will tend to disappear and that eventually a most economical size and type will be evolved, on the basis of range and Total Striking Power [i.e. bomblift] factors".[8] The second paper was produced by the Operational Requirements branch in the Air Ministry, and was concerned with the defence of light bomber aircraft, with a putative speed of 300 mph and a range of 800 to 1,000 miles, against single-seat fighters with a speed in excess of 300 mph and eight fixed machine-guns. The author concluded that, whilst the margin of superiority of the fighter over the light bomber in terms of speed was narrowing, the increasing range and, therefore, depth of penetration into hostile territory had a compensating effect which favoured the defence. He dismissed the concept of a high speed bomber evading the defences on the familiar ground that, "as aircraft design advances so quickly, especially in war…a type which may fulfil this function at one period may be hopelessly outclassed six months later".[9] He also concluded that no light bomber could be given both sufficient performance and defensive firepower to enable it to beat off attacks from an eight-gun fighter. The reaction of the Chief of the Air Staff to these papers was to agree, on 26 January 1936, "on the gradual elimination of the light bomber for European warfare".[10]

This pronouncement did not, however, mean that speed, as opposed to defensive armament, did not feature prominently in the RAF's consideration of future types. In the deliberations and discussions which took place in June 1936 around what became Specification P13/36, which led to the Avro Manchester and ultimately the Lancaster, both the Command and Group Headquarters

---

7 Minute, Chief of the Air Staff [Ellington] to Deputy Chief of the Air Staff [Courtney], 8.11.1935, NA AIR2/2718,

8 Ibid. Minute Deputy Director of Plans [Harris] to DCAS [Courtney] on "Bomber Squadrons Policy", dated 16.1.1935 [*sic* the date of the minute was clearly January 1936]. Harris, of course, was to become the most famous wartime C-in-C of RAF Bomber Command.

9 NA AIR2/2718, Paper on "The Defence of Light Bomber Aircraft", 15.1. 1936.

10 Ibid, minute CAS [Ellington] to DCAS [Courtney], 26.1. 1936.

expressed a preference for speed over armament.[11] Wilfrid Freeman, meanwhile, had taken over from Hugh Dowding as the Air Member for Research and Development on the Air Council in April 1936, and, therefore, had only limited influence on these early discussions. Nevertheless, in March 1937, Freeman, prompted by an approach from the Gloster Aircraft Company—not, we should note, de Havilland—rekindled the debate over the unarmed bomber. The chief designer at Gloster, George Carter, had only recently left A.V. Roe, where he had been a principal member of the team which designed the Avro Manchester to Specification P13/36, and he now claimed that by eliminating the rear gun turret, and reducing the bombload, he could produce an aircraft 50 mph faster than Avro's design. Freeman told Carter that the Air Staff would not consider an aircraft without rear defence, and the design was then amended to include a prone-position gun turret which the gunner was to reach by sliding along the fuselage on rails. Freeman sent this revised design to Group Captain Oxland, the Deputy Director of Operational Requirements (DDOR) in the Air Ministry.

The subsequent exchange between Freeman and Oxland is of interest for several reasons. It shows that the idea of the unarmed bomber was one which attracted a great deal of interest amongst designers from several companies, not just de Havilland. In his letter to Oxland, Freeman mentions that "[Arthur] Gouge [of Short Bros] made a similar proposal some time ago", and Oxland in his reply referred to a previous proposal by H.P. Folland, Carter's predecessor at Gloster, for an unarmed bomber.[12] Neither Oxland nor Freeman mentioned de Havilland. Oxland's reply also shows that the Air Staff were not at this stage totally supportive of the concept of a fast bomber: Oxland interestingly foresaw its greatest utility as being "for use at night, when all defensive armament is comparatively useless to the bomber, [and] we may be quite justified in sacrificing armament to speed".[13]

Here, however, another factor came into play, and one which was to form a consistent stumbling block to the concept of the unarmed bomber. This issue was the need to concentrate production on as few types as possible in order

11 Colin Sinott, *The Royal Air Force and Aircraft Design* (London, Frank Cass, 2001), 165. This study is the most comprehensive on the subject of the RAF and pre-war aircraft design and the only study to recognise that the Mosquito was not a purely private venture design by de Havilland.

12 Letter Freeman to DDOR [Oxland] 12.3.1937, NA AIR20/180. Freeman's letter to Carter and the original Gloster proposal are not on the file in the National Archives. However, all the details of the proposal and Freeman's reply to Glosters are contained in his letter to Oxland.

13 Letter Freeman to DDOR [Oxland] 12.3.1937; Minute DDOR [Oxland] to Freeman. The copy on this file is dated simply 3.37, but a further copy at NA AIR20/9 is dated 30.3.37.

ILLUSTRATION 9.2 Mosquito aircraft production line showing fuselages under construction—note the visible wooden plys.
CREDIT: AIR HISTORICAL BRANCH (RAF)

to maximise the use of industrial capacity (Illustration 9.2). One of the earliest problems identified by the Air Staff, when expansion was first considered, was the excessive diversity of types in production. In 1930, when the bomber squadrons of the Metropolitan Air Force numbered 24, there were no less than 11 different types of aircraft in service, of which only one type contributed more than three squadrons, and five types contributed only one squadron each. This situation was the result of the Air Ministry's inter-war policy of spreading its exiguous orders between sixteen or so competing aircraft firms in order to retain as many design teams, and as much production capacity, as possible. Oxland was quick to point out to Freeman that producing an unarmed bomber "purely for night work…means reverting to the old policy of specialised night bombers, and I think we are all agreed that we must aim for as much standardisation as possible in the future".[14] The need to make maximum use of industrial resources was to become an ever more prominent factor in Air Ministry thinking as rearmament got underway.

14 NA AIR20/180.

It is clear that Freeman did not entirely accept these views, and there is some evidence that he was working quietly behind the scenes to ensure that the concept was not allowed to wither simply through lack of enthusiasm on the part of the Air Staff. The vehicle which he used to advance the concept further was provided by a letter he received in May 1937 from Frederick Handley-Page. Handley-Page enclosed a memorandum on the subject of the unarmed bomber from his chief designer, G. R. Volkert, which argued that a twin-engined bomber designed to Specification P13/36, but omitting all defensive armament, would be able to carry three times the bombload, with far fewer crew members, at speeds up to 20 mph faster than the P13/36.

It is quite clear from the minute sheet to this file that Handley-Page had not simply submitted this memorandum out of the blue. In fact, R. H. Verney, the Director of Technical Development at the Air Ministry, had discussed it with Captain Liptrot, an aerodynamics specialist on his staff, and had asked Handley-Page to produce the scheme "as a basis for discussion". Both Verney and Liptrot were ultimately responsible to Freeman as Air Member for Research and Development (AMRD), and it is equally clear that Freeman was aware of the approach, and had presumably encouraged it himself, since his first response to Handley-Page's letter was to send it to Verney with a minute which read in part, "I believe you have already started investigations along the same lines as Volkert…"[15] Freeman was charged with the responsibility of maintaining contacts with industry, and was, therefore, quite justified in his encouragement of such exchanges.

The "Volkert's Proposal" was subject to intense scrutiny by Freeman's technical staff, principally Liptrot and the Chief Superintendent of the Royal Aircraft Establishment, A. E. Hall. Liptrot argued that the unarmed bomber saved on manpower and training, and gave a cleaner, smaller, better shaped, and more manoeuvrable aircraft—one which was more economical in structure weight. He calculated that such an aircraft with a bombload of 1,000 lbs., and a range of 1,000 miles, would cruise at 309 mph and have a maximum speed of 355 mph, "over [i.e. greater than] that of the Spitfire, which is the fastest fighter of today, so that it should be quite immune from attack by any single seat fighter". He concluded that, "The above arguments and those presented by Volkert give strong support to the view that the elimination of defensive armament ought to be given serious consideration…"[16] Hall also pronounced himself favourably impressed by the proposal, though Verney was a little less sanguine.[17]

15 Minute, Director of Technical Development [Verney] to Deputy Director of Operational Requirements [Oxland], 21.5.1937, and Minute Freeman to DTD 20.5.1937, NA AIR2/2059.

16 Ibid. Minute RDA3 [Liptrot] to DTD[Verney] 15.7.1937.

17 Ibid. Minute Chief Superintendent RAE [Hall] to DTD [Verney] 15.7.1937 and minute DTD to DDOR [Oxland] and AMRD [Freeman] 26.7.1937.

All these deliberations were, however, confined to the Technical Staff under Freeman as AMRD, and they could not, and did not, work in a vacuum. It was, therefore, necessary at an early stage to bring in the Operational Requirements Staff under the CAS, to consider the aircraft's operational utility and its possible place in the RAF's plans for conducting a war. Thus, on 6 August 1937, Freeman sent the file to the DDOR, Group Captain Oxland, which indicated that he was himself probably in broad agreement with the generally favourable reaction of the Technical Staff. By 1 September, the proposal had reached the CAS, who asked for a *précis* of the papers and comment from Freeman and the Deputy CAS, Air Vice-Marshal Richard Peirse.[18] Matters were further delayed whilst Handley-Page produced revised performance figures (which are not in the surviving files), and it was not until November 1937 that Oxland produced a paper entitled "A Consideration of the Possibilities of the Undefended Bomber", in which he compared the "Volkerts" and the P13/36.

Oxland attributed the following advantages to the unarmed aircraft: it would be quicker, cheaper, and easier to produce; it could be operated by fewer personnel requiring less training; it would be more economical in unrestricted and gas warfare operations, as well as at night, or in bad weather. Its disadvantages he categorised as: being very uneconomical and tactically limited against strongly defended targets in fine weather; the need to avoid hostile aircraft would be bad for crew morale (a rather odd conclusion one might think!); and it would require an increase in the number of different types of aircraft in service. He nevertheless concluded that "while it is the policy to have several types of Bomber in the service it is likely that the Volkert will show great economy of force over existing types in use under certain conditions".[19] He stated that these advantages in unrestricted, night, or bad weather operation, together with its ease of production and of economy of trained personnel, justified practical investigation, and he recommended the construction of a prototype.[20]

---

18 NA AIR2/2059. Minute PS/AMRD to DDOR 6.8.1937 and minute CAS [Newall] to DDOR [Oxland] 1.9.1937.

19 Ibid. Paper "A consideration of the possibilities of the unarmed bomber" by DDOR, nd but November 1937.

20 Ibid. Covering minute DDOR to AMRD 16.11.1937. Volkert's paper had referred to "unrestricted" warfare by which he meant a recognition that in future wars unrestricted bombing of civil populations was highly likely and he therefore argued that a bomb aimer was unnecessary because precision would not be required. He was partly right, but the eventual adoption of "area" rather than "precision" bombing was more a pragmatic recognition of reality and did not, in any case, mean accuracy and thus bomb aiming/aimers became entirely superfluous.

Freeman, to whom Oxland sent this paper on 16 November 1937, was surprisingly and uncharacteristically dilatory in his response, and it was not until 24 February 1938 that he at last minuted Newall, via Peirse. Its tardiness notwithstanding, Freeman's response was a trenchant and concise statement of the potential advantages of the unarmed bomber. He wrote:

1. The advantages of the bomber described in these papers are:
   (i) very high operating speed.
   (ii) economy in trained personnel.
   (iii) economy in equipment and its maintenance.
   (iv) simplicity of construction.
   (v) low first cost.
2. The outstanding performance upon which the success of the proposed type depends can be realised only if the whole design reaches the highest standard of streamlining… The requirements of existing specifications combine to preclude the attainment of such a standard.
3. I recognise fully the disadvantages… Defensive armament is reduced to the absolute minimum, or possibly eliminated. Accurate bombing from high altitude may not be possible.
4. At the moment it is by no means clear how far types now being designed suffer, in effect, from the same disadvantages. But they are undoubtedly extravagant in all the respects in which the proposed type is economical.
5. The potentialities of the type are such that I consider that we should explore them further by building prototypes. I propose, therefore, to prepare specifications.[21]

This minute shows clearly that not only was Freeman in favour of exploring the concept further, which is hardly surprising since he had probably encouraged the initial approach to Handley-Page, but also that it was the unarmed bomber's economy which was of particular attraction. Freeman was at that time struggling with the burdens of the increasing demands for aircraft and the unmistakeable signs of strain within the aircraft industry. However, what these exchanges also show is that the Air Staff, far from being irreconcilably opposed to the concept of the unarmed bomber as has been suggested, not least by the Official History quoted above, were in fact perfectly prepared to support Freeman and his technical staff in exploring the concept further and in building a prototype.

21 NA AIR2/2059. Minute Freeman to CAS [Newall] through DCAS [Peirse] 24.2.1938.

By April of 1938, the file having again been delayed by "more urgent matters", presumably the crisis over the Austrian *Anschluss*, Freeman's proposal had been approved by both Peirse and Newall. At this point, at the suggestion of Sholto Douglas, then Assistant CAS (ACAS), Freeman sent the file to Air Chief Marshal Sir Edgar Ludlow-Hewitt, the Commander-in-Chief at RAF Bomber Command.[22] Ludlow-Hewitt was to prove a staunch supporter of the projected bomber, though, paradoxically he did not favour the removal of all the defensive armament because of the possibly deleterious effect on crew morale. Nevertheless, he foresaw the need for just such an aircraft, and, in a letter to the CAS, he pointed out the type's utility for sustained bombing of industry and of communications by small formations or single aircraft; photographic reconnaissance, and, possibly, as a long-range fighter.

Ludlow-Hewitt thus identified three of the major roles eventually undertaken by the Mosquito, and for which the existing and putative designs were largely unsuitable. He also pointed out in his letter to the CAS that the current designs for "big bombers" were "frightfully expensive" and presciently stated that they might prove unable to defend themselves against modern fighters in which case "we may easily find ourselves with a huge force of bombers unable to reach the Ruhr or at least unable to maintain sustained attacks on the Ruhr".[23]

In the meantime, there had been a curious lack of action to put into effect Freeman's proposal to prepare specifications for a prototype. Although the reasons are unclear, it was Ludlow-Hewitt, not Freeman, who pressed the Air Staff on the issue. The former was apparently stung into action when, despite his support for the idea of a fast bomber following Freeman's initial approach, the subject was excluded from the agenda for discussion by the Air Ministry's Bombing Committee in May 1938. Ludlow-Hewitt wrote a formal letter to the Under-Secretary of State at the Air Ministry explaining the need for a fast, lightly-armed bomber with reasonable range.[24] The immediate Air Staff response was a paper, again written by Group Captain Oxland, but which appeared to retreat substantially from the position he had adopted nine months earlier. Oxland now concluded that any increase in speed at the expense of armament would not be sufficient to enable the aircraft to avoid interception. He

22 Ibid. Minutes by DCAS [Peirse] and CAS [Newall] to Freeman 16.3.1938 and 11.4.1938; and letter Freeman to Ludlow-Hewitt 20.4.1938, in file NA AIR14/251.

23 Letter Ludlow-Hewitt to CAS [Newall] 3.12.1938, NA AIR14/251. In the letter Ludlow-Hewitt mentions a meeting the previous Tuesday at which Newall had given the C-in-C the impression "that you are not disposed to consider production of a speed bomber".

24 Ibid. Letter Ludlow-Hewitt to USofS 19.7.1938. A copy is also on file NA AIR2/2059. The formal nature of the letter indicates Ludlow-Hewitt's desire to have his views put on record.

also argued that the speed bomber would not show any substantial economy over the standard bomber in terms of production, maintenance, flying personnel, or wastage, and that it did "not enjoy sufficient advantage over the standard bomber in any material respect, which would compensate for its limited operational role".[25]

Whatever the reason for Oxland's *volte face,* one might have thought this opinion from the Air Staff would have been sufficient to kill the proposal stone dead. By now, however, operational factors were but one aspect of the problem. Political, financial, and production problems began to impinge to an increasing degree. The successive rearmament programmes proposed by the Air Council as the international situation progressively worsened, and the rate of Luftwaffe expansion increased, eventually exceeded the available resources. Initially, the problem was one of finance: The Air Ministry programmes as formulated at the end of 1937 required £650 million over four years, yet the massive rearmament vote and loan of March 1938 only planned for RAF expenditure of £500 million.[26] Sir Thomas Inskip, the Minister for Co-ordination of Defence, explicitly challenged the Air Staff's Expansion Scheme J in December 1937, stating amongst other things that "the general suggestion I want to make is that, without loss of efficiency… we might substitute a larger proportion of light and medium bombers for our very expensive heavy bombers".[27]

The major problem, however, was one of productive capacity, not finance. In April 1938, the Cabinet, their minds newly focussed by Hitler's annexation of Austria in March, approved the Air Council's revised Expansion Scheme L, which provided for the production of 12,000 aircraft over two years, but, more significantly, also marked the end of the purely financial restraints on rearmament.[28] The production problems were not, on the whole, related to factories or plants, which had been adequately expanded under earlier forward-looking schemes, but, instead, were linked to more intractable shortages of raw materials, and, most critical of all, skilled manpower.[29]

25 Paper by DDOR on "The Speed Bomber" 7.8.1938, NA AIR 2/2059.

26 For a concise account of the arguments of the programme known as Expansion, Scheme J., see H Montgomery Hyde, *British Air Policy Between The Wars 1918–1939* (London, Heinemman, 1976), 407–413.

27 Sir Thomas Inskip's *Aide-Memoire* of 9.12.1937 is reproduced in its entirety in Sir Charles Webster and Noble Frankland, *The Strategic Bombing Offensive Against Germany* (London, HMSO, 1961), Vol 4, Appendices, 96–98.

28 Hyde, 415–417. The Air Council had produced a revised Scheme K post-Inskip, but this was initially rejected.

29 The Minister of Labour told the Cabinet in March that he found it difficult to envisage where the labour to fulfil Scheme L could be found. Smith, 203.

It is perhaps in this light that we must interpret the Air Staff response to Oxland's paper. Sholto Douglas, whilst agreeing that the arguments adduced in the paper were "cogent", pointed out that the "ideal bomber" specification, which had been the subject of much consideration by the Bombing Committee, was going to result in an aircraft of 50,000–60,000 lbs. It would thus "be an expensive article and far from easy to produce; and I doubt if we shall be able to have a large number of them on account of the expense and difficulties of production". He went on to suggest that an attempt should be made to produce smaller more easily constructed aircraft, with a smaller crew, and he considered that "an aircraft of the type proposed by the AOC-in-C [Ludlow-Hewitt] might fit the bill".[30]

Douglas, having been in part responsible for the Bombing Committee's lack of interest in the "Speed Bomber", as it had come to be called, now suggested that the committee should again consider it. In point of fact, the Bombing Committee does not appear ever to have done so, perhaps, because, by October 1938, international developments were beginning to impinge on Air Staff thinking to an ever increasing degree. Within days of the Munich settlement, Douglas was setting forth, in a minute to the CAS, four reasons in support of building a prototype. Firstly, he questioned the wisdom of building only large heavy bombers of around 50,000 lbs. Secondly, he stated that "political considerations" might force parity to be considered by numbers of aircraft, and not, as the Air Staff wished, by "parity of bombload [lifted]". Here he may have been influenced by the Munich Settlement, the ink on which was hardly dry, and which he obviously believed might lead either to a reduction in rearmament, or to continuing political pressure simply to obtain parity. If such equality were measured by numbers of aircraft, he argued that, "it is open to doubt if we shall be able to afford the expense of having our force … composed entirely of heavy bombers". Thirdly, he pointed out (optimistically) that Battles and Blenheims would disappear within a year, leaving no aircraft suitable for direct support of the Army—a role for obvious reasons generally ascribed to lighter bomber aircraft. Lastly, he admitted the force of Ludlow-Hewitt's pleas concerning flexibility and variety.[31]

Newall apparently remained unconvinced, however, and was only prodded into reluctant agreement to "preliminary discussions" in the Bombing Committee by the long and forceful letter from Sir Edgar Ludlow-Hewitt at Bomber Command in early December, which implicitly criticised the Air Staff for its

30 Minute Douglas [ACAS] to Peirse [DCAS] 13.8.1938, NA AIR2/2059.

31 NA AIR2/2059. Minute ACAS [Douglas] to CAS [Newall] 4.10.1938.

continuing vacillation.[32] If Ludlow-Hewitt entertained the hope that rapid results would follow, he must have been sorely disappointed; for although Sholto Douglas instructed Robert Saundby, Oxland's successor as DDOR, to prepare a paper on "The Need for a Light Bomber for use in the Royal Air Force", it was to be a full six months before it was ready.

The reasons for this lengthy delay are not obvious from the files relating directly to the unarmed bomber, but it seems reasonable to infer that there were two factors involved. Both illustrate very clearly the diverse pressures which affected decisions over which aircraft designs were to go into production, so it will be as well to consider them before going on to a more detailed description of the contents of the paper itself. The first factor was the further deterioration in the international situation in early 1939. The continuing slide towards war certainly had a direct political effect on the Government's military policy, and this in turn impinged on the Air Staff's policy debate on the unarmed bomber. The most fundamental changes in policy which resulted were the successive increases announced in the strength of the Territorial Army and the introduction of peacetime conscription, together with a vastly increased commitment to send military forces to the Continent of Europe in the event of war. The previous policy had envisaged the despatch of only four divisions in eleven months, but the Government performed a truly remarkable *volte face* in deciding to increase this to two corps within thirty-three days of the outbreak of war. This force was also to be provided with an RAF Air Component.[33]

Where the Army had previously been the "Cinderella" service during rearmament, the Government had now not only promised "Cinders" a new gown, but had also issued an invitation to the Ball. The result was seen when the much delayed Saundby paper on the light bomber finally saw the light of day. Whilst it was in many ways yet more inimical to the unarmed bomber than Oxland's of ten months previously (as will become clear), it ended by saying "direct support [of the Army in the field] is a definite commitment of the Royal Air Force, and it is clear that a light bomber is the most suitable aircraft for this task. It is recommended, therefore, that requirements should be drawn up for a light bomber primarily suited to this role".[34]

The second factor behind the delayed appearance of the Air Staff paper was the parallel, and from the RAF's point of view, more fundamental debate on what was referred to as the "Ideal Bomber". This discussion concerned the

32 Ibid. Letter Ludlow-Hewitt to CAS [Newall] 3.12.1938.

33 L F Ellis, *The War In France and Flanders 1939–1940* (London, HMSO, 1953), 6.

34 NA AIR2/2059. Air Staff paper "Consideration of the need for a light bomber for use in the Royal Air Force", 23.6.1939.

type of bomber aircraft which even supporters of the unarmed aircraft recognised was to provide the great bulk of the air striking force, namely the heavy bomber. The debate here was focused on whether these aircraft were to be truly heavy, i.e. four-engined heavily armed bombers of perhaps 50,000 lbs. plus weight, such as the Short Stirling, or smaller twin-engined aircraft of around 30,000 lbs. such as the Vickers Wellington.

The Air Staff based all its arguments on the single criterion of equivalent bomb-lift, just as it did for the purposes of assessing air parity with Germany. For a given bomb-lift, four-engined bombers, although individually more expensive, were shown to be more economical overall. The opening paragraph of an “Air Staff Note on the Size of Bomber Aircraft” prepared for the Secretary of State for Air in December 1938 read, in part,

> At first sight, especially to the non-technical mind, the medium size types of aircraft appear to offer several advantages as compared with the larger bombers. Each aeroplane is cheaper and easier to produce than one of the larger type, from which it is sometimes inferred that this will make for the easier replacement of losses in time of war, and the fact that the smaller bomber carries a smaller crew also seems attractive at first sight in view of the fewer casualties in each aircraft lost. The operation of a larger number of small bombers also appears to offer the advantage that there would not be too many eggs in one basket and thus that individual losses would be less serious.[35]

The paper, having adduced these apparent advantages, then proceeded to demolish them one by one. The arguments themselves are outside the scope of this paper (though it has to be said that the Note seriously skewed the arguments in favour of the Stirling), but their object, and especially their timing, had a great bearing on the simultaneous battles over the unarmed bomber.

The issue of the Ideal Bomber had as weighty a political as a military dimension. Thus, the subject of the Ideal Bomber was discussed by the British Cabinet on 7 November 1938, wherein the Air Staff’s proposal to build heavy bombers met with fierce opposition, including that of Prime Minister Neville Chamberlain, largely on the grounds of expense, and the Secretary of State for Air, Kingsley-Wood, had been forced to agree to a re-examination of the proposal. This had occasioned the Air Staff paper just quoted, and it was on the basis of this paper that Kingsley-Wood wrote to the Prime Minister in December 1938 re-stating the case for the “very heavy” bomber. In this letter, it was

35 “Air Staff note on the size of bomber aircraft”. 22.12.1938, NA AIR8/257.

claimed, amongst other things, "that the large bomber, because of its better defensive armament, higher speed (sic), ability to operate singly and lessened vulnerability would be a more effective weapon than the smaller bomber".[36]

Paradoxically, several of these same arguments were, of course, deployed by proponents of the speed bomber, and it would obviously have severely weakened the Air Staff case with the Cabinet had it become known, particularly in the Treasury, that substantial debate was taking place within the Air Ministry over the merits of a small, and therefore relatively inexpensive, fast bomber. Hence Newall's lukewarm reaction to Ludlow-Hewitt's December letter, which arrived in the middle of the crucial battle with the Cabinet. It is interesting to note that Air Vice-Marshal Arthur Tedder, then serving as Director General of Research and Development (DGRD), prepared a note for Newall in November in which he argued that "evasion based solely on speed cannot be relied on", but that speed was a vital requirement which was in part a function of size.[37] Tedder stated explicitly that Freeman was in entire agreement with this line of argument. At this point, because of the essentially political need to present a united front to the hostile politicians, both the Air Ministry technical and Air Staffs, including apparently Freeman, were talking only in terms of heavy bombers. Only the dissenting voice of Ludlow-Hewitt at Bomber Command continued to cry forth in favour of the "speed bomber".

When Saundby's paper finally did appear, its hostility to the unarmed bomber was based on familiar arguments, but now clearly slanted in tune with the prevailing politics: the evanescent nature of technical advantage would lead to obsolescence in two years, and thus "the costs of replacement would offset any economies which it could hope to achieve in other directions". Saundby, like Oxland before him, was sceptical of the claimed economies, and of the technical feasibility of producing a bomber with a significant speed advantage over contemporary fighters, even in the short term. Production factors also loomed large—the paper took as its starting point the need to reduce the number of different types to the minimum consistent with operational requirements—and, whilst Saundby conceded that speed and manoeuvrability would be useful assets in a night bomber, he stated firmly that it would be "undesirable to produce distinct types of bomber for day and night bombing".[38]

36 NA AIR8/257. Letter Sir Kingsley Wood, Secretary of State, to Prime Minister Neville Chamberlain ?.12.1938. Probably sent 9.12.1938.

37 NA AIR8/257. DGRD [Tedder] to CAS [Newall] 22.11.1938 enclosing a note on "The relation between aircraft performance and aircraft size as regards bombers".

38 Air Staff paper "Consideration of the need for a light bomber for use in the Royal Air Force". 23.6.1939, NA AIR2/2059.

A week after Saundby's paper appeared, Douglas wrote to Ludlow-Hewitt at Bomber Command, enclosing a copy, and inviting the AOC-in-C to comment. In his covering letter, Douglas made it clear that he did not favour such a design, and, despite his earlier promise, he now stated that discussion in the Bombing Committee would be "a waste of time". He was forced to admit, however, that the increase in the size of the British Army Field Force so recently announced now meant that the RAF would be forced to take its responsibility for providing air support for the Army far more seriously than it had previously, and that this meant producing a light bomber.[39] Ludlow-Hewitt accepted neither Douglas's strictures, nor the substance of the paper, and, when two letters to Douglas failed to alter the situation, Ludlow-Hewitt took it to ministerial level by writing an official letter to Harold Balfour, Under-Secretary of State for Air at the Air Ministry.[40]

It was at this point that Freeman's Department re-entered the fray. In a minute to Saundby of 14 August 1939, Douglas revealed that Air Commodore R.H. Verney, the Director of Technical Development, and effectively number three in the Air Member for Development and Production's Department after Freeman and Tedder, was considering a proposal from Blackburn's for a speed bomber, and that he wished to build a prototype to explore the concept.[41] In addition, a senior member of the Air Staff, Air Vice-Marshal Peck (Director of Operations), also thought that there was merit in Ludlow-Hewitt's strictures on too much reliance being placed on heavy bombers. Faced with these pressures, Douglas again backtracked and instructed Saundby to re-draft his paper yet again, and submit it to the Secretary of the Bombing Committee with a request to call a meeting as soon as possible.[42]

Even before the revised paper was ready, however, Douglas came under further pressure from within Freeman's technical staff. Early in 1939 it had been agreed that Tedder as DGRD, and Douglas as ACAS, would hold regular liaison meetings to co-ordinate the technical research programmes of the Air Ministry with the latest Air Staff thinking. At the meeting which took place on 28 August 1939, Tedder proposed that certain firms, not currently employed on

39 NA AIR2/2059. Letter ACAS [Douglas] to Ludlow-Hewitt 29.6.1939.

40 Ibid. Letters Ludlow-Hewitt to ACAS [Douglas] 11.7.1939 and 28.7.1939 and letter to Under Secretary of State for Air 7.8.1939.

41 23.6.1939, NA AIR2/2059. Minute ACAS [Douglas] to DOR [Saundby] through Director of Operations [Peck] 14.8.1939. Freeman's formal title was changed from Air Member for Research and Development [AMRD] to Air Member for Development and Production [AMDP] on 1st August 1938.

42 Ibid. Minutes DofOps [Peck] to DOR [Saundby] and ACAS [Douglas] 16.8.1939. Minute ACAS to DOR 21.8.1939.

definite specifications, should be given design projects approved by the Air Staff, in order to keep their design teams in work.

The first such project proposed by Tedder was a "relatively small very high speed bomber", which it was proposed to give to Blackburn, Armstrong Whitworth, and Avro. Whilst Douglas did make it clear that the Air Staff's views on air support for the Army were "not yet settled", the meeting nevertheless agreed to the high speed bomber prototypes going ahead. It is important to note here that this decision to go ahead with high speed, though not necessarily unarmed, bomber prototypes was thus taken eleven days before the outbreak of war and that it was apparently principally driven by production considerations, i.e. the need to retain design capacity intact, along with Army requirements.[43]

Three days after this meeting, Saundby produced his revised paper, which differed little in its conclusions from earlier drafts. He stated that the B1/39 four-engined bomber specification should provide the bulk of the striking force, but the greatly increased importance of the Army air support requirement was reflected in his parallel conclusion that a light bomber optimised for this task would be needed. Perhaps influenced by his knowledge of the ACAS/DGRD meeting, he also recommended that the Director of Technical Development be instructed to develop a bomber with the highest possible speed, carrying a reasonably powerful armament covering a small arc of fire astern.[44]

Douglas's capacity for blowing hot and cold on the project was perennial, nevertheless, and within a week he was again minuting Saundby, postponing discussion of the speed bomber "for the present".[45] The lessons of war are seldom susceptible to the whims of Air Staff procrastination, however, and Douglas's minute continued:

> As a result of operations during the past week, [i.e. the first week of the war] a requirement for a "Speed" aircraft for reconnaissance purposes has been raised by D[irector]G[eneral] of Ops [Operations]. It has become apparent that we shall need a really fast light aircraft for long distance high altitude reconnaissance [a requirement Ludlow-Hewitt had identified more than a year before].[46]

43 Minutes of 8th Monthly ACAS/DGRD Liaison Meeting on 23.8.1939, conclusion 12. NA AIR 20/425.

44 AIR 2/2059 Air Staff Paper "Consideration of the need for light and/or fast bombers for use in the Royal Air Force" 31.8.1939.

45 Ibid. Minute ACAS [Douglas] to DOR [Saundby] 9.9.1939.

46 AIR 2/2059, Minute ACAS [Douglas] to DOR [Saundby] 9.9.1939.

Douglas asked for an Operational Requirement to be produced, but he himself thought that "most if not all the bombload" could be sacrificed.[47] As we shall see, it was increasingly the reconnaissance requirement which drove the matter of the high-speed aircraft forward, and, by the time of the 9th Monthly ACAS/DGRD Liaison Meeting, on 4 October 1939, it was already beginning to subsume the bomber requirement. The item was included in the future experimental aircraft programme discussed at this meeting, under the heading "High speed photographic reconnaissance and bomber aircraft". The remarks column of the minutes contained this significant statement:

> The Air Staff ...thought that their most probable use would be as high speed photographic reconnaissance aircraft rather than bombers; the Air Staff would not be averse to provision for carrying bombs so long as this provision did not detract from the value of the aircraft for reconnaissance duties, but expressed the view that for acceptance as bombers, EFFECTIVE REAR DEFENCE ARMAMENT WOULD BE ESSENTIAL.[48]

In the meantime, however, Sir Wilfrid Freeman had received a letter from Geoffrey de Havilland containing the first formal outline of a proposal which was to have a profound effect on the equipment of the Royal Air Force in the Second World War. The letter, dated 20 September 1939, indicated in its second sentence that Geoffrey de Havilland had already consulted with Freeman. At that point, de Havilland wrote, "From former [i.e. previous] conversations with you [i.e. Freeman], and using the experience we have gained in very quickly producing types which have to compete with others from all over the world, we believe that we could produce a twin engine bomber which would have a performance so outstanding that little defensive equipment would be needed".[49] The proposal was for a wooden twin Merlin-engined bomber with a crew of two, maximum speed of 405 mph, a bombload of 1,000 lbs., and a range of 1,500 miles. Shrewdly, de Havilland emphasized production factors: the aircraft's simplicity and its consequent rapidity of construction, initially, and in full scale production. He also stressed the fact that it utilised raw materials and classes of skilled labour not required by other aircraft manufacturers.[50]

---

47 AIR 2/2059, Minute ACAS [Douglas] to DOR [Saundby] 9.9.1939.

48 Minutes of 9th ACAS/DGRD Liaison Meeting and attached "Notes of a Conference held on 4th October 1939, to discuss the future Experimental Aircraft Programme", NA AIR 20/425. Emphasis added.

49 Ibid.

50 Ibid.

The proposal was scrutinised by the Air Ministry's technical experts who, as usual, expressed some scepticism over the performance figures quoted. The similarity of the proposal to the high speed bomber/reconnaissance project already approved was obvious, however, and it is difficult to believe that Freeman had not had this in mind in his earlier discussions with de Havilland, as the reply to de Havilland, written by Tedder, reveals. Tedder wrote, "Your letter… proposes a form of aircraft which we have had in mind and on which we have already been in touch with several firms [i.e. presumably Blackburn, Avro, and Armstrong Whitworth] upon preliminary layouts for the construction of experimental prototypes". Tedder went on to say that the Air Ministry had proposed a 2,000 lbs. bombload and a 2,000 mile range, and that the complete absence of defensive armament "is not generally accepted. Its acceptance would depend very much on the additional speed that it would permit as demonstrated in tests on the prototype".[51] Note, however, that he did not rule the absence of defensive armament expressly out of court. He also intimated that the Air Ministry felt the designers were over-optimistic in their claims.

In view of the ebullient and optimistic tenor of de Havilland's initial letter, the response from the Company to Tedder's letter was surprisingly hesitant. The Company requested more time as the project was not "fully evolved", and they also asked for a meeting between the firm's design staff and Air Ministry technical experts.[52] This took place at the Grand Hotel, Harrogate, on 14 October 1939, and, although no record of exactly what passed at this meeting survives, it is clear that the Air Ministry representatives did not entirely accept de Havilland's view that they could build a bomber faster than contemporary fighters, even though the firm stressed that this would be achieved in part by reducing the time lag between design and production to about nine months. The Air Ministry representatives appear to have emphasised, to at least some degree, the prevalent Air Staff view that some form of rear defence was required for bomber aircraft.[53] There is also evidence to suggest that the Ministry was pressing for a four-gun tail turret, and that de Havilland—willing to compromise to secure a production order, but also anxious not to go so far as to render the design no more than mediocre—began to push the idea of a twin-gun position in the tail.

By early November, however, de Havilland's appear to have bowed to Ministry pressure and adopted a four-gun tail turret, although they did attempt to

51 Ibid.

52 NA AVIA 15/4. Letter C C Walker of de Havilland to Tedder 5.10.1939.

53 See Ibid. Letter from C C Walker to DDRD [H Grinstead] dated 19.10.1939, which refers to the discussions on the 14th October.

retain a high performance by asking for Griffon rather than Merlin engines.[54] The de Havilland proposal was apparently not considered at the ACAS/DGRD meeting of 1 November 1939, where instead Douglas asked for a "reasoned statement" as to why the Beaufighter could not be converted into a bomber.[55]

Whilst these largely technical deliberations were proceeding, the Air Staff came under renewed pressure from the indefatigable Ludlow-Hewitt. On 14 November 1939, he again raised the subject of the speed bomber, specifically the degree of progress made with his earlier suggestion of converting Beaufighters to bombers, with Air Vice-Marshal Peck, then Director General of Operations. Peck, quite rightly, referred the problem to Douglas, pointing out that "it [i.e. the speed bomber] did not appear to be covered in the [production] programme which I understood was being adopted". He continued: "I think that the representations of C-in-C Bomber Command should be considered at once and the matter settled so far as may be practicable, so that we may put in train for the future development of the Air Force the best means which can be found for meeting this need".[56]

Three separate strands were now beginning to coalesce and to force the hand of those members of the Air Staff who remained unconvinced. First was Ludlow-Hewitt's unremitting pressure; second was the early realisation that the Blenheim was unsuitable as a vehicle for strategic reconnaissance; and third was the desire of the technical staff of the Air Ministry, from Freeman downwards, to build prototypes to prove or to disprove the various contentions concerning high speed photo-reconnaissance/bomber aircraft, and the continuing interest of several aircraft firms in building such an aircraft.

On 15 November, the day after Peck's minute to Douglas, the twelfth ACAS/DGRD meeting took place, and there was a lengthy discussion on the question of the high speed bomber. The minutes of the meeting are worth quoting at some length, since they reveal much about the existing attitudes in both the Air Staff and the technical branches of the Air Ministry. The discussion was opened by Douglas who referred to the fact that Ludlow-Hewitt was still pressing for the rapid provision of such an aircraft, a clear reference to the latter's scheme for converting Beaufighters. Captain Liptrot argued that stripping the B1/39 "Ideal Bomber" of all armament, and thus having a smaller fuselage, would give a fast aircraft with good crew/bombload ratio, but it would take time to produce and would by then have been overtaken by contemporary

54 See NA AVIA 15/4, exchange of minutes between DDRDT [Grinstead] and DDRDE 10-16.11.1939.

55 ACAS/DGRD Liaison Meetings Minutes of 11th Meeting 1.11.1939. NA AIR 20/425.

56 Minute DGOps [Peck] to ACAS [Douglas] 14.11.1939 NA AIR 2/2059. Ludlow-Hewitt had apparently raised the matter verbally.

fighters. Consequently, the participants decided that "the best approach to this problem would be through the development of a high speed photographic reconnaissance aircraft for which it had been decided to place orders for two prototypes from three separate firms".[57] The discussion continued and Douglas raised de Havilland's suggestion that they be given an order forthwith for 25 high-speed bombers with Griffon engines, a four-gun tail turret, and a top speed of 350 mph; a proposal contained in a letter from Geoffrey de Havilland to Tedder dated 8 November.[58]

It was at this point that production factors began to exert a decisive influence. A suggestion that the proper way to proceed would be to invite tenders from a number of firms, because de Havilland's scheme was only one of many, was rejected firmly on the grounds that de Havilland's were free from other design commitments. Further, the firm had great experience in wooden construction, and any effort to build the design in metal would "trespass upon facilities required for the production of other aircraft".[59] Finally, Tedder did not believe that de Havilland's could produce a prototype in less than eighteen months, "by which time production facilities for the airframe and the Griffon engine should be available." Therefore, the participants decided that de Havilland should be given an order to proceed with two prototypes carrying 1,000 lbs. of bombs and a three-man crew, including a gunner in a four-gun turret, over a range of 1,500 miles.

Five days after this meeting, Douglas sent a minute to Freeman indicating that Ludlow-Hewitt had been agitating for the speed bomber, and that Newall had accepted the need for such an aircraft. Douglas asked Freeman to consider an aircraft with 1,500 mile range, 1,000 lbs. bombload, and a top speed not less than 300 mph. He mentioned both the de Havilland and the Blackburn designs as possibilities, and referred to them both as having four-gun rear turrets—a policy with which he agreed.[60] On the same day, Tedder wrote to Geoffrey de Havilland indicating that orders would be placed for two prototypes, each with the tail turret.[61] It is clear from these two documents, and from the minutes of ACAS/DGRD liaison meeting, that both the Air Staff and the technical staff, the latter probably influenced by the former, were at this point in favour of giving the aircraft reasonably powerful rear armament. It seems likely that de Havilland, anxious to secure a production order, were therefore prepared to accept the heavy tail turret provided they could use more powerful Griffon engines.

57 ACAS/DGRD Liaison Meetings Minutes of 12th Meeting 15.11.1939. NA AIR20/425.
58 Ibid. For de Havilland's letter to Tedder of 8.11.1939 see NA AVIA 15/4.
59 NA AIR20/425.
60 Minute ACAS [Douglas] to AMDP [Freeman] 20.11.1939.
61 Letter Tedder to Sir Geoffrey de Havilland 20.11.1939, NA AVIA 15/4.

Sir Wilfrid Freeman, however, was more inclined to accept the views of the firm than he was those of the Air Staff. Within two days of receiving Douglas's minute he was attending a meeting with de Havilland representatives at which he proposed some radical changes. On the morning of this meeting he had written to Douglas outlining his views. The letter contained a concise review of the situation, and ran, in part, as follows:

> First I think it just as well to say that I do not believe either the Beaufighter or the new Blackburn machine will fill the bill; the former because it will not have the speed, the latter because it will take too long to mature … my first reaction is rather as follows. You want a bomber so fast that it will have the legs of an enemy fighter. Having decided on that you decide that it will not be fast enough to get away from the enemy and will therefore make it slower by adding defensive armament; and it is probable that you will achieve a compromise which will not fit the bill in either of these two characteristics. It seems to me that we must rely on engine development to give us what we want, and I am inclined to think that we should aim at getting a new bomber into the air each year, sacrificing everything for speed. If it has not got the speed of the enemy then it is too bad: if it has got the speed of the enemy then it is likely to retain that for a limited period of say 9 to 12 months.[62]

Freeman did not envisage large scale production of one type, but rather a whole series of aircraft based on the concept, the first being a twin-engined Merlin aircraft of 350 mph, to be followed by a twin-Griffon model of 370–380 mph, and then a model with twin-Sabre engines in 1942 with a speed of over 400 mph. His letter continued:

> If we bank on the compromise we will have to turn out in 1941 a bomber with a speed of 350 mph with two Griffon engines and a four-gun turret. This machine cannot be either quickly designed or quickly manufactured, since the turret and hydraulics complicate matters to an extraordinary extent. Moreover, in this case, I very much doubt whether the Griffon engines would be available.[63]

---

62 Letter AMDP [Freeman] to ACAS [Douglas] 22.11.1939. NA AIR 20/2059. In the letter Freeman stated explicitly that he had not discussed this matter in detail with either Tedder or Farren[Deputy Director Research and Development for Aircraft], and that the views that followed were a first reaction.

63 NA AIR 20/2059. Freeman had personal experience of the tremendous problems associated with trying to integrate complex hydraulic turret designs with modern aircraft.

Having despatched this letter, Freeman then went into a meeting between his own technical staff and representatives of de Havilland, and whereas the thrust of all previous communications from the technical staff to the firm had been to argue for armament, Freeman reversed the Air Ministry line in favour of de Havilland's original suggestion stating that "the introduction of the turret tended to defeat the object of the form to produce a bomber which could outpace the contemporary fighter". He also went further than Tedder in meeting the firm's request for a production order. He made a series of far-reaching proposals, which must have stunned the de Havilland representatives given the Ministry's earlier lukewarm reaction to the aircraft. Freeman stated that if the firm would guarantee to produce an unarmed Merlin version within nine months, start building a Griffon-engined successor—again within nine months—and subsequently begin work on a further Sabre-engined version, then "he would be prepared to recommend it to the Air Council purely as a gamble".[64] Freeman had obviously carried his technical staff with him, as by the next day W. S. Farren, his Deputy Director for Research and Development (Aircraft), was minuting him re-iterating the main arguments, and stating that the firm were anxious to continue development and had the productive capacity and thus "it remains only to discuss it with the Air Staff and get their agreement".[65] Farren did, however, suggest to de Havilland the same day that they base the design on the long range reconnaissance requirement with some provision for bombs.

At this juncture, Ludlow-Hewitt, who as we have seen had been a moving spirit in ensuring that the concept remained alive, now began to exert a less positive influence. He took issue with Freeman's letter to Douglas of 22 November, partly because he obviously, and correctly, believed that building an aircraft from scratch would impose further delays to a matter which he considered so pressing. He urged instead his alternative scheme of adapting the Beaufighter.[66]

The point had been reached where it was considered necessary to bring together all the interested parties in the saga to thrash out an agreed requirement, and a meeting was consequently arranged for the 12 December 1939 at which Bomber Command, the Air Staff, the Director of Operations, and the technical staff were all represented. Before entering this meeting, however, Freeman, Tedder, and Farren met with representatives from de Havilland led

64 "Minutes of a Meeting at the Air Ministry Department ZA on 22 November [1939] to discuss a new de Havilland aircraft". NA AVIA 15/4.

65 Minute DD/RDA [Farren] to Freeman 23.11.1939, NA AIR 2/2059.

66 Ibid., Letter Ludlow-Hewitt to Douglas 25.11.1939.

by the Chief Designer, Lee Murray. The latter expressed the view that final decisions on general questions covering the design of the aircraft must rest with the designer. Freeman was quick to disabuse him of this notion, and stated firmly that, while the detailed requirements would finally be laid down at the meeting later in the day, "the firm were to look to him [i.e. Freeman] as the authority on requirements, and were not to take instructions from anyone else".[67]

Freeman also stressed to those present that "speed of production was of first importance", and his view that the aircraft should be an unarmed two-seater twin-Merlin model, with the highest possible speed (Illustration 9.3).[68] There is no doubt, therefore, that Freeman entered the later meeting with Douglas, Ludlow-Hewitt, *et al*, with his own mind clear, and with the technical staff, including Tedder, all brought firmly into line. The later meeting was chaired by Douglas who opened the proceedings by stating that its purpose was to analyse the high speed bomber proposals put forward by Bomber Command, and that previous analysis had concluded that any such aircraft would be a poor compromise. Freeman then put forward his well-rehearsed arguments, and

ILLUSTRATION 9.3 Mosquito production line showing aircraft in various stages of completion.
CREDIT: AIR HISTORICAL BRANCH (RAF)

67 NA AIR 2/2059. Note of Meeting 12.12.1939.
68 Minute DD/RDA [Farren] to Freeman 23.11.1939, NA AIR 2/2059.

stressed that speed versus armament was the crux of the question, and that, from the production point of view, the provision of armament would delay production and disrupt the present programme. He was opposed by Ludlow-Hewitt, who felt that some form of armament, whether fixed or moveable, was essential for morale. Tedder, unsurprisingly, supported Freeman's argument, and the two men worked more or less in tandem to persuade the meeting that, on production and technical grounds, the aircraft required should be unarmed, and that the aircraft proposed by Bomber Command would have insufficient speed and ineffective armament. Freeman also made it clear that it would be necessary to build two distinct types: a high speed unarmed bomber, and a photo-reconnaissance version, and that the former would have to be re-engined each year to maintain its ascendancy over contemporary fighters. After further discussion, and a contribution from J. S. Buchanan, Deputy Director General of Production, who stressed the production advantages in time, raw materials, and labour of using wood, the meeting agreed to pursue the unarmed de Havilland option. In addition, work was to be started on a large, high-speed bomber.[69]

The Mosquito's future was sealed at this meeting, and it was Freeman, ably supported by Tedder, who succeeded in getting his vision of the aircraft accepted by a doubtful Ludlow-Hewitt, and an unconvinced Sholto Douglas. A draft specification for the aircraft was telephoned to R. E. Bishop at de Havilland that same afternoon, a mock-up conference for the aircraft followed on 29 December, and an Air Ministry Letter of Intent to order 50 aircraft was sent on 23 January 1940.[70]

All these complex interactions show very clearly that the notion that the Mosquito was put into production by a combination of a brave air marshal and an independent minded aircraft company as private venture, largely against the opposition of an Air Staff wedded to the concept of the heavy bomber, is simplistic. Freeman's interventions were indeed crucial to ensuring that the initial design conformed to the original notion of a speed bomber, i.e. a fast, unarmed aircraft intended to reach its target by speed and evasion. However, Freeman's support of the aircraft was based firmly on his understanding of the crucial production issues—in other words, that it would utilise labour and raw materials not needed for the wider aircraft programme; that it had to be developed quickly; that the inclusion of a turret was inimical to this necessity, and to the operational concept; and that its long-term future depended on rapid development utilising successively more powerful engines.

69 Minutes of Conference held at King Charles Street Whitehall, 12.12.1939. NA AIR 14/251.

70 File NA AVIA 15/4 contains details of the letter of intent.

Even then, the first production order probably would not have gone ahead had there not been a clear operational requirement for a fast long-range photo-reconnaissance aircraft. The Bristol Blenheim had proved entirely inadequate in that role. Further, there was a clear recognition of the increased requirement to provide air support to the Army, which pointed to the need for a "light bomber".

Ludlow-Hewitt must also take some credit for his constant pressing of the Air Staff on the issue, though ultimately he had to concede to Freeman on the question of defensive armament. The Air Staff blew hot and cold on the matter. Cost was a significant factor here, initially in appealing to the Air Staff, but at some points conflicting with their nervousness over the political opportunity it presented to the Treasury and to politicians to argue against the more expensive four-engined heavy bombers. At different times during the gestation of the concept, cost factors were therefore used to argue both for and against its adoption, largely dependent on external politics.

During the War the RAF exploited the complementary characteristics of both the heavy bombers and the Mosquito to the undoubted benefit of overall operational capability—thus confirming the views of those who had argued against placing "all our eggs in one basket". The Mosquito itself, proved a peerless aircraft in the photo-reconnaissance, bomber, night-fighter, fighter-bomber, maritime strike, and weather reconnaissance roles. Yet, Freeman himself recognised initially that the aircraft was "a gamble". The gamble paid off spectacularly, and whatever else it was the aircraft proved not to be "Freeman's folly".

CHAPTER 10

# The Impact of Institutional Context: Anglo-American Naval Fire Control

*Corbin Williamson*

On 11 December 1941, Japanese aircraft flying from airfields near Saigon sank the British battleship HMS *Prince of Wales* and the battlecruiser HMS *Repulse*, inflicting one of the most serious defeats of World War II on the Royal Navy. The loss of these two capital ships capped the period from September 1939 to December 1941, during which dozens of Royal Navy warships fell victim to air attack. Efforts to explain these losses in the early years of the war have often focused on the comparatively inferior anti-aircraft (AA) equipment available on British warships during this period.

Scholars such as Stephen Roskill contended that the Royal Navy's inferior anti-aircraft capabilities in 1939 reflected the interwar Admiralty's failure to understand the threat posed by aircraft.[1] Comparisons of the AA fire control systems in use in the British and American fleets at the start of World War II highlighted the superiority of U.S. Navy long range, high-angle AA fire control equipment. American and British naval officers observing one another's ships in 1940 and 1941 repeatedly commented on this discrepancy. However, a simple comparison of fire control technical performance gives insufficient consideration to the wide array of other factors that can shape a military organization's employment of technology: finance, national and service policy, organizational culture, administration, and commercial innovation. An analysis of the development anti-aircraft fire control in the British and American navies from the 1920s through 1942 illustrates the critical importance of organizational context in understanding technological development in the military.

Sailors employed World War II-era naval guns as part of a larger fire control system that could include some or all of the following: a director, a basic mechanical computer, and a stabilizing element. The director looked like a small

1 Jon Tetsuro Sumida, "British Naval Procurement and Technological Change, 1919–1939," in *Technology and Naval Combat in the Twentieth Century and Beyond*, by Phillips Payson O'Brien (London: Frank Cass, 2001), 144; Stephen Roskill, *Naval Policy Between the Wars, Volume II: The Period Reluctant Rearmament, 1930–1939* (South Yorkshire, UK: Seaforth Publishing, 2016), 334.

 | DOI:10.1163/9789004393301_012

turret located high in the ship's superstructure, though, without a protruding barrel. Inside the director, personnel used optical devices to measure the range, location, and movement of a target, relative to the ship. The data collected in the director was passed to the plotting room, through either mechanical or electronic means, where an early computer calculated the projected position of the target and the appropriate firing solution for the ship's guns. The firing solution was relayed to the gunnery crews and consisted of the train (horizontal rotation) and elevation (vertical rotation) for the gun. For aerial targets, the firing solution also included the time delay to be set in the shell's fuse, since direct hits on aircraft were rare.

Measuring an aircraft's range, altitude, speed, and course was a tall order in the early years of World War II when radar was not commonly found in warships. Taking these measurements was even more challenging given that a ship could be moving in the X, Y, and Z axes simultaneously, and smoke and clouds could obscure or hide the target aircraft. The effectiveness of the entire system depended on a number of factors such as the training of the ship's personnel, the accuracy of the measurement devices, the speed of data transmission within the system, and the speed of changing the gun's orientation.[2]

## 1 Interwar Development

At the end of World War I, the Royal Naval Air Service stood pre-eminent with over 55,000 personnel, over 3,000 planes, and four aircraft carriers and seaplane tenders.[3] The armistice overtook the Grand Fleet's plan to launch an attack on the German fleet in port using torpedo-carrying aircraft.[4] Still, the planned operation highlighted the potential challenge posed by modern aircraft to warships. Two questions in particular vexed naval planners and leaders in the years after the war: How much of a threat did aircraft pose to warships, and what was the best means of countering this threat?

One of the first tests designed to answer these questions occurred in July 1921 when the U.S. Navy conducted bombing tests against several retired warships

2 David Mendell, *Between Human and Machine: Feedback, Control, and Computing before Cybernetics* (Baltimore, MD: Johns Hopkins University Press, 2004), 27.

3 Geoffrey Till, *Air Power and the Royal Navy, 1914–1945: A Historical Survey* (London: Macdonald and Jane's, 1979), 29–30; Geoffrey Till, "Retrenchment, Rethinking, Revival, 1919–1939," in *The Oxford Illustrated History of the Royal Navy*, ed. J.R. Hill (Oxford, UK: Oxford University Press, 1995), 320.

4 Mike Farquharson-Roberts, *A History of the Royal Navy: World War I* (New York: Palgrave Macmillan, 2014), 153.

off the Virginia coast, culminating in the sinking of the former German battleship *Ostfriesland.* While air power advocates, such as Brigadier General William Mitchell, claimed the tests represented the triumph of air power over the battleship, naval officers pointed out that the ships were unmanned, did not maneuver to avoid the attacks, and did not shoot back. The U.S. Navy's General Board studied the data from the tests closely and made a number of recommendations that helped improve the navy's anti-aircraft capabilities. For example, while the 1922 Washington Naval Treaty limited the size of battleships, the American negotiators succeeded in inserting a clause that allowed a slight expansion of battleship displacement to defend against air and submarine attack, based in part on the Board's recommendations.[5]

The *Ostfriesland* tests also shaped the Royal Navy's views on the threat posed by aircraft to ships at sea. The tests came too late to impact the 1920 interim report of the Admiralty's Naval Anti-Aircraft Gunnery Committee which recommended the development of tachymetric fire control (capable of tracking and aiming at targets moving in three dimensions) and the deployment of large numbers of small AA weapons.[6] However, studies ordered by the Imperial Defense Committee after World War I on defense against aircraft did consider the American tests. These studies concluded in 1926 that aircraft, equipped with the simple bombsights then available, attacking small targets such as ships, would need to fly straight and level at a constant speed. This conclusion served as the basis for the development of the British High Angle Control System (HACS), the fleet's primary anti-aircraft fire control system for vessels of cruiser size and larger.[7]

Introduced in 1930 and produced by the Vickers Corporation, HACS went through a number of iterations as British engineers made a series of gradual improvements in the system, though with only marginal improvements in performance. By 1935, British cruisers, aircraft carriers, and battleships carried HACS Mark 1. However, the system suffered from a number of limitations that hindered its capabilities before the introduction of radar. Within the director,

5 John Kuehn, "The Ostfriesland, the General Board of the Navy, and the Washington Naval Treaty: A Relook at a Historic Sinking," in *New Interpretations in Naval History: Selected Papers from the Sixteenth Naval History Symposium Held at the United States Naval Academy, 10–11 September 2009*, by Craig Felker and Marcus Jones, U.S. Naval War College Historical Monograph Series (Newport, RI: Naval War College Press, 2012), 81–83.

6 James Goldrick, "For Want of a Nail … Royal Naval Anti-Aircraft Gunnery 1919–1939," *Journal of the Royal United Services Institution of Australia* 6 (April 1983): 62–63.

7 H.W. Pout, "Weapon Control in the Royal Navy, 1935–45," in *The Applications of Radar and Other Electronic Systems in the Royal Navy in World War 2*, ed. F. A. Kingsley (Baskingstoke, UK: Palgrave Macmillan, 1995), 87.

personnel had to track the target and compensate for the ship's motion manually, which necessarily introduced error into the firing solution calculation.[8] The rangefinder used to determine the range to the target aircraft proved problematic, which necessarily led to errors in setting the fuses on the AA shells. Furthermore, the HACS designers' assumption that the target would fly straight and level at constant speed proved to be inapplicable to the high performance dive bombers and torpedo bombers of World War II. HACS designers had recognized this premise as a problem, but believed that the ship's crew would be able to correct for errors in the firing solution as the engagement progressed.[9] However, as aircraft speed increased due to the improved engine technology of the 1930s, AA engagements became shorter and shorter, preventing the crew from gradually adjusting their fire onto the target plane.[10] In short, the planning assumptions used to establish technical requirements for British naval AA systems were not borne out by events, leading to combat capabilities ill-suited for fighting high-performance aircraft.

In addition to these technical issues, British anti-aircraft fire control performance suffered from stringent budgets. Throughout the 1920s and up to 1932, the British government budgeted on the assumption that a major conflict would not come for at least ten years.[11] This rule limited the funds available for experimental fire control systems as the Admiralty struggled to maintain the existing fleet with tight finances. British designers considered a tachymetric system for HACS, but rejected it as too expensive.[12]

Financial restrictions also affected readiness and training. Limits on ammunition expenditure in AA exercises, delays in introducing high-speed AA exercise targets, and restrictions on live-fire exercises all stemmed from monetary concerns, and all reduced the skills and capabilities of AA gun crews. During a particularly telling anti-aircraft exercise in 1937, the Mediterranean Fleet failed to register a single hit on a target drone flying straight and level at low speed over the fleet for an hour.[13]

---

8 As was the case with intercontinental ballistic missile (ICBM) accuracy in the Cold War, the key to improved AA performance was to identify and eliminate sources of error in the system.

9 Pout, "Weapon Control in the Royal Navy, 1935–45."

10 Ibid., 87, 90, 92; Colin Sinnott, *The RAF and Aircraft Design, 1923–1939: Air Staff Operational Requirements* (London: Frank Cass, 2001), 38–39.

11 Joseph Moretz, *The Royal Navy and the Capital Ship in the Interwar Period: An Operational Perspective* (London: Routledge, 2002), 69.

12 Pout, "Weapon Control in the Royal Navy, 1935–45," 50.

13 Till, "Retrenchment, Rethinking, Revival, 1919–1939," 320; David Brown, *Nelson to Vanguard: Warship Development, 1923–1945* (Annapolis, MD: Naval Institute Press, 2000), 209;

In wartime, these training problems compounded technical deficiencies, especially in the large number of British destroyers built in the late 1930s. British naval architects, preparing destroyer designs for these ships, relied on the deliberations of the 1931 Naval Anti-Aircraft Gunnery Committee. The Committee's report downplayed the effectiveness of attack by dive bombers as the Royal Air Force (RAF) held that the difficulty of performing diving attacks required purpose-built planes of little utility in other roles. Due to the narrow utility of the specialized planes required for dive bombing, the RAF did not expect dive bombers to be widespread in a future conflict.[14]

This assessment reflected the state of aviation in 1930, but in a few years purpose-built dive bombers were introduced by the American and Japanese navies.[15] The unfortunate timing of the Committee's work led the Admiralty to underestimate the need for weapons capable of firing on targets directly over a ship. Furthermore, the Committee argued that destroyer armament should be designed to engage aircraft flying past the destroyers on their way to attack the more valuable capital ships at the center of a formation. British planning and the Royal Navy's own naval aviation doctrine suggested that destroyers did not need high-angle guns. British planners assumed that air attacks would focus on high value targets such as battleships. In particular, in an engagement with the Imperial Japanese Navy presumably conducted at a distance from land, each side would be limited to the use of aircraft carried to the engagement in aircraft carriers. Since the numbers of aircraft available would be limited and multiple aircraft were required to successfully strike a capital ship, fleet commanders should concentrate these limited aerial assets on high-value targets.

This doctrine was precisely what the British planned to do with the Fleet Air Arm. In this threat environment, heavy warships would need high angle anti-aircraft guns, but putting these guns on escorts was a lower priority. Taken together these conclusions led British naval architects to equip British destroyers such as the *Tribal*-class with medium-caliber AA guns incapable of elevating above 40 degrees, leaving these ships with little ability to defend themselves against dive-bombing attacks that came from almost directly above.[16]

---

Roskill, *Naval Policy Between the Wars, Volume II: The Period Reluctant Rearmament, 1930–1939*, 333.

14 Till, *Air Power and the Royal Navy, 1914–1945*, 142–143.

15 Thomas Wildenberg, *Destined for Glory: Dive Bombing, Midway, and the Evolution of Carrier Air Power* (Annapolis, MD: Naval Institute Press, 1998), 80–82; David Evans and Mark Peattie, *Kaigun: Strategy, Tactics, and Technology in the Imperial Japanese Navy, 1887–1941* (Annapolis, MD: Naval Institute Press, 1997), 307.

16 Brown, *Nelson to Vanguard: Warship Development, 1923–1945*, 208–209.

As aircraft performance steadily improved in the 1930s the Royal Navy took what steps it could to improve the fleet's anti-aircraft capabilities. Successive versions of HACS corrected some of the system's technical problems. The 1937 HACS Mark 3 compensated for some of the ship's movement, but still relied on human input at a number of points. However, by 1937, the Royal Navy was in the midst of a major shipbuilding program in response to the government's decision to begin rearmament in the face of the growing German threat. The decision to rearm invoked the iron law of mobilization according to Norman Friedman: "produce what is on hand – you do not wait for something better."[17] The long lead times required for warship construction precluded designing a new fire control system that could be produced in large numbers in time to be fitted to the ships being ordered. Thus, the Royal Navy went to war in 1939 with an AA fire control system with serious limitations.

On the other side of the Atlantic, the U.S. Navy's approach to interwar anti-aircraft fire control reflected the fleet's unique strategic context and organizational culture. In contrast to the British, the U.S. Navy planned to operate at a considerable distance from friendly bases as the fleet moved into the western Pacific to engage the Japanese, the assumed enemy during the 1920s and 1930s. As a result, the Americans could not rely on fighter support from shore bases and needed a native anti-aircraft capability that could travel with the fleet. Fighters operating from aircraft carriers provided part of this capability in combination with ship-board AA guns, which together formed the fleet's air defenses.[18]

The U.S. Navy also benefited from an organizational culture that emphasized decentralized oversight of corporate and scientific naval work, overlapping lines of research, and a willingness to accept satisfactory equipment as opposed to pursing even more advanced equipment at the risk of diminishing research returns.[19] This approach to research, development, and production was highlighted in a series of reports produced by British technical missions

17 Norman Friedman, *Naval Anti-Aircraft Guns and Gunnery* (Annapolis, MD: Naval Institute Press, 2013), 10.

18 Edward Miller, *War Plan Orange: The U.S. Strategy to Defeat Japan, 1897–1945* (Annapolis, MD: Naval Institute Press, 1991); Norman Freidman, "Mark 37 Direction and Fire Control," *Naval History* 25, no. 4 (August 2011): 28–30.

19 H. W. Pout, "Part 1 of Report of the British Naval Fire-Control Technical Mission (Commercial) to the United States of America, 1949 (Fire-Control, Computers, and Components), A.S.R.E. Monograph 861" (Admiralty Signal & Radar Establishment, June 1950), 119–121, ADM 220/189, National Archives of the UK; H. W. Pout, "Part 2 of Report of the British Naval Fire-Control Technical Mission (Commercial) to the United States of America, 1949 (Guided Missiles), A.S.R.E. Monograph 863" (Admiralty Signal & Radar Establishment, January 1951), 27, ADM 220/189, National Archives of the UK; Commander M. J. Ross, RN,

that visited the United States during World War II and the early Cold War to study American industrial and scientific practices. The missions found that in contrast to the British approach, where the Royal Navy stated its material requirements with great precision and closely oversaw research and engineering, "the Americans give the manufacturer a much freer hand in development than is the case here," according to the report of Dr. Cecil Dannatt, a British electrical engineer who headed one of these missions. Dannatt's report also noted that "the [U.S.] Navy Department functions mainly to translate Fleet requirements to the manufacturer, to select the best solution ... and to overlook production."[20] He argued that "by contrast, the British method is to wet nurse the manufacturer," by not providing sufficient latitude for corporate initiative.[21] Commander M. J. Ross, RN, a British gunnery expert, commented on American acceptance of satisfactory, rather than perfect, fire control equipment: "they have been content to progress steadily one step at a time, whereas we have tried to take several steps in one stride."[22] H. W. Pout, a noted British engineer, highlighted the good working relations between naval officers and U.S. companies after surveying American fire-control work: "the relationship between BuOrd [Bureau of Ordnance] project engineers and the U.S. firms seemed very good, more detailed, and rather less formal than in this country."[23]

This amicable relationship stemmed in part from the administrative structure of the Bureau of Ordnance (hereafter, BuOrd). BuOrd used line officers in addition to staff officers to procure equipment for the fleet. Line officers who rotated between sea duty and staff positions could not gain as much experience in weapon acquisition as full-time staff officers. As a result, BuOrd line officers relied heavily on engineering companies, giving them broad mandates rather than detailed guidelines.[24] Together, these methods of drawing upon the expertise of American industry allowed the U.S. Navy to deploy an

"American A.A. Fire Control Systems, Report of a Visit to U.S.A" July 1945, 31–35, B15 52, HMS Excellent (courtesy of Dr. Jon Sumida).

20 Dr. Cecil Dannatt, "Report of Dr. C. Dannatt on American Fire Control Practice and Developments (Visit of U.S.A. and Canada – July, 1943)" 1943, B15 20, HMS Excellent (courtesy of Dr. Jon Sumida).

21 Dr. Cecil Dannatt, 12.

22 Commander M. J. Ross, RN, "American A.A. Fire Control Systems, Report of a Visit to U.S.A," 31.

23 Pout, "Part 1 of Report of the British Naval Fire-Control Technical Mission (Commercial) to the United States of America, 1949 (Fire-Control, Computers, and Components), A.S.R.E. Monograph 861," 121.

24 Mendell, *Between Human and Machine: Feedback, Control, and Computing before Cybernetics*, 29.

effective fire control system for its dual purpose (weapons capable of engaging ship or aerial targets) 5-inch guns.[25]

A number of corporations contributed to this system. The Ford Instrument Company supplied directors and fire-control computers; General Electric provided automatic data transmission systems; and the Arma Corporation contributed a stable element, a gyroscope to allow the system to compensate for the ship's motion.[26]

The Ford Instrument Company (no relation to Henry Ford) already produced fire-control equipment for warships engaging surface targets and, in the late 1920s, Ford began developing the Navy's first anti-aircraft director.[27] In 1939, the Navy adopted the Mark 37 director, the latest in a series of directors produced by Ford. The Mark 37 could track targets moving up to 400 knots horizontally and 250 knots vertically, which meant it was capable of tracking modern dive bombers, a task with which British systems struggled. As avid users of dive bombers, the Americans emphasized dive bomber defense in their technical requirements for naval anti-aircraft systems.[28] By 1940, 300 Mark 37 directors had been produced, and were being installed on U.S. warships, especially the large number of *Sims*, *Benson*, and *Gleaves*-class destroyers coming out of American shipyards.[29]

The Mark 37 worked in conjunction with the Mark 1 computer, also produced by Ford, and located in the plotting room below decks.[30] Moving the computing element below decks reduced the weight of the director elements installed in the ship's superstructure which in turn increased the director's ability to rapidly turn to track targets.[31] Housed in a cabinet, the Ford Mark 1 computer was roughly the size of an oversized office desk. The computer electrically received constant range, elevation, and bearing data from the director, ship speed information from navigation, and pitch and roll information

25 Mendell, 50, 55, 60.

26 *Ibid.*

27 "Ford Instrument Company," *Journal of Jet Propulsion* 26, no. 5 (May 1956): 24–25.

28 Bureau of Ordnance, "No. 79, Fire Control (Except Radar) and Aviation Ordnance" 1946, Fiche 3, 145–149, U.S. Naval Administration in World War II; Microform Room, U.S. Naval Academy Library, Annapolis, MD; Friedman, *Naval Anti-Aircraft Guns and Gunnery*, 138.

29 Mendell, *Between Human and Machine: Feedback, Control, and Computing before Cybernetics*, 61.

30 A. Ben Clymer, "The Mechanical Analog Computers of Hannibal Ford and William Newell," IEEE *Annals of the History of Computing* 15, no. 2 (1993): 25; Burford Rowland and William Boyd, *U.S. Navy Bureau of Ordnance in World War II* (Washington, D.C.: Department of the Navy, 1953), 377.

31 Bureau of Ordnance, "No. 79, Fire Control (Except Radar) and Aviation Ordnance," Fiche 3, 147.

(known as level and cross-level) from a gyroscope, and calculated the future position of the aircraft relative to the ship and the necessary firing solution. The computer transmitted the solution, consisting of the train (horizontal) and elevation (vertical) settings for the gun and the fuse settings for the shell, to the gun platform where hydraulic motors automatically shifted the gun to match the firing solution train and elevation, and an automatic fuse setter adjusted the time delay fuse on the shell.[32] At each step in the process, automation served to reduce the need for data to be inputted manually or set by hand, reducing the delay between the director's crew reporting the target's position and the guns actually firing.

British observers who inspected the Mark 37 during visits to American warships in 1940 and 1941 praised the system for its capabilities. Captain Hugh Faulkner, RN, observed a Pacific Fleet destroyer anti-aircraft exercise off Hawaii in September 1940. "The destroyer firings were the most impressive that I have witnessed ... From our position the bursts seemed to be continually around the target."[33] Faulkner praised the system's ability to "nearly always succeed in getting at least one hit" during each practice run in the exercise.[34] An August 1941 general survey of the U.S. Navy's capabilities, circulated throughout the Admiralty, noted that American long range anti-aircraft fire "continues to be much better than British," especially on destroyers, due to superior fire control.[35]

The U.S. Navy's anti-aircraft fire control capabilities on the eve of World War II reflected both the Navy's management of industrial research and development, and an appreciation of the threat posed by aircraft to warships. The U.S. Navy's approach to developing weapons took advantage of the relative strength of American scientific research capabilities, both civil and military. A 1945 British report lamented that "the lavish staffs and equipment and the drive behind such American organizations as the Massachusetts and California Institutes of Technology....have no counterpart in England."[36] Furthermore, the development of American naval aviation, particularly its offensive capabilities,

32 Baldwin Bridger, "An Earlier Kind of Computer," *IEEE Industry Applications Magazine*, October 2007, 8–10, https://doi.org/10.1109/MIA.2007.902394.

33 Captain Hugh Webb Faulkner, RN, "Visit to the U.S. Fleet Based at Honolulu" September 29, 1940, 5, AVIA 10/1, National Archives of the UK.

34 Captain Hugh Webb Faulkner, RN, "Practice Results" September 17, 1940, 1, AVIA 10/1, National Archives of the UK.

35 Rear Admiral Herbert Pott, RN, Naval Attaché, Washington to Director of Naval Intelligence, Admiralty, "N.A. 501, Observers with U.S. Fleet," August 30, 1941, ADM 178/223, National Archives of the UK.

36 Admiral James Somerville, "Memorandum, 'Comparison between British and American Ships and Equipment'" July 3, 1945, 42, ADM 1/19308, National Archives of the UK: Public Record Office.

made the U.S. Navy aware of the need for advanced anti-aircraft capabilities, especially if the fleet was to successfully operate deep in the Western Pacific with little support from shore bases.

## 2 Wartime Experience

While the early months of World War II witnessed few air attacks at sea, the German invasion of Norway in April 1940 forced the Royal Navy to operate for extended periods within range of German aircraft. The result was a number of attacks on British warships that highlighted the threat posed by modern aircraft. The fall of France in June gave the Germans airfields within range of a number of British naval bases as well as the convoy routes through the English Channel and the Western Approaches. France's defeat also changed London's attitude toward a standing U.S. Navy request to place observers on British ships. In the summer of 1940, the two navies agreed to exchange observers. The first British officers arrived in September 1940 as part of the Tizard Mission, which travelled to the United States to improve scientific cooperation between the two nations. American observers began arriving in London in July and August 1940.[37]

By the end of 1940, the Royal Navy had suffered significant losses due to German air attacks. Between January and December 1940, 12 British destroyers were lost to German air attack, primarily off Norway and Dunkirk and in the English Channel. In early 1941, the scene of action shifted to the Mediterranean where the carrier HMS *Illustrious* was heavily damaged by German dive bombers in January. Later in the spring of 1941, the withdrawal of British forces from Greece to Crete and then off of Crete again placed the Royal Navy under threat of constant German air attack. Although the Mediterranean Fleet succeeded in evacuating thousands of British and Commonwealth soldiers from Crete, the Navy took heavy casualties from German aircraft. During the April and May 1941 Crete evacuation, the Royal Navy lost 4 cruisers and 6 destroyers, while a carrier, 2 battleships, 4 cruisers, and 2 more destroyers were damaged, and rendered out of action for months. American observers, who witnessed a number of these engagements, made a number of critical remarks about British anti-aircraft equipment and doctrine.

American observers typically found British anti-aircraft systems to be technically inferior to American systems. One observer noted that British range

37 Brian Hussey, "The U.S. Navy, the Neutrality Patrol, and Atlantic Fleet Escort Operations, 1939–1941" (Annapolis, MD, 1991): 29, ADA245396, Defense Technical Information Center.

finders did not give accurate readings when elevated above 45 degrees due to design problems. Upon reading the report, the senior gunnery officer for the U.S. Pacific Fleet wrote "if British rangefinders do this it is because of poor design ... we licked this problem eight years ago."[38] Lieutenant Commander John Opie, USN, spent considerable time with the Mediterranean Fleet in the spring of 1941. He observed that in "general British material is sturdy and simple."[39] However, he found British anti-aircraft fire control systems lacking. Specifically, the fire control equipment on British cruisers and battleships was not tachymetric and was not fully stabilized.[40] Consequently, Commander Clark, an American observer with the Mediterranean Fleet, concluded that British high-angle anti-aircraft fire "was not particularly impressive."[41]

These observations were made by officers with greater technical education than their British counterparts. Indeed each navy's approach to officer

---

38 Ltr, Commander Willard A Kitts III, USN, Fleet Gunnery Officer to Commander in Chief U.S. Fleet, 14 April 1941, RE: Intelligence Report, Serial F-1, O-26; Gunnery Comments on, enclosure to Office of Naval Intelligence, F-1 to Commander in Chief U.S. Fleet, "Serial F-1, O-26, Miscellaneous Information," March 25, 1941, Folder A8-2, ONI Intelligence Reports, 5 Apr 1941-26 May 1941 (2 of 2); Box 3; Entry P 88, CINCPAC Secret and Top Secret General Administrative Files, 1941–1944; Records of Naval Operating Forces, Record Group 313, National Archives at College Park, College Park, MD.

39 Office of Naval Intelligence, F-1 to Commander in Chief U.S. Fleet, "Serial F-1, O-11, Great Britain - Navy - Operations," February 14, 1941, 11, Folder A8-2, ONI Intelligence Reports, 1 Jan 1941–4 Apr 1941 (2 of 3); Box 3; Entry P 88, CINCPAC Secret and Top Secret General Administrative Files, 1941–1944 [hereafter CINCPAC S and TS Files]; Records of Naval Operating Forces, Record Group 313 [hereafter RG 313], National Archives at College Park, College Park, MD [hereafter NACP]; Commander in Chief, U.S. Fleet, "A8/01884, Intelligence Bulletin No. 35," November 30, 1940, 2, Folder #27, A16-3, Miscellaneous Ship Operations, 1942 (2 of 2); Box 5; Entry P 97, Southwest Pacific Force (Red 164) Secret Administrative Records, 1942; Records of Naval Operating Forces, Record Group 313, National Archives at College Park, College Park, MD.

40 Office of Naval Intelligence, F-1 to Commander in Chief U.S. Fleet, "Serial F-1, O-11, Great Britain - Navy - Operations," February 14, 1941, 11, Folder A8-2, ONI Intelligence Reports, 1 Jan 1941–4 Apr 1941 (2 of 3); Box 3; Entry P 88, hereafter CINCPAC S and TS Files; Records of Naval Operating Forces, hereafter RG 313, hereafter NACP; Commander in Chief, U.S. Fleet, "A8/01884, Intelligence Bulletin No. 35," November 30, 1940, 2, Folder #27, A16-3, Miscellaneous Ship Operations, 1942 (2 of 2); Box 5; Entry P 97, Southwest Pacific Force (Red 164) Secret Administrative Records, 1942; Records of Naval Operating Forces.

41 Naval Attaché, London to Office of Naval Intelligence, "Serial 1887, Great Britain - Navy - Operations," August 14, 1941, 21, Folder A8-2, ONI Intelligence Reports, 10 Sep 1941–17 Oct 1941 (3 of 3); Box 5; Entry P 88, Records of Naval Operating Forces, Tony DiGiulian, "The British High Angle Control System (HACS)" (Naval Technical Board), http://www.navweaps.com/index_tech/tech-066.htm#7.

education played a role in the development of anti-aircraft capabilities. In comparison with American junior officers, the average Royal Navy junior officer had far less training in engineering and significantly more experience as a ship driver. In the Royal Navy, an officer who specialized in engineering was not considered a "proper" officer and was not eligible to command a warship or fleet.[42] Thus, British equipment was designed for use by officers with less technical training than in the United States Navy.

This difference in technical backgrounds can be demonstrated by the comments made by British officers who visited American warships. One observer in 1941 reported that American "gunnery and electrical equipment is generally more complicated than that fitted in British warships."[43] After viewing a new American radio receiver, one Canadian officer wrote, "being quite a new USN electronic gadget it goes without saying that it is very complicated, technically."[44] The result of this difference in technical training was that British anti-aircraft equipment was seen by American observers as "sturdy and simple," lacking in some of the technical sophistication found in the U.S. Navy. Diverging approaches to officer technical education shaped the manner in which each navy designed and employed its weaponry.

## 3 Long-Term Results

The reports of their observers in the early years of World War II had long-term impacts on each navy's perception of the other. Observer reports led portions of the U.S. Navy to conclude that the Royal Navy could not defend itself against an air attack, feeding into a narrative in which the British were substandard in carrier warfare.

This attitude was most common among U.S. Navy aviators. Vice Admiral Lloyd Mustin, who became the Commander, Amphibious Force, Atlantic Fleet in the Cold War, later described British naval anti-aircraft performance as

42 Jock Gardner, January 2015, Author's personal collection; Admiral James Somerville, "Memorandum, 'Comparison between British and American Ships and Equipment,'" 43.

43 Lieutenant Commander Colin Colenso Martell, RN to Naval Attaché, Washington, April 25, 1941, ADM 178/223, National Archives of the UK.

44 Commodore Horatio N. Lay, RCN, Naval Member, Canadian Joint Staff (Washington) to Naval Secretary, "Report of a Visit to NAS Norfolk by Staff Officer (Air)," April 30, 1952, 3, File 1225-193-139 Part 2, "Visits and Reports, Naval Member, Canadian Joint Staff, Washington, 1949–1964," Volume 33487, RG24-D-1-c, Library and Archives Canada.

follows: "they had no anti-aircraft capabilities. Their capabilities were pitiful."[45] Rear Admiral Harold Miller, a USN pilot during World War II, later recalled that the British "just really didn't know how to operate aboard their carriers."[46] During his time with NATO in the early 1950s, Vice Admiral Herbert Riley, USN, remembered forcefully telling a British senior admiral: "I don't think that there's anything that the Royal Navy or the Fleet Air Arm can teach American naval aviators about operations in combat." The reports on British anti-aircraft capabilities filed by American observers in 1940 and 1941 played into this narrative; they highlighted the success of German air attacks on British warships, attacks that, to naval aviators, ought to have been countered by carrier-based fighters.

British observer reports on the U.S. Navy led many British officers to conclude that while the Americans had outstanding equipment, they did not know how to effectively use it. One British officer who observed an American carrier in 1952 concluded: "the US Navy has good equipment for providing information from all sources, but doesn't know how to use it."[47] A Canadian officer who observed U.S. airborne ASW practices in 1950 wrote:

> the United States Navy ... reminds me of a 1950 "Golden Jubilee" Packard with ultramatic drive. It is an extremely fine piece of machinery with loads of power under the hood. With a mighty bellow, gears whirring, lights flashing, and bells ringing it is away at a slow crawl. There seems to be just a bit of trouble in getting all that power to the wheels. The USN has a generous supply of the latest and in most cases the best equipment but the training given is most elementary.[48]

Observer reports could have a major impact on long-term attitudes; however, these reports rarely considered the full context within which each navy made its decisions. Each navy existed within a unique native environment. This environment consisted of internal characteristics such as doctrine, administration,

45 Reminiscences of Vice Admiral Lloyd M. Mustin, USN, interview by John T. Mason, Jr., 2003, 1:378, U.S. Naval Institute Library, Annapolis, MD.

46 Reminiscences of Rear Admiral Harold B. Miller, USN, interview by John T. Mason, Jr., 1995, 116, U.S. Naval Institute Library, Annapolis, MD.

47 Lieutenant Commander Robin Buller, RN and Lieutenant Commander David Seely, RN to Commander in Chief, Home Fleet, "Report on Experiences in Exercise Mainbrace," October 6, 1952, 2, ADM 1/23452, National Archives of the UK.

48 Report, Lieutenant (P) Frederic G. Townsend, RCN, Enclosure A to Commodore Kenneth F. Adams, RCN, CO HMCS Magnificent to Naval Secretary, "A/S Training in the USN," November 23, 1950, File ACC 1225-1, "Flag Officer Atlantic Coast – Visits and Reports – General, 1950–1957," Volume 11189, RG24-D-10, Library and Archives Canada.

and personnel as well as external ones such as national strategic policy, physical features of the expected area of operations, and national resources: fiscal, industrial, and demographic. Unless observers spent more than a year on observation duties, they rarely conveyed this context in their reports. As a result, technological comparisons were made without sufficient consideration given to the framework in which these choices had been made. British anti-aircraft gun technology in 1939 reflected a constrained fiscal environment, the Navy's assumptions about aviation at sea, the influence of the RAF on carrier design, unlucky timing in setting requirements, and the position of engineers in the fleet. American anti-aircraft capabilities grew out of the U.S. Navy's experience with naval aviation, a preference for more rather than less technical education for officers, and the Navy's management of American industrial and scientific resources.

As military organizations, the Royal Navy and the U.S. Navy were avid users of advanced technology. However, their decisions about which specific systems to design, purchase, install, and employ in their ships and aircraft reflected their own strategic, organizational, fiscal, and administrative contexts. These contexts formed the background for each navy's technology, a background from which observers found it difficult to separate themselves.

To a greater degree than land forces, the capabilities of naval forces flow from their technology, as is true of air forces. As a result, naval history often focuses on the design and employment of specific ships or weapon systems. The prominent role of technology in naval warfare can serve as a powerful catalyst for innovation, though the maritime environment plays a role as well. The dangers to life and limb inherent in sea voyages encourages a certain naval reluctance to abandon proven designs and methods.

Thus, naval historians ignore technology at their own peril. However, technology must be understood within a broad historical context that takes into account national interests, societal structures, and organizational culture. Furthermore, war is fundamentally a human endeavor in which people employ weapons, rather than the other way around. This does not mean that human will can inevitably triumph over advanced weaponry, but that we must understand the wielder as much as the weapon.

CHAPTER 11

# Strategy, Technology, and Timing: Aircraft, the Mediterranean Air War, and the Turning of the Tide in the European Theater

*Robert S. Ehlers, Jr.*

## 1 Why Wartime Airpower Technologies Mattered in the Mediterranean Theater

Were an interested person to seek a nearly ideal case study of the interactions between war and technology in the context of industrial-age warfare, it would be difficult to find a better one than the role of airpower in the Mediterranean Theater during World War II. Given the theater's underappreciated strategic importance, geographic complexity (see Maps 11.1–11.3), and calamitous impact on the Axis war effort, it is well past time to examine it with greater care.[1]

If, as Douglas Porch asserts, the Mediterranean Theater was a pivotal (although not a decisive) one, a critically important Allied learning laboratory, and one of the great graveyards of the Axis war effort, then we must seek to understand why and how this was so.[2] Along with a series of disastrously misguided grand-strategic and military-strategic choices—and failures to choose—on the part of the Axis Powers, especially Germany, the conduct and course of the air war emerge as vitally important aspects of the Axis defeat. Combined-arms efforts on the Allied side, including a far more effective use of air assets in cooperation with land and sea power, proved central to operational victories and the prevention of strategic disasters in a theater comprised predominantly of desert and ocean—two mediums in which airpower proved particularly effective. These combined-arms efforts turned the tide of the war in the theater firmly toward the Allies and denied the Axis a series of strategically vital accomplishments that were well within their reach.[3]

---

1 Robert S. Ehlers, Jr., *The Mediterranean Air War: Airpower and Allied Victory in World War II* (Lawrence, KS: University Press of Kansas, 2015), *passim*.

2 Douglas Porch, *The Path to Victory: The Mediterranean Theater in World War II* (New York: New York: Farrar, Straus, and Giroux, 2004), 10, 12.

3 Ehlers, *Mediterranean Air War, passim;* Porch, *Path to Victory, passim.*

 | DOI:10.1163/9789004393301_013

MAP 11.1 (West Point Atlas Map 33) – The Mediterranean and Middle East Theater covered an area of more than 3 million square miles. Its geography demanded a capable and complex interplay between land, sea, and air forces. The side that held air superiority and learned more effectively to employ combined-arms operations within the theater, based on sound grand and military strategies, would have a major advantage. (USMA)

The positive and "negative" strategic benefits that the Axis powers could have derived by winning in the theater—with "negative" successes defined here as the denial of something vital to the British Empire's and larger Allied war effort—were substantial and would have changed the course and duration of the war. Capturing British oilfields in Iraq and Iran would have done severe damage to the Empire's efforts in the Middle and Far East. Whether or not the Germans and Italians could have repaired whatever damage the British did to the oilfields as they withdrew, and then begun production in their own right, is debatable. However, available evidence suggests strongly that they could have done both. The Germans' dismantling of the Port Jerome refinery in France in anticipation of rebuilding it in the Caucasus Mountains after Russian oilfields there fell to German forces is a case in point. Once it became clear that they would not be able to reconstruct the facility in the Caucasus, the Germans rebuilt it at Moosbierbaum in Austria. The process took just over a year, and

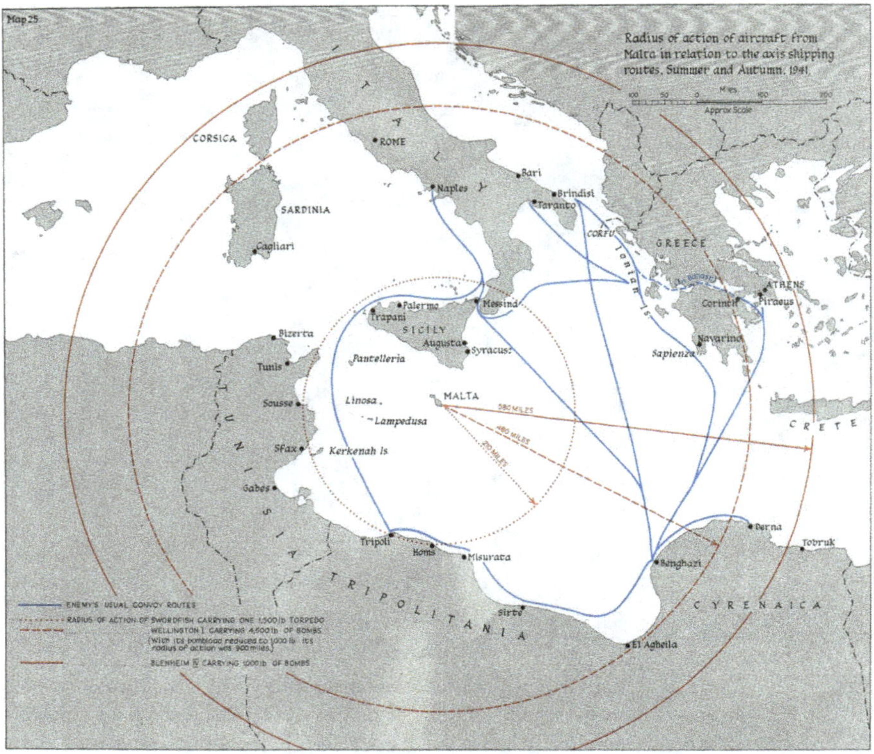


MAP 11.2 Malta aircraft ranges and major Axis convoy routes, 1941. Malta was the logistical hinge upon which all else ultimately turned in the Mediterranean Campaign. The Axis failure to maintain air superiority over the island, much less capture it, was one of the major blunders of the war. When combined with larger British air-superiority and combined-arms efforts in North Africa and throughout the theater of operations, the successful defense of Malta set conditions for Allied victory in the theater. (HMSO)

oil production was about to begin when American heavy bombers from the Foggia airfields in Italy raided the refinery in the first of nine attacks. It never produced useable quantities of product, but it was up and running in a year—a useful gauge for thinking about what German engineers and construction firms might have been able to do in Iraq and Iran, even if on a more modest scale, with secure sea and rail lines behind them after a military victory in the Mediterranean.[4]

4 Combined Strategic Targets Committee report, "Combined Strategic Targets Committee, Working Committee (Oil Production), Weekly Bulletin No. 25," Dec 19, 1944, U.S. Air Force Academy Special Collections Branch, Manuscript Record 16, Papers of Brigadier General

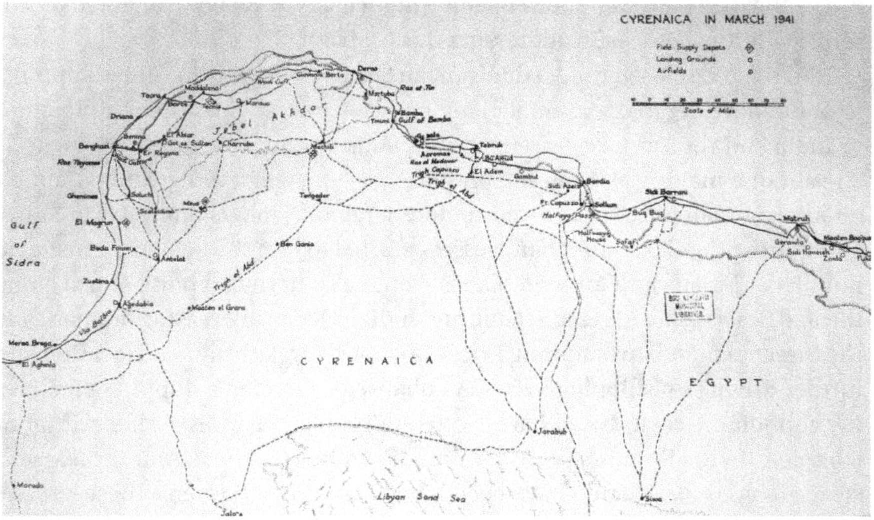


MAP 11.3 The Cyrenaican Hump. Note the heavy concentration of airfields here. Whichever side controlled these airfields tended to have the upper hand in the "battle for the SLOCs" since the RAF had trouble reaching key convoy areas unless it could engage in offensive operations from both Malta and these airfields in Cyrenaica. Every major British ground offensive gave the highest priority to capturing airfields so the RAF could play its full role in the combined-arms effort, whether direct support to the army or ship-attack missions. (HMSO)

Even without substantial Axis petroleum production, damage to the Allied war effort would have been severe, particularly given the major oil-tanker shortage the British and Americans suffered as a result of German U-boat actions. U-boat captains targeted oil tankers specifically given the importance of the resource they carried, and once Operation *Paukenschlag* (Drumbeat) began against American shipping in the Gulf of Mexico and along the Eastern seaboard, there were simply not enough tankers left to make good from American sources the 15 million barrels a year produced in Iraq and Iran. The United States produced one billion barrels of oil by 1940 and was the world's leading producer, but it could not deliver more than a tiny fraction of this oil to Great Britain, and nowhere near enough to compensate for the loss of the Middle

George C. McDonald, Section 5, Box 8, File 7, 2–3 and 7. British photointelligence analysts found Moosbierbaum, near Vienna, and confirmed that it was an oil refinery and a processing plant for a high-grade distillate used to boost aviation fuel octane ratings. It was confirmed by Allied Central Interpretation Unit reports to be the same one dismantled by the Germans at Port Jerome in France. Records obtained after the liberation of France reconfirmed this. Estimated output was 150,000 tons per year.

East oilfields. There were not enough American Type 2 and 3 large tankers in service until 1944 to begin addressing this problem.[5]

Had the Axis closed the Mediterranean Sea permanently to Allied shipping, it would have created extremely long and slow merchant shipping pipelines to India and the Far East—assuming the Axis did not overrun those areas as a result of a major victory in the Middle East—placing even greater strain on an already scarce resource a year before large oil tankers and Liberty Ships, along with other large merchant vessels, started to join the fleet in appreciable numbers. U-boats and torpedo planes would also have had ports of call along the Red Sea, Persian (Arabian) Gulf, and Indian Ocean, increasing dramatically the threat to merchant shipping in this previously (relatively) safe location and further driving up shipping losses. A collapse of merchant shipping pipelines in the Indian Ocean would have led to additional collapses, either partial or complete, in the Empire's positions in India and the Far East. American assets, scarce in 1941–1942, would have been inadequate to help reverse these losses.[6]

If the Allies had lost control of the Trans-Persian Railway, they would have been unable to deliver Lend-Lease to the Russians. Over 40 percent of total Lend-Lease, including a very large proportion of the vehicles, radios, landline cable, and other technologies necessary to give the Russians their own form of effective maneuver warfare, traveled to the Persian Gulf and then along that railway. Had this route been lost—the sea approaches and the railroad—the Americans could not have delivered nearly as much materiel to the Russians,

5 In 1942—a crucial year in the Mediterranean Theater and the last, best Axis opportunity to win there—the Americans lost 99 oil tankers. The British lost another 78. Given heavy British losses in previous years and the full range of delivery requirements to multiple theaters, the tanker force could not meet existing requirements in addition to delivering an additional 15 million barrels a year of American oil to replace oil lost in case of an Axis capture of the Middle East oilfields. During World War II, American tankers made 6,500 voyages, carrying 65 million tons of oil and gasoline from the U.S. and the Caribbean to American and allied forces. This amount was the equivalent of 260 million 55-gallon barrels, which weigh 500 pounds each. They supplied 80 percent of all the fuel used by Allied aircraft, vehicles, and vessels during the War. If we take the signing of the Lend-Lease Act on 11 March 1941 as the starting point for these deliveries, and VJ Day on 2 September 1945 as the ending point, the monthly delivery rate would have been 4.8 million barrels, with an annualized rate of 57.6 million barrels. British oil tankers had about 20 percent of the American capacity, meaning that total deliveries per year to all fronts would have been about 70 million barrels. However, the monthly averages would have been much lower in 1942 and 1943 pending the arrival of large numbers of American T2 and T3 tankers. During these two years, available tankers could not have delivered more than a small percentage of the 15 million barrels of oil produced in the Middle East had the Axis powers captured it. See http://www.usmm.org/tankers.htmla and https://www.battleships-cruisers.co.uk/merchant_navy_losses.htm.

6 Ibid.

given tonnage limitations at Murmansk and Archangel, the harsh and dangerous climate of the seas through which ships had to travel, and the high threat from U-boats operating along these routes. The fate of convoy PQ-17 and others in the PQ series made clear how dangerous this route was for shipping. The impacts of such a major decrease in deliveries on Red Army operational effectiveness would have been highly detrimental.[7]

Taken together, these positive and "negative" Axis strategic gains would have done grave harm to the Allied cause, extended the length and costs of the war, and perhaps even tilted its long-term course in favor of the Axis. In other words, rather than simply increasing the time and costs required to defeat the Axis, the Allies might have found themselves trying to survive, rather than win, at least until the successful detonation of atomic weapons. All of these points are counterfactual. They are also important and well worth discussing in terms of why and how a superior Allied air and combined-arms effort, executed within a clear and achievable grand strategy, prevailed over a less effective Axis effort that robbed Hitler and Mussolini—and Tojo for that matter—of what could have been a major military-strategic and even grand-strategic victory. Turkey would very likely have joined the Axis alliance had Germany and Italy won in the Mediterranean, and Spain might have done the same. Turkey's entry into the war would have given German forces a direct pathway into the Caucasus and southern Russia and secure lines of logistics behind them—a flanking effort that could have paid huge dividends for them and created a disaster for the Red Army and the larger Allied cause.[8]

## 2 Defining Airpower Technologies

To understand the dynamics involved in the complex interplay of strategy, combined-arms operations, and the employment of air assets in the high-stakes

7 Richard Overy, *Russia's War: A History of the Soviet War Effort, 1941–1945* (New York: Penguin, 1998), 193–198.

8 Gerhard L. Weinberg, *A World at Arms: A Global History of World War II* (Cambridge, UK: Cambridge University Press, 1994), *passim;* Major-General I. S. O. Playfair, *The Mediterranean and Middle East, vol. I, The Early Successes against Italy* [*to May 1941*] (Uckfield, UK: Naval and Military Press, 2004), 125–130. This is the British Official History and is referred to as "BOH" (plus volume number) in subsequent footnotes. See also Gerhard Schreiber, Bernd Stegemann, and Detlef Vogel, *Germany and the Second World War Volume III: The Mediterranean, South-east Europe, and North Africa, 1939–1941*, trans. Dean S. McMurry, Ewald Ossers, and Louise Willmot (Oxford: Clarendon Press, 1995), 127, 132–135, 153–162, 195. This is the German Official History and is referred to as "GOH" (plus volume number) in subsequent footnotes.

effort to control the Mediterranean Basin and surrounding areas, it is necessary to make clear what we mean by the term "technology." We must do so in terms of its effect on the air war and, by extension, the larger efforts of the warring parties.

At the most fundamental level, geographical realities drove a "Mediterranean system of armed conflict" just as it had during the great conflict between Christendom and the Ottoman Empire in the sixteenth and seventeenth centuries, and as it had during the rise and predominance of Rome.[9] The Mediterranean's many islands, narrow sea lines of communication, and shallow seas made it exceptionally well-suited for the operation of land-based airpower, which reached maturity just in time for the war. What airpower gained in this environment, sea power lost, as shipping and warships alike proved shockingly vulnerable to land-based air attack, in particular by dive bombers and torpedo planes. On land, the desert coast of North Africa, where the struggle for dominance in the Middle East played out, offered great advantages to the side with air superiority and terrible difficulties to the side without it.

Airpower magnified or diminished gains by the combatants. The British Army's four victories were amplified and, indeed, facilitated by air superiority, and its three major defeats did not lead to the outright destruction of the Western Desert Force—later Eighth Army—precisely because the RAF was able to maintain control of the air over the battle space as British troops retreated and regrouped. The Germans suffered the obverse of these circumstances. They endured constant air attacks on ground columns, resulting from failures to gain air superiority, before and especially after they outran their own, much less ground-mobile air units. Consequently, they were unable to turn tactical and operational victories on the ground into strategic ones.[10]

Given the importance, and in certain cases predominance, of airpower in the theater, one would think that the warring parties would internalize this reality. Only the British did to the necessary degree. The Italians failed despite the courage of their pilots and the effectiveness of their small torpedo-bomber force. The Germans fared little better despite having learned a great deal—at least in theory—about the RAF during the Battle of Britain, and in spite of their effective airframes and aircrews, and their tactical acumen. Why was this so? To the degree that it was a technological failure along with a strategic and operational failure on the part of the Axis, we must consider what comprised the

9 John F. Guilmartin, Jr., *Gunpowder and Galleys: Changing Technology and Mediterranean Warfare at Sea in the 16th Century,* revised edition (Annapolis, MD: Naval Institute Press, 2003), 16–21, 73–97.

10 Ehlers, *Mediterranean Air War, passim.*

airpower technologies employed by the warring parties, and how the relative success or failure to understand all elements of these technologies contributed to victory or defeat in the theater.

To do this properly, it is first necessary to develop a holistic view and understanding of technology. A seminal work on technology notes that "Three layers of meaning of the word 'technology' can be distinguished. First, there is the level of physical objects or artifacts...Second, 'technology' may refer to activities or processes...Third, 'technology' can refer to what people know as well as what they do..."[11] This tripartite definition is a useful starting point that allows us to discuss "technology in terms of a 'systems' metaphor," another key insight discussed regularly by historians of technology.

Further, any analysis of the role of technology and of technological systems in the conduct of the Mediterranean conflict must take full account of the range of contextual factors at play, including, but not limited to, history, geography, climate, and socio-cultural predilections. As Dr. John F. Guilmartin, Jr., a preeminent historian of armed conflict in the Mediterranean, noted: "[T]echnological factors cannot be analyzed outside their social, cultural, and economic context without inducing serious distortions...the crucial turning points hinged more often upon cultural, social or economic factors than upon narrowly technical ones."[12] In other words, as others have indicated, we must remain focused on "breaking down the distinction between human actors and natural phenomena" if we are to understand fully the impacts and limitations of technology in war.[13] Indeed, one of the greatest pitfalls the historian faces is to default to a technologically driven explanation of events. As scholars deeply conversant with this conundrum have admitted, "Who among us would deny that it is easy to be drawn into technology-driven explanations of cultural and historical processes? The frustration we experience with technological determinism and technocratic thinking is as much with ourselves as it is with the larger societies in which we live. As moths to a flame, we find ourselves continually attracted to its alluring but dangerous glow."[14]

This warning, along with the others relating to our fixation on technology, must inform our efforts going forward, not just for this particular topic, but for all aspects of history. Dr. Guilmartin pondered these issues deeply as he

11 Wiebe E. Bijker, Thomas P. Hughes, and Trevor Pinch, eds., *The Social Construction of Technological Systems* (Cambridge, MA: Massacusetts Institute of Technology, 1989), 4.

12 Guilmartin, *Gunpowder and Galleys*, 21.

13 Bijker et al., 4.

14 Merritt Roe Smith, "Technological Determinism in American Culture," in Merritt Smith and Leo Marx, eds., *Does Technology Drive History? The Dilemma of Technological Determinism* (Cambridge, MA: Massacusetts Institute of Technology, 1994), 35.

developed the following working definition: Technology, n. (1): "the application of knowledge to achieve a physical effect by means of an artifact, object or thing (distinct from art, which involves the application of knowledge to create aesthetic effect). (2): the artifact itself; the class of artifacts to which it belongs. (3): the knowledge needed to design, manufacture, operate, sustain and logistically support the artifact or thing and its user(s)."[15] Guilmartin's rigorous examination of the role of technology in war, particularly at the level of grand and military strategy, concluded with the following cautionary assertion:

> And so this analysis of the technological limits of strategy ends on a cautionary note about the essentially human nature of war…success in war [is] determined not just by strength and technological advantage alone, but by the effective application of those qualities in human context…understanding of policy goals presides over our application of military force. Whatever the technology, war remains as Carl von Clausewitz characterized it, a test of will and faith. Do not lose sight of that reality.[16]

Once again, we have a clear warning to value technology properly and to avoid assigning to it excessive credit for victory or defeat in war. Based on Guilmartin's observations and those of other prominent scholars, we thus arrive at three basic principles regarding technology. First, technology is much more than the artifact—we must address it holistically. In fact, technologies, including aircraft, are best addressed as "systems," and complex processes such as combined-arms warfare as "systems of systems." Unless we look comprehensively and subtly at what comprises a technology and a larger technological system, or system of systems, we cannot understand its impact on a given set of events. Second, intellectual and professional qualities are part of any technology. Put another way, technologies are "learning systems" in a very basic and profound way. To emphasize this point, Guilmartin noted regularly that "Intellectual infrastructures undergird technological ones."[17] Third, context, culture, cognition, and changes over time all influence uses of technology. These principles are central to understanding the Mediterranean air war and its role in the outcome of the larger conflict.

15 John F. Guilmartin, Jr., "Technology and Strategy: What Are the Limits? (1994), in *Two Historians in Technology and War*, 7.

16 Ibid., 25.

17 Discussions between Guilmartin and Ehlers, 2002–2005.

## 3 Strategy and Airpower Technologies

These admonitions and insights leave us with a clear understanding that airpower is much more than airplanes, but how much more? A great deal, as it turns out. Any discussion of Allied and Axis airpower technologies must begin with a discussion of grand and military strategy—the larger political and military ends for which the warring parties employed airpower. Of the major powers in the theater prior to November 1942—the British, Germans, and Italians—only the former had clear grand and military strategies that gave firm direction to their combined-arms effort.

Despite early defeats in France, Greece, and North Africa—and perhaps because of them—the British had the will and most of the means necessary to pursue their strategic aims. These included the three vital Mediterranean choke points—Gibraltar, Malta, and Suez—as well as access to the Middle East oilfields. Although the Americans played an increasingly important role in securing the Empire's position—most of all with the provision of aircraft—sound strategy set the conditions for ultimate success. Winston Churchill said, pointedly, that the British would "fight to the last inch and ounce" for the Middle East," that it was "the peg upon which all else hung," and that short of a German invasion of the British Isles, he could not conceive of a greater disaster than the Axis conquest of the Middle East. Consequently, he and his civilian and military advisors developed a realistic grand strategy that accepted a long war of attrition and an indirect approach to defeating the Axis powers that would first "close the ring" around Italy and only then turn fully to Germany.[18]

In contrast to British strategy, Italian high-level political and military aims were as inconsistent and muddled as its military was unprepared for a major war. Benito Mussolini's stunning statement that "I will declare war but not wage it" gives us some idea of the lack of direction his military chiefs received before and during Italy's involvement. As numerous scholars have observed, Italy lacked the strategic vision, military capabilities, and will to fight effectively

18 Michael Simpson, "Superhighway to the World Wide Web: The Mediterranean in British Imperial Strategy, 1900–1945," in *Naval Policy and Strategy in the Mediterranean: Past, Present, and Future,* ed. John B. Hattendorf (London: Cass, 2000), 51; John Strawson, "The Shape and Course of the Mediterranean War, 1940–43," in Royal Air Force Historical Society, *The End of the Beginning: Bracknell Paper No. 3, A Symposium on the Land/Air Co-operation in the War, 1940–43,* ed. Derek Wood (Bracknell, UK: Royal Air Force Historical Society), 20 March 1992, 14. Churchill's focus on the Mediterranean was in large part due to his determination not to lose sea lines of communication to, or control over, India.

in a long, attrition-based, great-power war. Italy's effort became a "strategy of expediency" dependent upon continuing German primacy in the conflict.[19]

German grand and military strategy were clear in the abstract and over the long term, but erratic and inconsistent during the struggle in the Mediterranean Theater. The Wehrmacht's "campaigns of opportunity" were designed to be short and sharp, allowing the military to refit before the next campaign in the larger quest for *Weltmacht*. This system worked well until the failure in Russia and the increasingly major involvement in the Mediterranean began consuming German human and material resources at the battle fronts faster than the Germans could make good the losses. More problematic for Germany was Hitler's view that the Mediterranean was an important but "secondary" theater that he could scoop up handily once he had disposed of Russia. Until then, it would be "Italy's theater." Hitler's priorities were to conquer the Soviet Union and in the process force the British to come to terms. Until those things happened, the Mediterranean would be a "defensive forefront and rampart" rather than a springboard for overrunning British possessions in the theater. However, had the Germans struck rapidly and with proper timing, concentration, and persistence in the air and on the ground, they could very well have defeated the British in North Africa with the relatively modest assets they had in place.

Either way, Hitler's long-term strategic approach did in fact focus on seizing Middle East oil sources, which produced as much crude oil as all of Germany's sources, including Ploesti, combined, but he failed to grasp the operational realities such a long-term plan entailed. He also planned for a division of Africa with Italy and Vichy France and a link-up with the Japanese in India once the British position collapsed. While a number of scholars have dismissed German efforts in the Mediterranean as both misguided and offering no strategic advantages, they might turn the question around and consider how destroying the British position there—which they came within a whisker of doing even with an ineffective air and sea effort—would have redounded to their advantage in ways both immediate and long-term. There were many potential benefits as noted earlier regarding "positive" and "negative" strategic advantages,

19 There are many works regarding Italian ambitions, policy issues, and military shortcomings. See Reynolds M. Salerno, *Vital Crossroads: Mediterranean Origins of the Second World War, 1935–1940* (Ithaca, NY: Cornell University Press, 2002), passim. Another crucial source is GOH III, 1–126. For Mussolini's longer-range ambitions, see Gerhard L. Weinberg, *Visions of Victory: The Hopes of Eight World War II Leaders* (Cambridge, UK: Cambridge University Press, 2005), 41–55.

but a failure of imagination and the improper use of the air weapon doomed the effort to failure.[20]

## 4 Operational Doctrine and Airpower Technologies

The operational doctrines each party used also played major roles in determining the kinds of airpower technologies they developed, and how they used them in the Mediterranean Theater. Germany's doctrine of *Bewegungskrieg* (mobile warfare) and the successive "campaigns of opportunity" it was designed to facilitate, drove the character and operations of the Wehrmacht's air arm. Every German combat aircraft was designed for take-offs and landings on grass or dirt strips. Early in the war, dedicated Ju 52 transports flew spare parts and highly skilled mechanics forward with the fighter, bomber, and reconnaissance aircraft that played a pivotal role in the rapidity of the Reich's early victories. Pilot training, though utterly inadequate for a major and prolonged war of attrition, as later events would underscore, was adequate for these short campaigns. The working assumption was that losses in men and machines could be made good during the pause between campaigns.[21]

20 Norman J.W. Goda, *Tomorrow the World: Hitler, Northwest Africa, and the Path toward North America* (College Station, TX: Texas A&M University Press, 1998), *passim.* Jak P. Mallmann Showell, Foreword, *Fuehrer Conferences on Naval Affairs 1939–1945* (London: Chatham Publishing, 2005), *passim;* abbreviated as "FCNA" in future citations. Raeder's conferences with Hitler from 26 September 1940 to March 18, 1941 (pages 141–186) are instructive regarding his arguments for a moderate but serious and sustained level of involvement in the Mediterranean to destroy the British position there, turn it into an Axis lake, and make it a springboard for colonial conquests stretching from Egypt to India and south into East Africa to conquer the Red Sea and dominate the Indian Ocean. For colonial aspirations, see GOH III, 279–301. Klaus Schmider, "The Mediterranean in 1940–1941: Crossroads of Lost Opportunities?" in *War and Society* 15, no. 2 (October 1997), 19. Walter Ansel, *Hitler and the Middle Sea* (Durham, NC: Duke University Press, 1972), 3. Norman Rich, *Hitler's War Aims: Ideology, the Nazi State, and the Course of Expansion* (New York: W.W. Norton and Company, 1992), 177.

21 Air Ministry A.D.I.(K), "Interrogation of German Officers from the Reichsluftfahrt Ministerium Conducted by Air Ministry A.D.I. (K) on Friday, 15th June 1945, 553.619, Air Force Historical Research Agency (referred to hereafter as AFHRA), 7. Air Ministry, *Supply Organisation of the German Air Force, 1935–1945* (East Sussex, UK: The Naval & Military Press Ltd., 2009), 48–51, 57, 59, 63–64, 68–71, 78–81, 199, 231–232, 237, reprint of 1946 original (original available at Air Historical Branch); Tedder to Portal, 23 January 1942, AIR 23/1396, The National Archives (referred to hereafter as TNA). Colonel-General Ernst Udet told Ehrhard Milch in 1938 that "The movement of squadrons must not be hampered by administrative work. Officers will not be dependent on engineers—such a situation would prejudice the whole morale of the Luftwaffe. All campaigns will be short and German aircraft production will be so tremendous that during such periods of operation no major

Italy's operational doctrine was neither formally written nor coherent. Mussolini and his generals counted on a very short war, taking advantage of German victories to claim as many territories around the Mediterranean Sea as they could. Italian military forces were woefully inadequate in almost every way to engage the British or any other great power, but *Il Duce* believed that he could seize Greece, Egypt, British Somaliland, Kenya, and perhaps part or all of the Vichy possessions in North Africa before the British or Vichy French could respond.

Unfortunately for Italy and its fighting men, the lack of a sound operational doctrine facilitated a series of military disasters that disabused Hitler and Churchill, among others, of any idea that a "New Rome" was rising in the Mediterranean. Interoperability between the military services was nearly nonexistent, training was substandard, logistics were insufficient, military hardware was for the most part obsolescent, and the lack of imagination senior military leaders showed in the employment of forces was as calamitous as it was disheartening to the rank and file. These problems were on full display in the Regia Aeronautica, which proved unable to contend with the RAF from the outset of hostilities between the two countries on 10 June 1940.[22]

British operational doctrine, on the other hand, was in a period of profound change throughout the war in the Mediterranean. This re-invention was true for the army and air force as a result of the disaster in France, and later for the navy as the lethality of land-based air attacks during the Greece and Crete campaigns made shockingly clear the impossibility of naval operations in an area where the British did not have air superiority. The RAF had already faced the Luftwaffe during the Battle of Britain and, unlike its foe in North Africa, learned to seek air superiority as the *first and most important step in every campaign*. Cooperation between the RAF and the army, nearly nonexistent in France, improved steadily during the Desert War, as did air-naval cooperation in the crucial battle for seaborne logistics, which was the single most important arbiter of success and failure in the Mediterranean Theater. The

---

repairs will be necessary. Damaged planes will be repaired and salvaged at home after the campaigns are won." See Air Ministry, *The Supply Organisation of the GAF*, 71.

22 GOH III, 62–98; Brian R. Sullivan, "Downfall of the Regia Aeronautica, 1933–1943," in Robin Higham and Stephen J. Harris, eds., *Why Air Forces Fail: The Anatomy of Defeat* (University Press of Kentucky, 2006), Chapter 5, *passim;* MacGregor Knox, *Hitler's Italian Allies,* 65–67, 163–167; Lucio Ceva and Andra Curami, "Air Army and Aircraft Industry in Italy, 1936–1943," in Horst Boog, ed., *The Conduct of the Air War in the Second World War* (New York, Berg, 1992), 106, 111; Hans Werner Neulen, *In the Skies of Europe: Air Forces Allied to the Luftwaffe 1939–1945* (Wiltshire, UK: Crowood Press, 2000), 23–24, 26, 37; Karl Gundelach, *Die deutsche Luftwaffe im Mittelmeer 1940–1945,* Vol. I, (Frankfurt am Mann: Peter D. Lang, 1981), 30–32. Referred to hereafter as "DLM."

widening gap in British and German operational doctrine in the combined-arms and air arenas gave the British Army time to evolve, survive, and evolve again—a cycle that ultimately resulted in a major turn of the tide at El Alamein. It would, however, be an error to say that the British merely copied and improved upon *Bewegungskrieg*. From the earliest days of their reforms, they developed a distinctive combined-arms warfare ultimately characterized by a level of air-ground cooperation greater and more effective than the German variant, and an air-sea effort far beyond what the Axis powers ever managed.[23]

## 5 Discerning the Elements of an Airpower Technology

Grand strategy, military strategy, and operational doctrine drove the ways in which the belligerents developed and used their air arms. However, there were many other factors at play, which we must examine in order to understand just how expansive, complex, and different the belligerents' airpower technologies and employment were. The feedback loops between these elements of airpower technologies, and from there back up to the development and conduct of strategy and operations, played vital roles in the conduct and course of the war in the Mediterranean Theater. In this sense, the British developed a much more integrated combined-arms system designed specifically to support strategic aims.

The first and most important element comprising the airpower technologies was what Guilmartin referred to as an *intellectual infrastructure*—the collectivity of cognitive, organizational, and procedural factors brought to bear in order to build, operate, and sustain an air arm. How effectively each side employed the key artifacts within the larger technology—in this case aircraft—depended to a great degree on how effectively they planned, organized, equipped, and employed them within the contextual and strategic realities they faced. The British proved adept here, as they built the equivalent of a metropolitan air force thousands of miles from the British Isles with a similarly high level of organizational, logistical, and infrastructural sophistication.[24]

23 GOH III, 31–32; BOH I, 35–37; Roderic Owen, *The Desert Air Force; an Authoritative History Published in aid of the Royal Air Force Benevolent Fund* (London: Hutchinson, 1948), 90. Referred to hereafter as "Owen, DAF"; Air Vice-Marshal Peter Drummond, "Appendix 'C' to R.A.F. Middle East Operation Plan: Directive to Air Officer Commanding, Egypt Group," 11 May 1939, AIR 23/781, TNA, 1–4.

24 Humphrey Wynn noted that "uniquely among the RAF commands, the Middle East had created a complete Air Force…" See Humphrey Wynn, "The RAF in the Mediterranean Theatre," in Wood, ed., *The End of the Beginning,* Chapter 5, 21.

Cognitively and organizationally, airmen and soldiers came together to work out the severe problems with coordination and control of assets that they had experienced in France. Similarly, airmen and sailors corrected many of the errors made during the German conquest of Norway, Greece, and Crete, during which the Royal Navy suffered severe losses. Combined-forces headquarters, forward headquarters, and innovations such as Air Support Control units all developed steadily and proved much more capable over time at executing combined-arms warfare than the Axis powers, who lacked anything like the British model.[25]

British commanders also understood that air forces are of little use without extensive infrastructural, logistical, mobility, communications, command-and-control, and other assets comprising distinct parts of the larger airpower technology. The RAF's superb tactical and operational mobility, which developed rapidly in North Africa, was front and center here. An increasingly large, capable, and specialized set of vehicles allowed aircraft, aircrews, ground crews, and supplies to "leapfrog" or "bound" rapidly forward and back as the fortunes of war on the ground dictated. Unlike the Luftwaffe, which invariably fell behind Rommel's advances, RAFME stayed up with the troops and fell back just ahead of them, delivering direct and indirect support at every step of every battle. A major building program led by the Army's Royal Engineers produced hundreds of new airfields and landing strips all over Egypt and Libya, facilitating operations within hours of the first aircraft's arrival. Convoys laden with fuel and spare parts, and divided into forward and follow-on echelons for every air unit from the squadron level up, delivered supplies, ground crews, anti-aircraft units, and many other assets, in many cases within an hour of the arrival of the first aircraft at an airfield, and on occasion even before the planes arrived.[26]

25 This topic receives extensive attention in Ehlers, *Mediterranean Air War, passim.* See also documents such as the following for insights into various aspects of combined-arms efforts: "The Middle East Campaigns," Vol. 1, AIR 41/44, TNA, 6, 25, 27–28; "Middle East: Strategy for Operations (August 1940–February 1941)," AIR 8/514, TNA; "Air Support," 47, AIR 10/5547, TNA; "Army Air Support," War Office (referred to hereafter as WO) 277/34, 40; Brad William Gladman, *Intelligence and Anglo-American Air Support in World War Two: The Western Desert and Tunisia, 1940–43* (New York: Palgrave MacMillan, 2009), 42; "Air Support," AIR 10/5547, TNA, 47–48; Army Air Support," WO 277/34, TNA, 39–40; "The Middle East Campaigns," Vol. 1, AIR 41/44, TNA, 7; Tedder, *Report*, 55–59; HQ RAF ME, "Paper on a Suggested Inter-Allied Inter-Service Communications Scheme," 30 January 1942, AIR 23/1293, TNA, 1–3; "Report on Visit to the United Kingdom by Chief Signals Officer, R.A.F. M.E., Chief Signals Officer, Air Formation Signals, and Chief Radio Officer, R.A.F. M.E.," 27 October 1941, AIR 23/1293, TNA, 1–5.

26 Ibid. The bureaucratic infighting and command relationships within the German armed forces, and between the Germans and Italians, exacerbated these problems. So did Germany's historical context as a land power in which the army had primacy.

Logistically, the British developed a superb capacity for supporting air and combined-arms operations while the Axis powers failed to do so. Expanded Mobility, high readiness rates, excellent salvage capabilities, and superb maintenance hubs around Cairo all underpinned combined operations. Transport aircraft eventually joined vehicles to facilitate the movement of RAFME combat assets, and especially single-engine fighters and fighter-bombers, very quickly across long distances. This "leapfrogging" or "bounding" gave the British major advantages in the air, on land, and at sea. Entire squadrons, wings, and, eventually, groups were able to move forward or back in a matter of hours, delivering direct support to advancing or retreating British Army forces even as they moved to new airfields. The Axis powers never developed anything like this capability, and their aircraft, including the superb Bf 109 fighter, constantly lagged far behind Rommel's troops as they advanced or, conversely, were overrun on their airfields as Rommel's forces retreated due to chronic Luftwaffe shortages of vehicles, fuel, and spare parts.[27]

Finally, British infrastructural achievements proved vital. Often overlooked given its decidedly unromantic nature, the physical infrastructure undergirding air operations proved extraordinarily important in all theaters of war, but nowhere more so than in the Mediterranean. The proliferation of airfields in the Western Desert gave the British major advantages across the theater. Italian efforts were also substantial in Libya as well as Italian East Africa, but the greatest distinguishing feature between British and Axis efforts was the great emphasis the former placed on developing the full logistical infrastructure required to sustain air operations in time of war. While Air Vice-Marshal Graham Dawson developed Egypt's repair, salvage, and supply capabilities, the Axis powers failed to do so in Libya. Because the Axis had no major aircraft depots in Africa, they had to fly planes back to Europe or send them by ship—an

27 BOH I, 64–65, 277–278, 281–282, 353–358; BOH III, 11–13; BOH IV, 216–217; Owen, *DAF*, 112; Robin Higham, *Bases of Air Strategy: Building Airfields for the RAF, 1914–1945* (London: Airlife, 1998), 124–125, 162–164; Longmore, "Despatch No. 1 on Middle East Air Operations, 13-5-40 to 31-12-40," 1 February 1941, AIR 23/808, TNA, 16; Tedder to Ludlow-Hewitt, 6 September 1941, Tedder Box 4, Air Historical Branch (referred to hereafter as AHB); Raymond Collishaw, "Brief Report on Royal Air Force Operations in the Western Desert From the Outbreak of War with Italy—The Capture of Cyrenaica to the Time of the Enemy Counter Offensive," 19 April 1941, AIR 23/6475, TNA, 14–16; HQ ME, "Report on Army Co-operation Carried Out in Connection or with the Land Operations in the Western Desert and Libya," 31 January 1941, WO 201/348, TNA, 1–3; HQ RAF ME, "Some Notes on Administrative and Organisational Difficulties Affecting the Royal Air Force in the Western Desert and Cyrenaica during the Period of Mobile Warfare—November 1941–February 1942," 30 March 1942, AIR 23/1048, TNA, 1–10; HQ RAF ME to All Subordinate HQs, "Mobility of Squadrons and Units," 6 April 1942, AIR 23/1048, TNA.

increasingly dangerous venture—for periodic depot maintenance and major repairs. Their spare-parts situation was also marginal. Conversely, Dawson and his subordinates conducted everything from major repairs to periodic depot maintenance, salvaged nearly every crashed RAF aircraft and repaired it or used it for spare parts, and developed sophisticated fuel and spare-parts delivery networks across the theater. Logistics, supply, and repair capabilities were as important as aircraft production and airfields because without enough of them, air forces were unable to engage in sustained or mobile operations across the huge distances involved in the Desert War.[28]

The second element comprising each side's airpower technology was the relative sophistication of its *tactical and operational employment*—namely, how each air arm employed its aircraft at the level of missions, battles, and campaigns. Airmen and soldiers took insights gained from France and the Battle of Britain to the Western Desert. Paramount among these was the overriding importance of holding air superiority, in general if possible, and at critical times and places at the very least (See Illustrations 11.1–11.3). Air Chief Marshal Sir Arthur Longmore and his better-known successor, Air Chief Marshal Sir Arthur Tedder, maintained an unstinting focus on air superiority—an effort that paid major dividends. Given the high quality of Luftwaffe fighters and pilots facing the generally less capable RAF machines and men, holding to this policy proved costly. *Jagdgeschwader* 27 pilots alone shot down nearly 1,300 aircraft, most of them fighters, from April 1941 to November 1942. The combined German-Italian kill total during this period was over 1,400 machines. British failures to train adequate numbers of pilots in the theater during the first year of war, or to introduce the Spitfire until March 1942, left the RAF at a serious disadvantage in terms of individual men and machines, but their more creative employment of both allowed them to maintain nearly constant air superiority.[29]

28 For the best single work on RAFME airbase planning and construction, see Higham, *Bases of Air Strategy.* For Dawson's key role in the logistical and maintenance efforts, see Mrs. Anthea M. Lewis, "A Tribute to Air-Vice Marshal GG Dawson, 1895–1944, CMSO ME & N Africa 1941–1944," AHB, *passim.* For general insights, see BOH II, 292–293, and Denis Richards and Hilary St. G. Saunders, *RAF 1939–1945, Vol. 2, The Fight Avails* (London: Her Majesty's Stationery Office, 1954), 162–167 and 227–228. Specific details regarding Dawson's maintenance organization are located in RAFME, "Maintenance Diagrams," June 1943, AIR 23/3417, TNA, Charts 1, 3, 14. A detailed discussion of British logistical and maintenance efforts and advantages over the Axis powers is available in Ehlers, *Mediterranean Air War, passim.*

29 Christopher Shores and Hans Ring, *Fighters over the Desert: The Air Battles in the Western Desert, June 1940–December 1942* (New York: Arco, 1969), 217; Werner Held and Ernst Obermaier, *The Luftwaffe in the North African Campaign, 1941–1943* (West Chester, PA: Schiffer,

ILLUSTRATION 11.1 Bf 109E over the southern Mediterranean, 1941. This fighter and its later F variant proved the most capable in theater until Spitfire Mark V's arrived, far later than necessary, in early 1942. Had OKL and its senior officers given Bf 109 pilots clear orders—or any at all—to make air superiority their first priority, the outcome of the campaign would likely have been very different. As it turned out, Bf 109s were tied mostly to bomber escort in support of advancing troops, robbing the Germans of this superb air asset's greatest strengths. (AWM)

Tedder and his operational air commander, Air Vice Marshal Sir Arthur ("Mary") Coningham, moderated the effects of these disadvantages by bringing experienced fighter pilots from the British Isles to teach RAF fighter pilots the latest tactics and counter-tactics employed over the Channel, by deploying mobile radar and radio units to give the RAF a warning network far superior to that of the Axis, and to allow subordinate airmen to improvise new defensive tactics such as loose escort and the defensive "flying circle" as they sought to survive German fighter attacks. British bomber raids on Axis airfields caused serious losses to aircraft, crews, and maintenance personnel. This unceasing assault on Axis airfields, which had no radar coverage until spring 1942 and

1992), 44; Tedder, *WP*, 75–76; Tedder to Freeman, 16 December 1941, AIR 23/1386, TNA; Tedder to Portal, 12 December 1941, AIR 23/1396, TNA; Tedder to Portal, 23 January 1942; Tedder to Portal (r) Pierse, 4 February 1942; both in AIR 23/1396, TNA; Portal to Tedder, 6 February 1942; Portal to Tedder, 26 February 1942; both in AIR 23/1396, TNA; Drummond to Coningham, 4 March 1942; Tedder to Drummond, 4 March 1942, AIR 23/1396, TNA; Coningham to Tedder, 14 January 1942, AIR 23/1391, TNA.

ILLUSTRATION 11.2 Hawker Hurricane I squadron in the Western Desert. These aircraft, along with the Hurricane II and the venerable Gladiator biplane, clinched air superiority during Operation Compass in December 1940, allowing bombers to concentrate heavily on airfields, vehicle convoys, and other high-value targets. However, the Hurricane was incapable of dealing with the Bf 109 and soon had to move into fighter-bomber and armed reconnaissance missions. (IWM)

relied entirely on spotters, proved lucrative and helped to ensure the RAF never lost air superiority where and when the combined-arms force needed it.[30]

30 R. Collishaw, "Brief Report on Royal Air Force Operations in the Western Desert From the Outbreak of War with Italy—The Capture of Cyrenaica to the Time of the Enemy Counter Offensive," 19 April 1941, AIR 23/6475, TNA, 1–3; "Report on Visit to the United Kingdom by Chief Signals Officer, R.A.F. M.E., Chief Signals Officer, Air Formation Signals, and Chief Radio Officer, R.A.F. M.E.," 27 October 1941, AIR 23/1293, TNA, 1–11; Tedder to Portal, 25 October 1941, Portal to Tedder, 25 October 1941, AIR 23/1395, TNA, Tedder to Portal, 30 October 1941, Tedder Journal, 21 November 1941, all in Tedder, Box 3, AHB; Tedder Journal, 21 November 1941, Tedder, Box 3, AHB; AHQ Egypt, Standing Operating Instruction No. 20 of 1942 (Replaced No.7 of 1942); "G.C.I. Control of Interceptions," 10 December 1942, AIR 23/808, TNA, 1; Basil Embry, *Mission Completed* (London: Quality Book Club, 1957), 209–216. Each GCI radar set provided coverage equal to 158 searchlights crewed by 3,667 men—a huge force multiplier. The fact that the Axis had no operational radar in North Africa until the spring of 1942 is thus even more telling in terms of their inability to defend airfields effectively against constant RAF raids.

ILLUSTRATION 11.3 Royal Australian Air Force (RAAF) P-40 Kittyhawks over North Africa, 1942. This aircraft, despite its major improvements over the P-40 Warhawk, was still inferior to the Bf109 in a turning fight. The consistent inferiority of their fighters and pilots relative to those of the Germans forced the British to pursue air superiority by raiding Axis airfields constantly and employing a range of defensive tactics in encounters with the Bf109. (Pineterest) http://photobucket.com/gallery/user/p40hawksnest/media/cGFoaDpSQUFGLzc2U3FuLotpdHR5aGF3aolJSXM3NlNxbjAwMWExLmpwZw==/?ref=

Bomber crews also did exceptionally well in the long and unglamorous, but critically important, battle against Axis sources of supply. They had already become quite good at navigating along coastlines as they hunted and destroyed nearly a third of the barges the Germans had gathered for the anticipated invasion of the British Isles. Bomber crews brought this expertise to bear in raids on Tripoli, Benghazi, and other port cities and towns in Libya. When combined with an increasingly effective RAF anti-shipping campaign in which aircraft surpassed submarines and surface ships as the foremost ship-killers, the logistical balance shifted steadily toward the British with a few notable exceptions. Predictably, those exceptions coincided with the arrival and employment of substantial Luftwaffe contingents in theater, when steady attacks on Malta and the SLOCs turned the logistical balance in the Axis powers' favor.[31]

Even as the long-term RAFME air-superiority effort bore fruit, Tedder and Coningham had to make do with single-engine fighters less capable than the Bf 109. Senior political leaders' insistence on keeping all Spitfires on the home islands even after the threat of invasion had long since passed drove Tedder to distraction as he sought to ensure air superiority. Tedder and his airmen had to make effective use of available aircraft. Increasing deliveries of British, and later American-provided Lend-Lease aircraft via the Takoradi Route across the Sahara, around the Cape of Good Hope to the Middle East, and to Malta during the desperate and costly but ultimately successful effort to defend it, gave the RAF enough fighters to hold the superior Bf 109 at bay. The Hurricane II was the best fighter the British had in theater early in the air war. With a capable pilot, it could hold its own against the Bf 109E but was badly outclassed by the BF 109F2, which began arriving in September 1941. Eventually, the American-built P-40 Warhawk and Tomahawk arrived, but these too were inferior to German fighters. Even the P-40 Kittyhawk, a modified Tomahawk with a Rolls Royce Merlin engine and heavy armament, could not hold its own with the Bf 109 in a turning fight. Tedder also made due with less effective pilots as the best ones remained in the UK with their Spitfires.[32]

Nonetheless, RAF senior leaders had the major advantages of a sound approach to the employment of air assets within the contextual realities of the theater. As a result of the debacle in France and the successful effort during the Battle of Britain, senior military leaders came to recognize the indispensability of air superiority as an enabler for other operations. This understanding

31 Air Ministry, RAF Summaries Nos. 328–834, 12 June 1940–31 October 1941, AIR 22/10–27, TNA; BOH II, 280–282; HQ RAF ME Tactics Assessment Office, "Air Tactics and Operational Notes on 201 Naval Co-operation Group, R.A.F.," N.D. (probably June 1943), Foreword and Chapter 1, AIR 23/1282, TNA, 1; *DLM,* 298; HQ RAF ME Tactics Assessment Office, METM No. 1, "Practice Low Level Attacks by Marylands on H.M.S. Carlisle (A.A.) Suez Eighth September 1941," October 1941, AIR 23/1281, TNA; USAAF/A2, "Signal Intelligence Service of the German Luftwaffe, Vol. X, Technical Operations in the South," 1945, 519.6314-10 V.9, AFHRA, 71–72. For the bombing campaign against German barges in 1940, see Robert S. Ehlers, *Targeting the Third Reich: Air Intelligence and the Allied Bombing Campaigns* (Lawrence: University Press of Kansas, 2010), 94–96.

32 For detailed discussions of these topics, see Ehlers, *Mediterranean Air War, passim.*

became clear in the Mediterranean Basin with its narrow sea lines of communication, shallow waters, and many islands suitable for air operations. The same basic principle applied in North Africa, where limited port facilities, roads, and railroads gave aircraft a particularly important role in the combined-arms effort. Early operations, including Army successes in Italian East Africa and Libya, and Royal Navy disasters around Greece and Crete, confirmed the pivotal importance of air superiority.[33]

Luftwaffe tactics were, on the whole, substantially less effective than those of the RAF. Most importantly, German airmen misused the relatively small number of Bf 109s they had in theater. The inimical practices of *Freie Jagd* ("free hunting"), close escort of bombers, and excessive control by the army all played their role in lessening the effectiveness of these superb German fighters and their veteran pilots. The failure to focus consistently, or even discernably, on winning air superiority theater-wide or even locally at the right times was a disastrous error. Airfield attacks were rare in comparison with the nearly non-stop RAF raids. Both before and during major ground offensives, the Luftwaffe generally failed to control the skies in the battle area. When they did, major tactical and operational successes followed, as evidenced in every one of Rommel's offensives prior to the battles around El Alamein. However, Luftwaffe air superiority was inevitably of very short duration, very local in nature, and confined to the immediate battle area due to the chronic and serious shortages of vehicles required to make Luftwaffe units highly mobile. As Rommel quickly outran his air cover, his advances slowed and then stopped in the face of heavy RAF air attacks. Further, German tactics changed very little from the *Schwarm*, finger-four formations and the unfortunate free hunting. The latter relegated wingmen to de facto roles as spotters for and protectors of their flight leads, who sought to increase their kills rather than considering carefully, along with their commanders, how they might better have gained air superiority within a larger combined-arms context.[34]

Because air superiority proved the enabler for all subsequent operations, single-engine fighters were important assets. The number of fighters on each side was important, and the Germans never sent enough. Even the ones they sent did not all stay long enough in theater, with most withdrawing before the Axis could turn tactical and operational victories into strategic ones. This issue

33 Ehlers, *Mediterranean Air War, passim.*

34 For an excellent account of the differences between Luftwaffe and RAF employment of fighters, Shores and Ring, *Fighters over the Desert*, 217–221. See also *DLM*, 119, 123; BOH II, 14–15; Arthur Tedder, *With Prejudice: The War Memoirs of Marshal of the Royal Air Force Lord Tedder G.C.B.* (Boston : Brown and Company, 1966), 184–190; Ehlers, *Mediterranean Air War, passim.*

of timing—arrival in theater, duration and persistence of contributions, and consistency of use in the air superiority role during the critical period of operations from March 1941 to August 1942—is one of the most important but generally overlooked reasons for Axis failure in the air, and by extension at sea and on the ground.[35]

The Germans relied on the Bf 109, which was ideally suited to the air-superiority mission. Inexplicably, the first Bf 109 unit, *Jagdgeschwader* 27, did not arrive until April 1941—two months after Rommel's first offensive. Even then, only a single *Gruppe* arrived, with the rest of the *Geschwader* deploying to Russia. A mere 30 single-engine fighters thus took on the air-superiority role. However, even this tiny number of aircraft was soon pulled in multiple directions, providing army support and close escort to sea as well as land supply convoys. Another 30 aircraft from a second *Gruppe* arrived in September and 30 more from a third *Gruppe* in December. The arrival of *I Gruppe*, JG 53, in January 1942 brought total Bf 109 strength to around 120 aircraft. The *II Jabogruppe* of JG 27 arrived in May 1942 for Rommel's Gazala Offensive and remained in theater until September, increasing the total to around 150—a high point from which numbers rapidly fell due to battlefield losses, a lack of replacements, and fuel and spare-parts shortages that kept fighters grounded much of the time. The *II Jabogruppe* JG 27 specialized in ground-attack missions but was fully capable of air-superiority missions. The staggered arrival of Bf 109 fighters and their veteran pilots, often too late to support Rommel's ground campaigns, represented an ineffective use of the best fighter in theater, particularly given the absence of Spitfires until early 1942. This deployment pattern undermined

35 See John F. Guilmartin, Jr., "The Aircraft that Decided World War II: Aeronautical Engineering and Grand Strategy, 1933–1945, The American Dimension," The Harmon Memorial Lecture in Military History, Number 44, United States Air Force Academy, 2001, 1, 6–8. Guilmartin discussed factors contributing to air superiority and the effective combined-arms operations that relied on it. He identified the Bf 109 as the single most important aircraft of the war not only because of vital role in gaining air superiority over France in 1940 and Russia in 1941, but because it was nearly the ideal airframe for gaining and maintaining air superiority at the operational and tactical levels. However, Guilmartin cautioned that the right aircraft types had to be in action at the right places and times, and for sustained periods, to provide the air superiority required to win combined-arms campaigns. Guilmartin noted, "...air power was critical to the conduct and outcome of the Second World War...because control of the air invariably provided an essential ingredient of victory in the battles and campaigns comprising World War II and...determined its flow and eventual outcome." This observation describes the differences in RAF and Luftwaffe approaches to air superiority and the general employment of airpower in North Africa. Regarding other factors at play in the German failure to gain air superiority, see Ehlers, *Mediterranean Air War, passim.*

whatever passed for an effort to gain battlefield air superiority—the very task for which the Bf 109 was designed.[36]

"Free Hunting" combined with a myopic focus on army support missions for bomber aircraft spared British airfields the same pounding that Axis airfields endured. From 12 June 1940 to 31 October 1941, RAF aircraft raided Axis airfields 1,078 times, compared to 259 Axis raids. Most of the latter were made by less than five aircraft, while British raids involved 20 or more. German fighter pilots shot down nearly 1,000 British fighters and killed many of their pilots, but never gained anything other than very local and temporary air superiority.

The number of planes destroyed in the air proved to be the wrong measure of effectiveness. RAFME leaders and their airmen understood that the more effective way to undermine an adversary's air and combined-arms effectiveness was to attack airfields without pause, causing maximum damage to aircraft and maintenance facilities while killing as many aircrew and ground crew as possible. Focused and effective attacks on aviation fuel deliveries, to include sinking Italian tankers, destroying Ju 52 transports, strafing vehicle convoys delivering it to airfields, and bombing fuel dumps at the operating locations, was an integral part of this effort.[37]

At the moment of contact in the tactical arena, nothing was more important than the quality of aircrews. German fighter pilots dominated here for most of the contest in North Africa. The quality of fighter pilots depended on an extensive training infrastructure designed to produce adequate numbers of well-trained airmen for the high-attrition air warfare that proved the norm in every theater of operations. The side that planned for high losses over sustained periods of time was bound to develop an advantage. While the British lost far more pilots than necessary given their policy of holding back the best ones in Great Britain, they nonetheless developed good Operational Training Units in the U.K. and in theater. The Germans failed to develop pilot-training programs for a long war of attrition, so as German pilot quality declined, British and American pilot quality improved, and, by summer 1942, allied pilots often outclassed their opponents.[38]

---

36 See *Jagdgeschwader 27* organization and deployment timelines at http://www.ww2.dk/air/jagd/jg27.htm. While this information is available in numerous secondary sources, it is set forth particularly clearly in this online document. The same is true for JG 53 at http://www.ww2.dk/air/jagd/jg53.htm. The episodic nature of the latter's involvement in the North Africa campaign becomes clear when reading this document. See also Guilmartin, "The Aircraft that Decided World War II," 6–8, and Ehlers, *Mediterranean Air War, passim.*

37 Air Ministry, RAF Summaries Nos. 328–834, 12 June 1940–31 October 1941, AIR 22/10–27, TNA.

38 Portal to Tedder, 19 June 1941, AIR 23/1395, TNA; Dawson to Tedder, "Summary of Wing Commander Messiter's Signals on American Projects," 30 October 1941; Lyttelton to HQ

The third key element of the airpower technologies at play in the Mediterranean air war was the *timing involved in massing and employing aircraft,* both to gain air superiority and to support ground operations. This element was one of the most important in Mediterranean air operations, and in the Axis defeat in the theater. It was particularly central in the case of the German Bf 109 single-engine fighter, which arrived too late, too episodically, and in insufficient numbers to make the critical difference it could have made in the combined-arms operations stretching from Malta to Mesopotamia. The ability to move entire *Luftflotten* (air fleets) to and from the Middle East and the Eastern Front had a paradoxical and largely negative impact on Axis air operations in both theaters, but especially in North Africa and over Malta, where Luftwaffe units arrived later than needed and in disjointed force packages, rather than as cohesive units. The move of single-engine fighters, torpedo bombers, and other critical aircraft types was neither prioritized nor synchronized to coincide with the start of major air operations over Malta or in support of Panzerarmee Afrika in Libya and Egypt. The process also worked in reverse with the bulk of single-engine fighters and other aircraft moving back to the Eastern Front just as the Axis had won major ground campaigns in North Africa and had gained a significant measure of air superiority over Malta. These errors in timing cost the Axis dearly in both the short and long-term, as they failed to turn hard-earned operational and tactical gains into enduring grand and military-strategic ones.[39]

The Axis habit of temporizing, defined here as being indecisive or evasive either to gain time or delay acting, something both the Italians and the Germans did repeatedly in the deployment and employment of airpower, was part of their bad timing. It was a product of unclear strategic thinking, priorities, and guidance. It was also a result of ineffective air and ground leadership in North Africa, with "ineffective" being defined here as a combined inability and unwillingness to learn how to employ air, ground, and sea assets effectively within a combined-arms context. The Luftwaffe and Regia Aeronautica did not coordinate their efforts to any significant degree. Nor did the air services work effectively with their ground and naval counterparts. The inconsistent use of air assets also resulted from the temporizing that occurred constantly from the highest levels of political and military leadership to the operational and tactical

ME, 10 December 1941; both in AIR 23/1315, TNA; Norman Franks, ed., *The War Diaries of Neville Duke 1941–1944* (London: Grub Street, 1995), 76, 134; Ehlers *Mediterranean Air War,* 109–110, 153.

39 The overarching factors related to questions of timing, tempo, persistence, and similar factors receive detailed attention in Ehlers, *Mediterranean Air War, passim.*

levels. Had the Axis air arms been employed properly, at the right times and places, gaining and holding air superiority in the Western Desert and Malta, and had the Axis high commands exploited this advantage with coordinated offensives in both locations, victory in the theater would very likely have gone to them. Instead, they failed, while the British devised new and constantly improving combined-arms efforts based on sound grand and military strategies.[40]

While the Luftwaffe did deliver effective support to Rommel's forces on several occasions, vehicle shortages prevented it from protecting Axis troops as they advanced. The one key advantage the Germans enjoyed in the air war—the Bf 109—fell prey, along with its pilots and ground crews, not only to badly timed arrivals in theater but also to a perennial vehicle shortage caused by inefficient Italian shipping processes and Rommel's penchant for commandeering Luftwaffe vehicles. Luftwaffe vehicle allocations were already at half strength as a result of large-scale transfers to units supporting Operation Barbarossa. These issues meant that after Rommel grabbed half of the Luftwaffe's vehicles, which were already at half strength, air units in North Africa had about one-quarter of the German vehicles they were supposed to have. Modest additions of Italian trucks did little to help the problem. This lack of ground transport made it impossible for air units to keep up with Rommel's advances and to escape intact when Rommel was forced to retreat. Consequently, Luftwaffe personnel could not concentrate air-superiority aircraft at critical times and places after ground actions, and Rommel's columns inevitably had to stop advancing as RAF aircraft attacked them with impunity on the few roads and desert tracks where they had to concentrate.[41]

Combined with a cadre of superb pilots, highly effective ground crews, and good battlefield maintenance practices at the unit level, even the modest number of Bf 109s could well have made the difference between victory and defeat had they been able to protect Rommel's forces once they broke through British positions. By extension, if they had been able—or allowed—to engage in operations at the right times and places, and been in place for long enough to allow ground forces to exploit their tactical and operational successes, the Bf 109s would have conferred major advantages on the larger Axis effort at critical moments. The operational, military-strategic, and potentially grand-strategic advantages flowing from such an employment of single-engine fighters, in coordination with other aircraft types, and with ground and sea assets, might have been sufficient to confer victory in the theater. However, the larger aspects of what we have determined comprises an airpower technology and

40 Ehlers, *Mediterranean Air War, passim.*

41 Ibid.

discussed previously, including depot-level maintenance; salvage capabilities; ground mobility and the vehicles as well as fuel required to facilitate operations; radar; and several other factors, were woefully inadequate on the Axis side (See Illustration 11.4). The Germans' small-scale, tardy, and episodic deployments of Bf 109 fighters to North Africa undermined their opportunity to win lasting gains.[42]

The operational and strategic consequences of the small numbers of Bf 109s, magnified by their misuse and limited operational mobility, were serious enough in their own right. Rommel's tendency to head into the desert without taking his Luftwaffe liaison officers or informing the senior Luftwaffe commander of his whereabouts made the problem even worse, and contributed to serious failures of coordination between the Luftwaffe and Panzerarmee Afrika. These persistent and increasing misuses of the most capable Axis fighter represented an inversion of priorities and, by extension, a lost opportunity to control the air in a theater of operations where the large proportion of desert and ocean made aircraft particularly effective technological artifacts.[43]

Another timing issue the Axis faced was the utter inefficiency of the Italian shipping and port infrastructures. Reliable and timely deliveries of replacement aircraft, engines, spare parts, ammunition, aviation fuel, and support vehicles were vital to the air and combined-arms efforts. The Axis powers were at a disadvantage here from the outset and never engaged in more than episodic efforts to turn this factor in their favor.

In contrast, RAFME's major depots and workshops in the Cairo area played central roles in this question of timely deliveries, given their proximity to the front. The British enjoyed another major advantage in that the Cairo maintenance and salvage complex, and in fact all aspects of logistical and maintenance support, were under Dawson's highly capable leadership and backed by secure, seaborne delivery pipelines. German and Italian reliance on ship and transport-aircraft delivery of spare parts and other key materials was a perennial weakness as the British attacked Axis SLOCs with a will whenever they

42 Ehlers, *Mediterranean Air War, passim.*

43 The Mediterranean Campaign: Signals and Directives concerning problems of supply and German-Italian co-operation April–June 1941," Trans. VII/128, Air Ministry A.H.B. 6, November 1953, AHB, 28–31,; "The Mediterranean Campaign: Signals and Directives concerning problems of supply and German-Italian co-operation June–October 1941," Trans. VII/129, Air Ministry A.H.B. 6, February, 1954, AHB, 5–6,; BOH II, 14–15; Helmuth Felmy, "The German Air Force in the Mediterranean Theater of War," AFHRA, K113.107–161, 73–74, 79–85, 257–259; B.H. Liddell Hart, ed., *The Rommel Papers* (New York: Harcourt, Brace and Company, 1953), 121; *DLM*, 135, 146; Franz Halder, Charles Burton Burdick, and Hans Adolf Jacobsen, *The Halder War Diary, 1939–1942* (Novato, CA: Presidio, 1988), 348.

gained the upper hand in North Africa with planes from advanced airfields that could strike Axis convoys.[44]

The fourth key element affecting airpower employment and effectiveness was the set of *larger-order organizational, bureaucratic, productive, and infrastructural components of the airpower technology that supported aircraft and air operations.* German aviation-fuel production, for instance, depended on inferior refining techniques, resulting in 92-octane fuel for fighters in the best case, and more often 89 or even 87-octane. The American and British use of 100-octane fuel, which was available regularly and in large quantities, unlike the often scarce and inferior German fuel, gave less capable Allied aircraft and pilots an important advantage in an otherwise disadvantageous air-to-air situation. German shortages of nickel, tungsten, chromite, and copper forced German aircraft manufacturers to use synthetic rubber hoses rather than metal tubing and fittings, which further decreased performance in the air and reduced serviceability rates as aviation fuel ate away at rubber parts. This reduced operational readiness rates and forced more frequent depot maintenance. However, because there were no major depots in North Africa, aircraft had to sail across to Italy aboard one of the increasingly endangered Italian merchant ships, make the flight over despite the current malfunction, or sit in the desert for lack of proper parts and maintenance expertise.[45]

At the start of the war, Luftwaffe units down to the *Staffel* level enjoyed their own field logistics support in the form of dedicated Ju 52 aircraft carrying spare parts, fuel, and ground crews. They also had adequate numbers of vehicles to allow for rapid movement behind advancing ground troops. This combination allowed fighters to gain air superiority over the battle area when needed, and it continued to work as long as the Germans waged the short, sharp campaigns they anticipated.

However, once the dual commitments to the Mediterranean and Russia stretched Luftwaffe capabilities to the breaking point, dedicated Ju 52 logistics

---

44 These factors are vital to understanding why the British gained an increasing advantage in the air and by extension on land and at sea, but they are very complex and beyond the scope of this paper. For a detailed look at each, see Ehlers, *Mediterranean Air War, passim.*

45 For inferior German aviation fuel and aircraft components, see Stephen L. McFarland and Wesley Phillips Newton, *To Command the Sky: The Battle for Air Supremacy over Germany, 1942–1944* (Washington, D.C.: Smithsonian Institution Press, 1991). For maintenance and operational-readiness issues, see Studiengruppe Geschichte des Luftkrieges (Karlsruhe, Germany), "The German Air Force: Aircraft Procurement," 1955, Sections 456–457, 466–468, 472–473, K113.107–170, AFHRA; Air Ministry, *Supply Organisation of the GAF*, 198. For propeller shortages, see Adam Tooze, *The Wages of Destruction: The Making and Breaking of the Nazi Economy* (New York: Viking, 2007), 598.

ILLUSTRATION 11.4 Damaged Bf. 109F fighters abandoned at their landing ground outside Tobruk. Note the missing propellers. RAF Bomber Command raids destroyed a number of propeller factories in Germany, making this the number-one Wehrmacht salvage item in 1942–43. Shortages were serious even earlier, and many planes captured on the ground throughout the campaign were missing propellers. The damage to these aircraft also highlights the constant airfield attacks made by British fighters and fighter-bombers. There were no German radar sets in North Africa until 1942, leaving airfields exceptionally vulnerable. The raids caused serious losses in aircraft, aircrews, ground crews, vehicles, aviation fuel, spare parts, and tools—all in short supply among Luftwaffe units in North Africa. (LOC)

aircraft disappeared. They were replaced by periodic deliveries of spare parts and replacement ground personnel to airfields, and later with centralized supply dumps that often contained too many of the wrong spare parts and too few of those most in demand, in particular propellers. In fact, these became the highest-priority salvage and replacement item for the entire Wehrmacht by the spring of 1942 because of the heavy wear they experienced in Russia and North Africa, and the damage propeller plants sustained as a result of Bomber Command raids on the Reich.

In addition, allied attacks hindered resupply. Replacement engines, large spare parts, vehicles for improved mobility, and other critical components of the larger airpower technology supporting Luftwaffe fighters increasingly ended up at the bottom of the Mediterranean or burnt up as a result of RAF attacks on the vehicle convoys bringing them forward or at the airfields where Axis logisticians had managed to deliver them. The Axis failure to develop effective convoying and other protective measures for their merchant-ship and vehicle convoys, itself a result of the ineffective use of airpower in a theater profoundly reliant on resupply over long distances and over water, thus exacerbated the tactical and operational shortcomings of Luftwaffe aircraft employment.[46]

The Germans furthermore relied entirely on overseas aviation-fuel shipments for their fighters, a problem that went from bad to worse as British naval and air assets targeted and hunted down Italian oil tankers. These had become an endangered species by the spring of 1942 and nearly extinct by the fall. The German airlift fleet, comprised of the venerable Ju 52 "Tante Ju" (Auntie Ju), did yeoman service under very dangerous conditions delivering aviation fuel to Luftwaffe units in Africa, but they could not deliver enough to allow for full employment of fighters except during short periods of time as Rommel's major offensives commenced. Finally, although Rommel captured huge stocks of vehicle fuel from the British Army, he found virtually no aviation fuel, which the highly mobile RAF logistics-and-supply organization transported away before Axis troops could overrun the fuel dumps. Excellent RAF tactical communications and warning capabilities played vital roles in this effort. In this sense, even the "negative" strategic effects of capturing British oilfields would have been well worth the effort given what that would have done to each side's relative ability to employ air, ground, and sea assets. The balance of fuel deliveries and availability would have turned steadily toward the Axis once they controlled the Mediterranean Sea and captured British oilfields and refineries.[47]

British logistical capabilities were superior to those of the Axis and extended to aircraft-salvage operations, which Air Vice-Marshal Graham Dawson placed at the top of his list of priorities (See Illustration 11.5). Salvage teams combed the desert in large numbers and with good information about the locations of downed aircraft, resulting in nearly 100 percent of repairable aircraft being sent back to Cairo and eventually returning to service. Salvage crews also stripped

46 "The Mediterranean Campaign: Signals and Directives concerning problems of supply and German-Italian co-operation April–June 1941," Trans. VII/128, Air Ministry A.H.B. 6, November 1953, AHB; "The Mediterranean Campaign: Signals and Directives concerning problems of supply and German-Italian co-operation June–October 1941," Trans. VII/129, Air Ministry A.H.B. 6, February, 1954, AHB; CSDIC (UK), "A Survey of the Supply System of the German Army 1939–45," 25 August 1945, RG243/E36/B187, NARA, 4–7; Air Ministry, *Supply Organisation of the GAF*, *passim.*

47 These topics receive detailed attention in Ehlers, *Mediterranean Air War*, *passim.*

ILLUSTRATION 11.5
Salvaged Hurricanes en route to the shops around Cairo for repair, 1942. British logistics and maintenance capabilities, all part of their larger aircraft technology, were exceptionally good. Salvage, repair, supply, airfield-construction, radar and early warning, photoreconnaissance, intelligence, and every other element of the supporting infrastructure proved effective. This facilitated a concerted and successful British bid for air superiority in North Africa, which remained in place virtually without interruption until the end of the campaign. (IWM)

every usable part from aircraft damaged beyond repair, aiding the repair and refurbishment efforts underway throughout the Nile Delta. Everything from the high prioritization of this effort to the quality of vehicles and manpower dedicated to it was very much in favor of the RAF. The Luftwaffe was forced to rely on far less numerous and capable Italian salvage vehicles to do what they could. By summer 1942, there were hundreds of salvageable German aircraft, including many fighters, littering the desert, while dozens of others sat idle on airfields waiting for spare parts and overhauls that sometimes never materialized. While British aircraft operational readiness rates fluctuated between 70 and 90 percent during the North African campaign, German rates hovered around 50 to 60 percent and Italian rates at or slightly below 50 percent.[48]

48 Mr. Humphrey Wynn, "The RAF in the Mediterranean Theatre," in Wood, ed., *The End of the Beginning,* Chapter 5, 22–23; Owen, *DAF,* 59; R. Collishaw, "Brief Report on Royal Air Force Operations in the Western Desert From the Outbreak of War with Italy—The Capture of Cyrenaica to the Time of the Enemy Counter Offensive," 19 April 1941, AIR 23/6475, TNA, 13–14; Longmore, "Despatch No. 2 on Middle East Air Operations, 1-1-41 to 3-5-41," 1 November 1941, AIR 23/808, TNA, 50–51; "Maintenance, Supply, and Salvage Organization, Western Desert, 1942," AIR 23/6493; Lewis, "Tribute to Air Vice Marshal Dawson," AHB, 21–24; B RAFME, "Maintenance Diagrams," June 1943, Charts 1, 3, 14, AIR 23/3417, TNA; Air

Organizational innovations also proved vital to British success in the air, on the ground, and at sea. Joint headquarters, dedicated army-air force and navy-air force units and innovations, a Chain Home-like air-defense capability both in forward areas and in the Nile Delta, and many other examples of the "intellectual infrastructure" to which Dr. Guilmartin often referred were in evidence across the larger British and Allied efforts, but much less so in the German joint effort. German-Italian coordination was so poor as to be nearly nonexistent. The technological components of the larger British (and later Anglo-American) combined-arms effort included landline and direct-broadcast radios, copious supplies of vehicles to facilitate mobility, and the full range of other artifacts discussed earlier. American aircraft deliveries and logistical infrastructure development beginning in March 1941 magnified these British advantages. Specific aircraft types such as the C-46 and C-47 air transports had impacts utterly out of proportion to their numbers as Tedder, Coningham, and Dawson learned to use them to speed and intensify the "leapfrogging" process.[49]

Most fundamentally, and to bring us full circle to the importance of the "intellectual infrastructure" underpinning aircraft technology, the RAF became a "learning organization." This development began with the Allied debacle in France. Paradoxically, the Luftwaffe appears to have learned relatively little about either the enduring importance of air superiority in a combined-arms context or just how capable and dangerous the RAF was becoming even in the summer of 1940. Any sober assessment of the German failure over the Channel and southern England should have yielded the conclusion that gaining and holding air superiority in North Africa and over the Mediterranean would be particularly difficult and thus require a sustained effort. The Italian fiasco in the air and on land there in 1940 underscored this point.

Nonetheless, the Germans committed far too few single-engine fighters, torpedo bombers, and other vital air-land and air-sea assets, did so too late, and then squandered them and their crews in missions that often had little impact on the course of events. The Luftwaffe's episodic presence, based largely but not entirely on the ebb and flow of operations in Russia, further reduced the time on station of *Luftflotten* and subordinate units. Finally, Axis command-and-control, early warning, logistics, mobility, and other critical components of

---

Ministry A.D.I.(K), "Interrogation of German Officers from the Reichsluftfahrt Ministerium Conducted by Air Ministry A.D.I. (K) on Friday, 15th June 1945, 553.619, AFHRA, 7; Air Ministry, *Supply Organisation of the GAF*, 231–232, 237; Tedder to Portal, 23 January 1942, AIR 23/1396, TNA; BOH II, 292–293.

49 Ehlers, *Mediterranean Air War, passim.*

airpower technology did not mature to any significant degree. Command relations and interoperability with the Italians in the air and at sea also remained rudimentary—a fatal weakness in a theater of operations defined by a complex geography that required careful coordination of sea, ground, and air assets.[50]

Even when British reverses in the region and further afield threatened their position in the Middle East, Axis efforts to capitalize were either nonexistent or ineffective. These events included Rashid al-Gailani's coup in Iraq; the Arab uprising proposed by the Grand Mufti of Jerusalem; German opportunities to operate from Vichy airfields in Syria and Lebanon; and the major diversions of RAF, Royal Navy, and British Army forces to Greece and the Far East. The diversions to Greece and the Far East were timed perfectly from the Axis perspective given that they occurred during periods of RAF and Army exhaustion and weakness following Operations Compass and Crusader. Rommel took full advantage of both in his ground offensives, but there was no commensurate Luftwaffe effort to gain and maintain air superiority at the very moments when RAFME was most vulnerable—and with it, the Empire's position in the Middle East.[51]

Axis air and ground commanders also failed consistently to think of major campaigns as encompassing the weeks of air operations that preceded ground offensives. This bifurcation of thought on force employment and on what a major campaign actually encompassed was disastrous for the Axis but not for the British, who treated major operations as combined air-ground-sea operations that began months before the Army advanced against enemy positions. This holistic view of campaigns and battles gave the British, and later the Americans, major advantages in terms of strategy development and implementation, planning and logistical preparation, and, execution and exploitation. The Allies carried this fundamental lesson forward with them into subsequent campaigns from Italy to Germany.[52]

Two specific instances of the very different approaches to and use of air assets—and particularly single-engine fighters—stand out within the much larger question of air employment during the three-year contest for control of the Mediterranean and North Africa. Two deliveries of Spitfire V aircraft and their experienced pilots to Malta by the USS *Wasp*, which was desperately needed in the Pacific Theater but nonetheless stayed back to make these crucial deliveries, epitomizes the British refusal to yield control of the air. Conversely, the Axis decision to cancel Operation Hercules—the carefully planned attack

50 Ehlers, *Mediterranean Air War, passim.*
51 Ibid.
52 Ibid.

on Malta—was based on a fundamental misunderstanding of just how quickly the RAF could and would recover both on Malta and in the Middle East. Axis political and military leaders clearly had no idea what that would mean for control of the SLOCs, deliveries of fuel and other supplies to Rommel's armies, and the loss of initiative—permanent this time, as it turned out—that would result in North Africa and the broader Mediterranean Theater. This proved deleterious to the course of Axis efforts in the theater and was clearly one of the most disastrous decisions Hitler and Mussolini made within the larger collection of bad decisions regarding the conduct of the war.[53]

## 6 Taking Stock

Any effort to gauge the importance of a technological artifact—in this case aircraft in general with a specific focus on the single-engine fighter—requires that it first be defined within a holistic and overarching approach to what comprises aircraft technology. From there, we must place this aircraft technology within the larger strategic, geographical, climatic, and other contexts in which it operated. Clear and ultimately achievable Allied grand and military strategies contrasted markedly with a muddled and ineffective Axis strategic approach that hamstrung every aspect of military operations, but particularly control of the air. This denied Rommel a range of major advantages that he otherwise would have enjoyed in his effort to defeat British forces and conquer Egypt and larger areas of the Middle East.

---

53 GOH VI, 664; BOH III, 159–162, 194–195; FCNA, 279; James Douglas-Hamilton, *The Air Battle for Malta: The Diaries of a Spitfire Pilot* (Barnsley, UK: Pen and Sword Aviation, 2006); Felmy, 246–255; Italian Air Staff, "Memorandum on the Malta Situation, 10th May 1942," Trans. VII/57, A.H.B.6, December 1947, AHB, 246–254. Robert Citino provides an informative discussion of the Malta agreement and Rommel's permission from Hitler to scrap it. He also reminds us that Rommel's impetus to advance and win a decisive victory rather than pause and wait for Operation Hercules was, like many things Rommel did, the result of the influence of "300 years of German military history." See Robert Michael Citino, *Death of the Wehrmacht: The German Campaigns of 1942* (Lawrence: University Press of Kansas, 2007), 193–195, 205. Vincent Orange argued that "Britain's army, navy, and RAF commanders in the Mediterranean could hardly have prevented a conquest of Malta had it been attempted with all the Axis power available before August 1942." Based on an assessment of the available evidence, his statement appears compelling. See Vincent Orange, "Getting Together: Tedder, Coningham, and Americans in the Desert and Tunisia, 1940–43," in Daniel R. Mortensen, ed., *Airpower and Ground Armies: Essays on the Evolution of Anglo-American Air Doctrine 1940–1943* (Maxwell AFB, AL: Air University Press, 1998), 18.

If air superiority were merely a matter of having the best single-engine fighter and pilots, the Germans would have dominated the skies over North Africa and Malta. However, their failure to send enough of these assets, the inevitable shifting of planes and pilots back and forth between Russia and the Mediterranean, serious misuse of fighters at the tactical and operational levels, and major failings across most of the other elements of their larger airpower technology robbed them of their advantage in this critically important arena. Deserts and oceans are particularly advantageous places to have air superiority, and the collective elements comprising airpower technology made this profoundly so by 1941. The British, despite the failure to send their best fighter in time, or to release sufficient numbers of good pilots, were able to make up for these major mistakes by excelling in virtually every other aspect of the air war. This excellence included the utter focus on maintaining air superiority at the right times and places, attacking Axis aircraft on the ground unceasingly, and waging an increasingly effective battle against Axis SLOCs. Superior logistics and communications systems—maintenance, salvage, depots, airfields, radar, command-and-control, inter-service cooperation, and integrated combined-arms doctrine—made up for other failures regarding their single-engine fighters.

British successes in the air and combined-arms arenas, including air-ground and air-sea coordination, were based on learning from their early mistakes and building on their early successes. This process, which we associate with "learning organizations" and the "intellectual infrastructures" underpinning them, was consistently evident. Among other things, it prompted them to compensate for their distinctly inferior fighters and pilots by attacking Axis airfields constantly and heavily, while deploying highly effective radar and other warning capabilities such as tactical signals intelligence units to give RAFME air assets time to react when Axis air raids did occur. They also dispersed their aircraft across a large number of airfields and landing strips; employed camouflage, concealment, and deception to the maximum extent; sent special-forces teams to attack Axis airfields to excellent effect; and mastered "leapfrogging" their fighters and other aircraft forward and back with extraordinary speed and efficiency. These efforts gave British forces increasing advantages in North Africa, which proved to be a highly effective learning laboratory for developing and employing new techniques in future Allied military operations from Sicily to Normandy and into the Reich.

Despite multiple opportunities to seize and hold air superiority, and to exploit the many advantages it conferred, the Germans failed in their timing, temporized to an excessive degree, and thus lost the opportunity to control the tempo and course of operations. The almost complete lack of focus on air superiority; constant dispersion of effort; and failure to consider when, where, and for how long they had to control the air in order to facilitate strategic rather than operational and tactical successes proved the single greatest weakness of Axis efforts in the theater. Consequently, operational and tactical victories remained unexploited, the positive and "negative" strategic advantages that would have accrued to the Axis remained unrealized, and the theater became one of the primary graveyards of the Axis war effort. In essence, the German military did not properly conceptualize or understand how to use airpower to produce effects at the level of grand strategy—a serious problem in a complex theater of war with a range of grand-strategic opportunities and risks.

Nearly 250,000 Axis prisoners—157,000 Germans and 87,000 Italians—fell into allied hands in Tunisia in May 1943. Tens of thousands more Germans and hundreds of thousands more Italians were captured, wounded, or killed from the Italian declaration of war in June 1940 to the collapse in North Africa. The Axis lost over 9,700 aircraft, representing nearly two-thirds of production for 1942–1943. Combined with the aerial resupply debacle at Stalingrad, the similar slaughter of transport aircrew during the final months of the war in North Africa proved the death-knell for Germany's air-transport fleet, which had given the Wehrmacht a substantial measure of operational and logistical mobility during the war's opening years. As Allied bombing raids increased during 1943, the German fighter force had no opportunity to refit or to make good the heavy losses of pilots in the Mediterranean, and it was drawn heavily and ultimately disastrously into an aerial battle of attrition over the day and night skies of the Reich. At sea, the Axis lost 762 merchant vessels. The Italian Merchant Marine was virtually annihilated and its large navy fell into Allied hands.[54]

Finally, lessons learned in the theater regarding the overriding importance of air superiority as an enabler for subsequent efforts and of combined-arms operations as a means for beating the Germans at their own game of *Bewegungskrieg*, gave the Allies major advantages, or at least minimized some of their previous disadvantages, in future campaigns. Mature Allied combined-arms doctrine and employment were in evidence by summer 1944 during the breakouts in Italy and France. Heavy-bomber operations from Italy joined those from the United Kingdom in doing severe damage to German transportation and oil assets, demotorizing the Wehrmacht and denying it any semblance of mobility. This situation occurred at the very moment that German defensive lines were shortening and German troops were fighting harder than ever—and inflicting ever-higher casualties on Allied forces despite their own lack of mobility—in late 1944 and early 1945. The Germans had already lost the war,

54 Ehlers, *Mediterranean Air War*, 286.

but lessons learned and advantages accrued in the Mediterranean Theater helped the Allies to win the war more quickly and at a lower cost in blood and treasure.[55]

## 7 Toward a More Nuanced View of an Aircraft Technology

Technology is greater than the artifact in question. We must address it holistically and place it within larger contextual factors relating to the armed conflict in which a given artifact appears. The character of the conflict, which is itself driven in large measure by contextual factors, and the ways in which the warring parties employ a technological artifact, are also of great importance. However, as Dr. Guilmartin warned, strategic acumen and choices still drive the character and course of armed conflicts. Technology plays its role, and often a disproportionate one as was the case of single-engine fighter employment in the Mediterranean Theater, but it does not determine the outcome of wars in its own right. The degree to which those using it develop sound and achievable grand and military strategies, operational concepts, tactical usages, and underlying doctrinal precepts will determine in very large measure how useful that technological artifact, and the larger technology within which it exists, prove to be in the course of the larger conflict.

Technological artifacts matter, but we must remember that they are themselves the representations of a collectivity of related elements of a larger and holistic technology underpinned by and in turn supporting related organizations and infrastructures. (See Table 11.1) In this sense, fighter and other aircraft were merely artifacts, if important ones, within a larger airpower technology. Further, this conception means they and the larger technology within which they reside are just one part, if an operationally and tactically crucial one, within larger technological "systems." Understanding the effects and effectiveness of a technological artifact becomes even more complex, but also more important, when we assess its role and impact within a larger technological "system of systems" such as combined-arms warfare structures. This reminds us that the intellectual and professional qualities at play within these larger systems and "systems of systems" determine in large measure how effectively each technological artifact comprising it will be used, and what the collective utility of these usages is to the overall effort. "Learning organizations" and the "intellectual infrastructures" underpinning them are at the heart of the effective or ineffective use of a given technology writ large, whether an air force, a combined-arms force, or a larger allied or coalition force.

55 Ehlers, *Mediterranean Air War, passim.*

TABLE 11.1 Airpower Technologies in Comparison in the Mediterranean and Middle East Theater of Operations

| Air Force Characteristic | Royal Air Force | *Luftwaffe* | *Regia Aeronautica* |
|---|---|---|---|
| Grand Strategy | G | R | R |
| Military Strategy | G | R | R |
| Operational Doctrine | G | Y | R |
| Tactical Effectiveness | G | G | R |
| Technological Development | G | G | Y |
| Intellectual Infrastructure | G | Y | Y |
| Organizational Structure | G | Y | R |
| Physical Infrastructure | G | Y | Y |
| Production Capacity | G | G | R |
| Logistics and Supply | G | Y | R |
| The Artifact (aircraft) | Y | G | Y |
| Artifact Operators (aircrew) | Y | G | Y |
| Training Infrastructure | Y | R | R |
| Force Apportionment | Y | R | R |
| Force Employment | G | Y | Y |
| Geography, Climate, Weather | G | G | Y |
| Historical Experiences | G | R | R |
| Aggregate Will and Capabilities | G | G | Y |
| Changes Over Time (e.g. attrition) | G | Y | R |

Green = effective constitutive or complementary element of airpower technology
Yellow = marginally effective constitutive or complementary element of airpower technology
Red = ineffective constitutive or complementary element of airpower technology

When we approach airpower as a technology writ large and comprised of the many facets addressed earlier, this question of technologies narrowly as "learning systems" and broadly as "systems of systems" takes on even greater importance. Because every war is contextually unique and features a distinctive collection of technologies and technological artifacts, they and the political and military structures directing their use become "learning systems" within these unique contexts. The warring parties thus have no more than a few opportunities to learn, evolve, and operate more quickly and effectively than their adversaries. How they choose to employ technologies is a critical element in this process. Their understanding of the kind of war upon which they embark—short or protracted, regional or global, symmetric or asymmetric, "limited" or "total" in terms of the level of effort and risk invested—is central to the proper employment of technologies within larger contextual realities.[56]

These basic propositions about technology—what comprises it, how senior leaders discern its constituent elements, whether and how well they guide its use based on clear and achievable strategic aims, and whether there are effective "learning systems" and intellectual infrastructures evolving along with available technologies—are all central to understanding the conduct and course of the Mediterranean air war and the broader conflict there. Technologies, timing, and temporizing come together with these larger factors and, in fact, drive them in various directions as warring parties seek to gain a continuing advantage and ultimately victory. Dr. Guilmartin's warning thus comes to the fore once again: Although technology is a key component of war, strategic context and acumen, combined with rapid and effective learning at the operational and tactical levels, are even more so. This wise insight remains our guide for considering the interplay between technology, strategy, and war.

56 John Lynn's discussion of the discourse and reality of war is particularly useful here. Military organizations that are flexible enough to modify their discourse regarding war—to include doctrine and force employment—by incorporating the often unexpected realities they encounter in war generally defeat less flexible opponents. See John Lynn, *Battle: A History of Combat and Culture From Ancient Greece to Modern America,* revised and updated edition (Cambridge, MA: Westview Press, 2003), 359–369.

# PART 4

## *War Since 1945*

CHAPTER 12

# The War Is Lost: Technological Surprise and the Collapse of Portugal's Colonial Airpower Strategy, Guinea 1963–1974

*Matthew M. Hurley*

The use of air airpower as a counter-insurgency (COIN) instrument has long engendered promise and disappointment, applause and condemnation; indeed, public debate on the subject dates back at least as far as 1917 and the aerial angst of T.E. Lawrence's "Arab Army."[1] Since before World War I, insurgents have both feared and despised the seemingly omnipotent and invincible flying machines at their enemies' disposal, while those enemies consistently sought to maximize their aerial advantage.[2] All the while COIN forces increasingly have attracted vitriolic scrutiny for "excessive" and "indiscriminate" air attacks, particularly when employing allegedly "inhumane" tactics and ordnance.[3] Despite this mounting criticism, however, air forces have played an increasingly significant role in nearly every COIN campaign waged by governments with the technological savvy and available resources to field them.[4]

1 Indeed, Lawrence complained bitterly and repeatedly about the "helplessness" of his forces in the face of enemy airpower, even concluding at one point that "everything was being wrecked by air-impotence." T.E. Lawrence, *The Seven Pillars of Wisdom: A Triumph* (New York: Anchor Books, 1991), 612, 615.

2 See, for example, Michael Paris, "Air Power and Imperial Defence 1880–1919," *Journal of Contemporary History* 24, No. 2 (April 1989): 209–225. Dr. Paris identifies French sorties against Moroccan tribesmen in 1913 as the first recorded instances of fixed-wing airpower used in operations that could, loosely at least, be considered COIN. The same label could be applied to analogous air operations in Spanish Morocco, German-occupied Southwest Africa, and the Netherlands East Indies the following year (Paris, 221).

3 The use of napalm by Western air forces is perhaps the most famous example, but more recently other munitions have risen to the public fore. By way of example, the Israeli Defense Forces endured a great deal of international criticism for their use of white phosphorus during operations in Gaza – an internet search for "Israel Gaza 'white phosphorus'" on news.google.com (9 March 2009) drew 145 hits to stories that were uniformly negative. Similarly, NATO was regularly derided for its use of depleted uranium ordnance during the 1999 Kosovo air campaign.

4 See, for example, three compendiums of air operations in a counter-insurgency role: James S. Corum and Wray R. Johnson, *Airpower in Small Wars: Fighting Insurgents and Terrorists* (Lawrence, KS: University Press of Kansas, 2003); Philip Anthony Towle, *Pilots and Rebels: The*

 | DOI:10.1163/9789004393301_014

Portugal's colonial wars in Africa were no exception. From 1961 to 1974, the Portuguese Air Force (*Força Aérea Portuguesa*, hereafter FAP) participated as a major – and increasingly central – element of Portugal's COIN effort in Angola, Mozambique, and Guinea-Bissau.[5] In Guinea-Bissau, especially, the FAP became the campaign's centerpiece, to its tactical and operational credit, but also to its strategic detriment. While the FAP consistently rendered crucial service in that theater, its value and effectiveness there led Portugal to a dangerous degree of dependence on airpower.[6] When insurgent forces upset the aerial equation with a new and unexpected weapon after nine years of escalating conflict, the FAP – and with it Portugal's imperial strategy – suffered a shock from which neither recovered. By this time, Lisbon's colonial enterprise in Guinea-Bissau had devolved to a situation described by one Portuguese commander as "hold on as best you can." With the loss of Portugal's sole remaining military advantage, even this meager objective was no longer achievable.[7]

Of Portugal's colonial conflicts in Africa, Guinea-Bissau consistently presented the Lisbon regime with its most vexing military and political challenge. Consequently, the small and economically insignificant province of "Portuguese Guinea" commanded Portugal's heaviest military commitment in Africa, proportionally speaking.[8] Airpower consistently played a key role in

---

*Use of Aircraft in Unconventional Warfare, 1918–1988* (London: Brassey's UK, 1989); and Joel Hayward, ed., *Air Power, Insurgency, and the "War on Terror"* (Cranwell, UK: Royal Air Force Center for Air Power Studies, 2009).

5 While under Portuguese sovereignty, Guinea-Bissau was generally known as "Portuguese Guinea." As insurgent elements began to attract more international support, and particularly after their unilateral declaration of independence in September 1973, the name "Guinea-Bissau" became more uniformly accepted. "Guinea" can be used to refer to either Guinea-Bissau, Equatorial Guinea, or Guinea-Conakry; or, it can refer to the general region of West Africa.

6 In this paper, as in US Air Force doctrine, "airpower" includes more than just the application of kinetic force from aircraft in flight. The term also incorporates roles more colloquially described as "support," such as reconnaissance or transport.

7 According to General Carlos Fabião, the last governor-general of Portuguese Guinea, "By mid-1973, with [Portugal's airpower] advantage gone, Portugal's 'strategy' amounted to little more than '*aguentar enquanto fosse possível*' (hold on as best you can)." Following the 25 April coup, Fabião oversaw the withdrawal of Portuguese forces from the territory. "A descolonização da Guiné-Bissau. Spínola: a figura marcante da guerra na Guiné," in *Seminário: 25 Abril 10 anos dopois* (Lisbon: Associação 25 de Abril, 1984), 310.

8 By 1973, over 32,000 Portuguese troops (and between 8,000 and 11,000 locally-recruited soldiers) were fighting in Guinea, or at least one soldier to every 17 Guinean inhabitants. By way of comparison, Portugal had only one soldier for every 74 inhabitants in Angola, and one for every 128 in Mozambique. Relatively speaking, then, Portugal had committed – at a minimum – over four times as many troops to Guinea as to Angola, and eight times as many as to Mozambique. Paulo Borges Coelho, "African Troops in the Portuguese Colonial Army, 1961–1974: Angola, Guinea-Bissau and Mozambique, *Portuguese Studies Review*,

that commitment, so much so that a 1968 Central Intelligence Agency (CIA) assessment described the FAP as "the decisive military factor" in Guinea-Bissau.[9] Indeed, relative to the size of the territory and its population, Guinea-Bissau hosted the greatest concentration of Portuguese airpower in Africa, and, for much of the conflict, the air effort drove Portugal's military strategy. The fortune of Portugal's campaign in Guinea-Bissau would consequently mirror the successes – and failures – of its air arm.

Objectively, Portugal's air commitment to Guinea-Bissau paled in numerical significance when compared to the French campaign in Algeria or the American effort in Vietnam. Unlike the vast air armadas employed in those earlier conflicts, the FAP's order of battle in its rebellious West African province comprised some 50 to 80 aircraft during most of the war, including helicopters – less than the complement of a modern American aircraft carrier.[10] These initially included eight U.S.-supplied F-86 Sabre fighters deployed to Bissau's Bissalanca Airport (known to the FAP as *Base Aérea* 12, or BA12) in 1961. Intense American diplomatic pressure, however, compelled Portugal to withdraw these aircraft from Guinea-Bissau in October 1964.[11] Oddly enough, American protests did not extend to the FAP's U.S.-built C-47 Dakota transports, some of them modified for night bombing, and armed T-6 Texan/Harvard piston-engined trainers. Rounding out the initial FAP commitment to Guinea-Bissau were a detachment of Dornier Do-27 and Auster D4/D5 light utility aircraft (the latter retired by 1965), and a handful of French-produced Alouette II helicopters, which were withdrawn in 1966.[12] By that time, the FAP had added Alouette III helicopters, additional Do-27s, and Fiat G.91 jet attack aircraft to

---

September 2002: 136; Norrie MacQueen, "Portugal's First Domino: 'Pluricontinentalism' and Colonial War in Guinea-Bissau, 1963–1974, *Contemporary European History* 8, no. 2, 1999: 226; Arslan Humbaraci and Nicole Muchnik, *Portugal's African Wars: Angola, Guinea Bissau, Mozambique* (New York: The Third Press, 1974), 18–21; and W.S. Van der Waals, *Portugal's War in Angola, 1961–1974* (Rivonia, South Africa: Ashanti, 1993), 247–250, 256.

9 Central Intelligence Agency, *The Guerrilla War in Portuguese Guinea*, Weekly Review and Special Report, 20 December 1968, 11.

10 Colonel Luís Alves de Fraga, Portuguese Air Force, *A Força Aérea na Guerra em África* (Lisbon: Prefácio, 2004), 122; Lieutenant Colonel José Fernando de Almeida Brito, Portuguese Air Force, untitled status report for Operational Group 1201, 1973, in the Amílcar Cabral Collection, Fundação Mário Soares Arquivo e Biblioteca, Lisbon, 1.

11 General José Lemos Fereira, Portuguese Air Force, "No ceú da Guiné," in José Freire Antunes, *A guerra de África (1961–1974)*, vol. 2 (Lisbon: Temas e Debates, 1996), 587; Duncan Curtis, *North American F-86 Sabre* (Marlborough, UK; The Crowood Press, 2000), 173; U.S. Department of State, *US Arms in Portuguese African Territories*, U.S. Department of State telegram, December 3, 1963, 1.

12 Fraga, 105.

its Guinean fleet. Total inventory peaked at 81 aircraft in 1970, between half and two-thirds of them operational at any given time.[13]

Throughout the conflict, the FAP principally flew in support of Portuguese surface forces operating against the *Partido Africano da Independência da Guiné e Cabo Verde* ("African Party for the Independence of Guinea and Cape Verde," abbreviated PAIGC). Air support to the surface forces was a determinant factor in the conduct of Portugal's campaign: aerial fire support, reconnaissance, transport, command and control, and casualty evacuation consistently conferred crucial advantages to Portuguese troops in Guinea.[14] These missions therefore drew the great majority of FAP sorties during the conflict, although the FAP's limited inventory remained a persistent problem.[15] However, the FAP was aided to an extent by geographical factors, most notably Guinea-Bissau's small size. From the FAP's primary base at Bissalanca, for example, Portuguese forces could arrive at any point in the territory within an hour by helicopter, and in just 15 minutes by jet aircraft such as the G.91.[16]

The expected and demonstrated effectiveness of airpower influenced Portuguese commanders in different ways over the course of the conflict. General Arnaldo Schulz, Portugal's Governor-General in Guinea from 1964–1968, preferred to unleash the FAP in direct attacks against guerrilla logistics, food supplies, and sympathetic villages in an effort to "starve the [PAIGC's] forces to death or better still force them to surrender or mutiny."[17] This effort proved sufficiently successful at the tactical and operational levels, for a time nearly paralyzing the PAIGC and forcing radical changes in its strategy.[18] Schulz's

13 Compilation and analysis of Commando Zona Aérea de Cabo Verde e Guiné weekly situation reports (SITREP), 1964–1973 inclusive, Arquivo da Defesa Nacional, Paço do Arcos; Fraga, 122.

14 Lieutenant General Jose M. Brochado de Miranda, Portuguese Air Force, "El empleo del arma aérea en la lucha antiguerrilla," *Revista de Aeronautica y Astronautica* No. 715 (July–August 2002): 596. Miranda's remarks echoed the opening pages of the FAP's COIN manual. That document groups air operations into four general categories: fire support, reconnaissance, transport, and "diverse" (i.e. "miscellaneous"). See Serviço de Instrução da Força Aerea, "Manual do Apoio Aéreo na Contra-Guerrilha," 17 February 1972, 2ª. Região Aérea Manuais Collection, Arquivo Histórico da Força Aérea, Lisbon, 1–5.

15 Al J. Venter, "Portugal's Forgotten War," *Air Enthusiast* 2, No. 2 (February 1972), 62.

16 Venter, "Portugal's Forgotten War," 60. During Operation "Barracuda" in February 1967, for example, FAP Fiat G.91s required a mere 10 minutes from take-off to time over target (TOT). Dr. José Manuel Coreia, "Os Fiat G.91 ao Ataque na Guiné: Operação Barracuda," *Mais Alto* 45, No. 367 (May–June 2007): 24.

17 Mustafah Dhada, "The Liberation War in Guinea-Bissau Reconsidered," *Military History* 62, No. 3 (July 1998): 582.

18 Mustafah Dhada, *Warriors at Work: How Guinea Was Really Set Free* (Niwot, CO: University Press of Colorado, 1993), 34–35.

1966–1967 air offensive, for example, compelled the insurgents to dismantle their painstakingly developed base areas and replace them with smaller, semi-permanent facilities. As PAIGC leader Amílcar Cabral later recounted, "We reduced the bases, dividing them into twos or threes," thus eliminating these lucrative targets "just in time." PAIGC cadres concurrently faced the unenviable task of trying, sometimes unsuccessfully, to persuade Guinean villagers to remain in the targeted "liberated areas" despite the risk of continued aerial bombardment, and even veteran PAIGC combatants began to question the liberation effort itself.[19]

All told, Schulz "played [the airpower] card to great effect," according to one historian,[20] and by the end of his tenure in Guinea-Bissau the Portuguese COIN effort was predominantly an aerial affair.[21] Although victory yet eluded Lisbon, Schulz' air campaign did present the greatest obstacle to the PAIGC's aims, leading the CIA to conclude shortly after Schulz' 1968 departure that Portugal's airmen and aircraft "have so far prevented the insurgents from exploiting many of their gains. As long as the Portuguese can bomb them at will, the insurgents probably will not try to hold fixed positions like towns."[22]

The ultimate defect in Schulz' strategy, however, lay in its exclusive reliance on a military solution to the problem of insurgency. By way of contrast, his successor Brigadier General António Spínola recognized that "Revolutionary war, the kind we are facing in Guinea, is primarily psychological, which necessarily exploits the motivations that are susceptible to influencing and conditioning attitudes and behavior."[23] Consequently, between 1969 and 1973, Spínola focused his efforts on improving social and economic conditions for Guinea-Bissau's population, conducting psychological operations, and pursuing negotiations to achieve a political, rather than a military, solution to the conflict.[24] The FAP played a significant role in this "psycho-social" strategy,[25] embarking

19 Dhada, "The Liberation War in Guinea-Bissau Reconsidered," 580, 583.

20 Dhada, *Warriors at Work*, 33.

21 Patrick Chabal, *Amilcar Cabral: Revolutionary Leadership and People's War* (Trenton, NJ: Africa World Press, 2003), 92.

22 Central Intelligence Agency, *The Guerrilla War in Portuguese Guinea*, 11.

23 Dutra Faria, "Governor Tells of Independence Party Activity," *ABC-Diario de Angola* (Luanda), 7 December 7 1969, translated and reprinted in Joint Publications Research Service, "Translations on Africa No. 867," JPRS 49903 (February 20, 1970), 68.

24 Spínola's overarching concept was embodied in a project called *Guiné Melhor*, or "A Better Guinea." For a full description of this program, see António de Spínola, *Linha de Acção* (Lisbon: Agência-Geral do Ultramar, 1971). An English-language synopsis is available in American Embassy Lisbon, *Portuguese Guinea and Its Governor*, U.S. Department of State Airgram, 19 October 1971.

25 In Portuguese, "*psicosocial*."

on such efforts as pamphlet delivery, loudspeaker broadcasts, presence operations, aid delivery, and medical evacuation for sick and wounded civilians.[26]

Spínola also recognized that his social and political focus "does not diminish the keenness of the military problem, since we cannot ignore the fact that although these wars are not won on the battlefield, they can certainly be lost there."[27] His dual-track strategy of psychological carrots and kinetic sticks subsequently earned the label "Smiles and Blood" from the PAIGC and, as under Schulz, the "blood" aspects of the program were delivered predominantly from the air.[28] Unlike Schulz, however, Spínola employed airpower "not as a show of brutal force, not as an instrument of tactical advantage in a given situation, not without an underlying strategy…but relentlessly, with precision, and always to exact maximum political damage."[29]

Spínola's airpower program intensified in 1969 after he withdrew Portuguese ground troops from large areas of Guinea-Bissau, where the PAIGC was particularly active and established aerial free-fire zones in those regions.[30] Subsequent operations assumed a consistent, and largely effective, pattern: FAP aircraft would begin by conducting air strikes, sometimes over several days, against PAIGC facilities and concentrations, as well as emerging targets encountered during armed reconnaissance sorties. These raids would be followed by helicopter-borne assaults deploying FAP paratroopers, escorted by fixed-wing attack aircraft and armed helicopters (so-called *helicanhãos*, or "heli-cannons"). After troop disembarkation, the helicopters remained available for casualty evacuation and personnel withdrawal upon completion of the mission.[31] Largely due to such methods, by 1972 Spínola had effectively stabilized a formerly losing situation in Guinea-Bissau, leading one visiting

26 General Tomás George Conceição Silva, Portuguese Air Force, "The Portuguese Air Force in the African Wars of 1961–1974," in *Memories of Portugal's African Wars, 1961–1974: Proceedings of a Conference, King's College London, 10 June 1997*, ed. John P. Cann (Quantico, VA: Marine Corps University Foundation, 1998), 122.

27 Faria, 67.

28 Luís Cabral, *Crónica da Libertação* (Lisbon: O Jornal, 1974), 365.

29 Dhada, "The Liberation War in Guinea-Bissau Reconsidered," 588.

30 American Embassy Lisbon, *Portuguese Guinea and Its Governor*, 2. To the FAP, these were known as Zonas *de Livre Intervenção da Força Aérea* ("Air Force Free Intervention Zones," or ZLIFA); see General Manuel Diogo Neto, "Voo em Três Frentes," in *A Guerra de África (1961–1974)*, vol. 1, ed. José Freire Antunes (Lisbon: Temas e Debates, 1996), 316–317, and Fraga, 106.

31 "Intensification of Armed Struggle," *Info Tchad* (Fort Lamy, Chad), 9 April 1973, translated and reprinted in Joint Publications Research Service, "Translations on Africa No. 1304," JPRS 58903 (May 2, 1974), 29. Portugal's use of this tactic in Guinea and elsewhere may have inspired Rhodesian "Fireforce" operations in present-day Zimbabwe.

journalist to observe "Though far from winning the war, his army and air force is [sic] no longer losing it."[32]

Some visitors to the PAIGC's "liberated areas" spoke of Guinean children singing a song featuring the lyrics "We have the land – the Portuguese have only the sky."[33] Beneath such public bravado, however, lurked the realization that the FAP posed a potentially existential danger to the guerrillas and their cause; a realization that resonated at all levels of insurgent deliberation. The PAIGC thus roundly condemned "the criminal action of the barbarian Portuguese Air Force against our people and against humanity" in any available international forum, more often and more vehemently than any other Portuguese activity.[34]

These complaints began almost immediately after the initiation of guerrilla operations in January 1963, when the PAIGC publicly protested FAP raids against villages in freshly "liberated" areas.[35] At one point Luís Cabral, who would later become the independent Guinea-Bissau's first president, remarked that Portuguese aircraft "caused great disturbances among our people, who were also suffering from the certainty of knowing that we could not counter the bombings with comparable blows."[36] PAIGC military commander João Bernardo ("Nino")

---

32 Venter, "Portugal's Forgotten War," 60. Said one Portuguese officer on the Guinean situation in 1971, "We may not be winning what is essentially an unwinnable war, but we have stopped losing" (Jim Hoagland, "Portugal's 'Unwinnable' War Heats Up," *The Washington Post-Times Herald*, 21 February 1971). As late as January 1973, according to the Chief of Staff of the Portuguese Armed Forces, "The military and subversive situation is at this moment under control" (Chefe do Estado-Maior General das Forças Armadas, "Visita do CEMFGA á Guiné, Janeiro – 73: Relatório," 22 January 1973, file F3/17/37/59, Arquivo da Defesa Nacional, Paço do Arcos, 43).

33 George M. Houser, No One Can Stop the Rain: Glimpses of Africa's Liberation Struggle (New York: The Pilgrim Press, 1989), 207.

34 "Guinea-Bissau: Advances on the Military Front," *Southern Africa: A Monthly Survey of News and Opinion* 6, No. 9 (November 1973): 15. By way of further example, during testimony before the U.S. House of Representatives Subcommittee on Africa in February 1970, PAIGC leader Amílcar Cabral frequently – and to the exclusion of *all* other military methods – decried the results of Portuguese airpower. See U.S. Congress House of Representatives, Subcommittee on Africa of the Committee on Foreign Affairs, *Report on Portuguese Guinea and the Liberation Movement* (91st Congress, 2nd session, 26 February 1970), 3.

35 *Qualques Crimes Perpetres par les Colonialistes Portugais contre Nosse Peuple en 1963* (Conakry, Guinea: Partido Africano da Independência da Guiné e Cabo Verde, 1964), 1. For later accounts, see Amílcar Cabral, *Aucune Manoeuvre ou Crime des Colonialists Portugais Ne Sera a Meme d'Eviter la Victoire Ineluctable de Notre People Africaine: Rapport Bref sur la Lutte in 1971* (Conakry, Guinea: Partido Africano da Independência da Guiné e Cabo Verde, January 1972), 13, and "War Communiqué," *PAIGC Actualités Quarterly English Edition* nos. 45–48 (1972).

36 Luís Cabral, *Crónica da Libertação*, 354.

Vieira[37] echoed these sentiments in 1971, observing that airpower "is the weapon that frightens the inhabitants the most. It is one of the weapons most used by the colonialists and at present it is almost their only offensive weapon."[38] Even PAIGC schoolbooks joined the chorus of protest. A Second Grade reading primer, for example, featured a reading exercise focused on the bombing of a PAIGC-dominated village, while another about a fisherman and his grandson highlighted the twin riverine dangers of crocodiles and Portuguese bombers.[39]

Such pervasive angst throughout the PAIGC's military, political, and social organs stemmed from the perception that FAP aircraft were omnipresent, invulnerable, lethal, and could come out of nowhere with precious little warning, if any. PAIGC units had learned early in the conflict how to choose surface targets with craftiness and care, and to evade superior Portuguese ground forces. Experience had taught them, however, that FAP aircraft would not grant them the opportunity to shun disadvantageous combat. In the air-to-ground fight, the aircraft dictated the location, timing, and tempo of operations, at least for the first ten years of the war.

Despite its impact, the FAP suffered from a number of inherent weaknesses that persistently limited its effectiveness in Guinea-Bissau. These disadvantages included an an obsolete aircraft inventory, a rudimentary aviation infrastructure, and chronic shortages of ordnance and ammunition.[40] Portuguese forces did construct 21 airfields throughout Guinea-Bissau in the late 1950s and early 1960s, and more afterward, but most of these featured improvised, unpaved surfaces suitable only to helicopters or to rugged STOL (short takeoff and landing) aircraft.[41]

Perhaps the FAP's greatest weakness, however, lay in the number of aircrew it could muster for operations in Guinea-Bissau. Although Portuguese troops in the territory numbered some 40,000 by 1971 – proportionately the largest Portuguese manpower commitment anywhere in its African empire – that number included only 50 to 70 pilots, including conscript enlisted aviators.[42]

37 Later President of Guinea-Bissau, assassinated in 2009 by soldiers avenging the murder of the Guinean-Bissau armed forces chief, General Batista Tagme Na Waie.

38 Pedro Machado, "Liberation Movement Discussed by Guerrilla Leader," *Etumba* (Brazzaville, Congo), September 11, 1971, translated and reprinted in Joint Publications Research Service, "Translations on Africa No. 1081," JPRS 54466 (November 11, 1971), 62.

39 "Morés" and "Aliu, o pescador," in *O Nosso Livro: 2ª Classe* (Regiões Libertades da Guiné: Serviços de Instrução do PAIGC, 1970), 28, 53.

40 Miranda, 594.

41 General Pedro Cardoso, Portuguese Army, "A Better Guiné," 95.

42 Fereira, 590.

During 1972 and 1973, the FAP complained that its aircrew roster in Guinea was half the number needed,[43] and, as a result, Portuguese pilots in the territory logged an extraordinarily high number of flying hours in an unusually diverse repertoire of aircraft.[44] The FAP's cadre of helicopter pilots, so essential to Portugal's counter-insurgency operations in Guinea, was particularly understaffed.[45]

Aircraft were in similarly short supply throughout the conflict, in part, due to U.S. policies prohibiting the use of American-manufactured weaponry in Portugal's colonies. This embargo did not prevent Portugal from using a variety of U.S.-built aircraft in Africa at various times, but it did result in American reluctance to provide new aircraft to the FAP.[46] To compensate, Portugal turned

---

43 Chefe do Estado-Maior General das Forças Armadas, "Visita do CEMFGA á Guiné, Janeiro – 73: Relatório," 22 January 1973, file F3/17/37/59, Arquivo da Defesa Nacional, Paço do Arcos, 18; Lieutenant Colonel Aniceto Afonso and Colonel Carlos de Matos Gomes, Portuguese Army, "Guiné: Maio de 1973 – O Inferno" (working paper, Avenida da Liberdade/Associação 25 de Abril, 22 February 2008), 1.

44 By way of example FAP personnel records show that, of the 41 pilots based at Bissalanca in January 1970, 16 flew at least three types of aircraft during that month's operations. These were not merely different models the same class of aircraft, but rather a mix of substantially different types such as propeller-driven aircraft and jets, or fixed-wing aircraft and helicopters. Two of those pilots flew *four* different types during that month alone. Of the 41, only two helicopter pilots and four fixed fixed-wing pilots flew a single type of aircraft. A similar situation prevailed in March 1973: of the 50 FAP pilots then assigned to Bissalanca, only fifteen flew a single type of aircraft for the month.

As of that same month, the average total accumulated flight time for company grade pilot officers (lieutenants and captains) in Guinea was an astonishing 2,731 hours – as much as most U.S. Air Force colonels and generals today. Pilots in the field grade ranks (major through colonel) averaged over twice that with 5,600 accumulated flying hours each. Even enlisted conscript pilots routinely tallied flight hours approaching the 1,000 mark, in one case even doubling that number. Comando da Zona Aerea de Cabo Verde e Guiné, "Envio de Relatórios Mensais (Março de 69)," 25 April 1969, 1ª Região Aérea, Comando da Zona Aérea Cabo Verde e Guiné – Relatórios Mensais do Comando, Anos 1968–1969 collection, Arquivo Histórico da Força Aérea, Lisbon, 1; Comando da Zona Aerea de Cabo Verde e Guiné, Base Aérea No. 12, "Mapa de Qualificação de Pilotos e Navegadores," January 1970, 1ª Região Aérea, Commando da Zona Aérea Cabo Verde e Guiné – Relatórios Mensais do Comando, Anos 1968–1969 collection, Arquivo Histórico da Força Aérea, Lisbon, 2–4; Comando da Zona Aerea de Cabo Verde e Guiné, Base Aérea No. 12, "Qualificação e Experiência de Pilotos e Navegadores," March 1973, 3ª Divisão–EMFA, 1ª Região Aérea, Relatórios do Comando 1972–1974 collection, Arquivo Histórico da Força Aérea, Lisbon, 1–2.

45 Secretariado-Geral da Defesa Nacional, "Acta de reunião da comissão conjunta dos chefes de estado maior em 6 Fevereiro de 1973 (Acta Nº. 1/73)," February 1973, F3/10/23/38, Arquivo da Defesa Nacional, Paço do Arcos, 5–6.

46 U.S. Department of State, *US Arms in Portuguese African Territories*, U.S. Department of State telegram, December 3, 1963, 1; National Security Council, *Implementation of Arms*

to France and Germany. Exploiting those nations' desire to utilize Portuguese facilities far behind Cold War frontlines, Portugal in the 1960s managed to acquire 76 Do-27 observation aircraft, 70 World War II-vintage T-6 training aircraft adapted for ground attack duties, and – most significantly – 40 Fiat G.91 jet fighter-bombers from Germany, as well as Alouette-III helicopters and additional T-6s from France.[47] In the end, though, these acquisitions simply injected a fresh supply of obsolete or second-hand aircraft into the FAP's inventory: it remained a 1940s force attempting to wage war into the 1970s (See Table 12.1). Such a fleet might have been sufficient to Lisbon's needs had the FAP's initial assumption that "there was no proper air opposition" proven correct.[48] In Guinea-Bissau, however, Portugal faced an adaptive, ingenious, and resilient foe, whose leader clearly understood the impact of government airpower, and correspondingly recognized the importance of air defense.[49] The PAIGC consequently expended Herculean efforts in its attempt to develop an effective air defense capability.

The PAIGC's armory was initially limited to small arms, machine guns, grenade launchers, and mortars provided from the Communist bloc, first acquired

*Embargo on South Africa and Portuguese African Territories*, National Security Council Memorandum 81, August 17, 1970, 1–2. This policy, enacted during the Kennedy administration and continued by President Johnson, was relaxed somewhat during the Nixon administration. To wit, in 1972 President Nixon permitted the long-delayed transfer of Boeing 707 airliners to Portugal, based on their presumed civilian nature. Unsurprisingly, these aircraft were soon put to use shuttling troops and military equipment between Portugal and its combat theaters in Africa. See for example Col.Edgar Pereira da Costa Cardoso, Portuguese Air Force, *História da Força Aérea Portuguesa, 3° Volume* (Lisbon: Cromocolor, 1984), 270. As explained by Portugal's foreign minister at the time, Rui Patricio, "If I buy an American car, can America tell me how I can use it? If I want to drive it in Africa, I will drive it in Africa. The Boeing is not an arm." Jim Hoagland, "Arms from East, West Used in Africa," *The Washington Post-Times Herald*, 4 April 1974.

47 Germany sought to construct a logistics and training base in the "European rear," while France planned to build a rocket tracking and instrumentation facility in the Azores. Although these deals included stipulations that the aircraft "will remain in Portugal and will be used in defense of NATO interests," Lisbon was left free to determine what constituted "Portugal," and whether or not the purported defense of Africa from encroaching communism served "NATO interests." In the end, Portugal interpreted these clauses to the advantage of its imperial venture and deployed such aircraft as it deemed necessary to Africa, including many of the recent German contributions. Ana Mónica Fonseca, "Dez anos de relações luso-alemãs," *Relacões Internacionais* 11 (2006): 54; Mario Canongia Lopes, "Portugal's Ginas," *Air Enthusiast* 36 (May–August 1988): 66; Fereira, 586–587.

48 Fereira, 584.

49 Amílcar Cabral, *A Acção Armada e os Métodos Militares* (Bissau: Departamenta de Informação, Propaganda e Cultura do C.C. do PAIGC, 1988), 4.

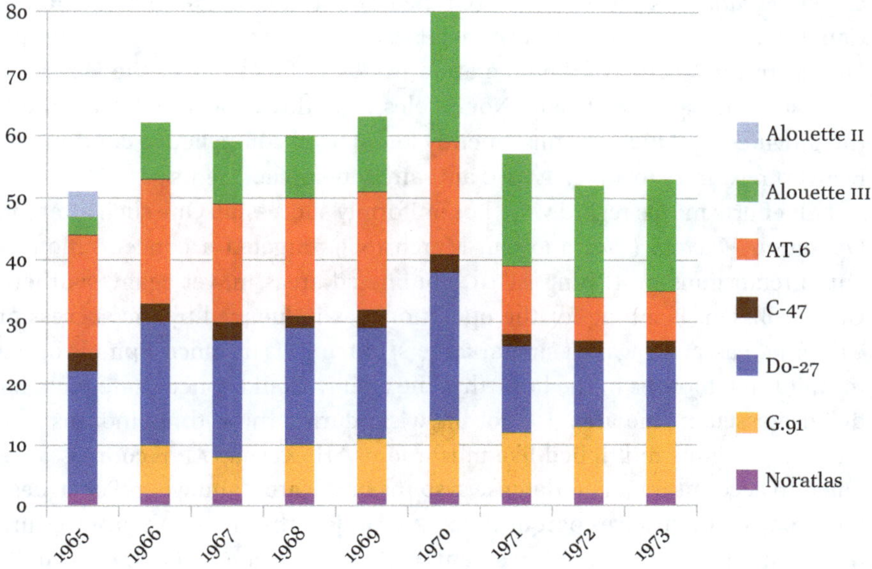


TABLE 12.1 Portuguese Air Force inventory in Guinea.

in late 1962.[50] However, as early as 1963 heavy anti-aircraft machine guns, including the 12.7mm Dyegtarov-Shpagin and the quad-mounted 14.5mm ZPU-4, began confronting Portuguese aircraft over Guinea-Bissau.[51] These soon took an inaugural toll on the FAP, claiming an F-86 and a T-6 that same year.[52] PAIGC anti-aircraft capabilities continued to expand with subsequent acquisitions, and by November 1965, FAP aircraft were taking 20mm antiaircraft fire in southern Guinea-Bissau, the guerrillas' stronghold.[53] The growing air defense threat resulted in rigorous Portuguese efforts to neutralize the PAIGC's expanding anti-aircraft capability. An intense two-day operation in December 1965,

50 These first arms shipments for the PAIGC originated in Morocco. Luís Cabral, "Prender Spínola," 536.

51 Coreia, 22; Venter, "Portugal's Forgotten War," 61; Fraga, 109.

52 António Lobato, *Liberdade o Evasão: O Mais Longo Captiveira da Guerra* (Amadora, Portugal: Erasmos, 1995), 33–40; Licínio Azevedo and Maria da Paz Rodrigues, *Diário da Libertação (a Guiné-Bissau da Nova África)* (São Paolo, Brazil: Versus, 1977), 103–109; Amílcar Cabral, *Communiqué* (Conakry, Guinea: Partido Africano da Independência da Guiné e Cabo Verde, June 1963), 1; Amílcar Cabral, *Communiqué* (Conakry, Guinea: Partido Africano da Independência da Guiné e Cabo Verde, March 11, 1963), 1; Fraga, 104, 118–121. António Lobato, author of the first cited source, was the pilot of one of two T-6s lost on a single mission, and his account indicates it was an indirect result of hostile fire (his wingman, while inspecting the damage to Lobato's T-6, inadvertently collided).

53 Fraga, 109.

called *Resgate* ("Rescue"), neutralized the PAIGC's anti-aircraft network in the Cantanhez region of southern Guinea-Bissau for a period of several months. Similar results accrued from Operation *Estoque* ("Stock") over the Quitafine Peninsula during August 1966. Nonetheless, the threat persisted throughout the province, and the FAP consequently found itself compelled to continuously divert resources to attack PAIGC anti-aircraft emplacements.[54]

FAP efforts in this regard were not uniformly successful. One such attempt, Operation *Vulcano* ("Volcano") in March 1969, targeted a concentration of anti-aircraft guns protecting PAIGC-dominated areas in, yet again, southern Guinea-Bissau. Planning for the operation was influenced by the success of earlier FAP assaults against similar targets, but, in this instance, Spínola opted to wait three months in the hope that the enemy would concentrate more air defense assets in the area. He got his wish: during those three months "the anti-aircraft guns multiplied like mushrooms." However, FAP reconnaissance failed to account for all air defense systems in the area, and when Portuguese paratroops – themselves part of the FAP – landed, they found themselves under intense fire, lacking reinforcement or air support due to the unexpectedly heavy anti-aircraft opposition.[55]

The Portuguese paratroops were ultimately forced to withdraw without achieving their objectives, but the PAIGC, nonetheless, concluded that its extant anti-aircraft guns would grow increasingly vulnerable to discovery and destruction. By 1970, insurgent air defense units had begun fielding more powerful 37mm anti-aircraft cannons, but these, too, proved insufficiently mobile and, thus, vulnerable to Portuguese air attack.[56] The PAIGC consequently began to look for more portable, flexible, and concealable weapons. During 1972, news reports from Vietnam suggested a solution: the Soviet *Strela-2* man-portable air defense (MANPAD) missile (NATO: SA-7 "GRAIL").[57] PAIGC leader

54 *Ibid.*, 109–111. The PAIGC air defense threat in 1965 and 1966 – and FAP efforts to counter it – are described in detail by Major General José Duarte Krus Abecasis, Portuguese Air Force, in his book *Bordo de ataque: memórias de uma caderneta de voo e um contributo para a história* (Coimbra, Portugal: Coimbra Editora, 1985); 489–494, 502–504, 557–562.

55 Silva, 119; Associação de 25 de Abril (April 25th Association), "Operação Vulcano," http://www.guerracolonial.org/index.php?content=396.

56 Comando-Chefe das Forças Armadas da Guiné, Quartel General, Repartição de Informações, "SUPINTREP Nº. 35 – Aditamento nº. 1 (evolução da actividade e da situação do IN no TO referente a 1970), 20 April 1971, file F2/SR007/SSR1/61/35, Arquivo da Defesa Nacional, Paço do Arcos, 10, 18.

57 For examples of American press coverage of SA-7 employment in Vietnam, see Sydney H. Schanberg, "Foe's New Missile Disrupts Air Tactics in South," *The New York Times*, 28 May 1972; "New N. Viet Missile Causes Changes in U.S. Pilots' Tactics," *The Washington Post, Times Herald*, 15 July 1972; and Malcom W. Browne, "New Aircraft Tactics Reducing Threat of Heat-Seeking Missiles in Vietnam," *The New York Times*, 30 July 1972.

Amílcar Cabral correspondingly made two visits to Moscow in July and in December 1972 to lobby his Soviet benefactors for delivery of this "precious weapon," which had publicly proven its worth earlier that year over the battlefields of Vietnam.[58] Cabral's second visit evidently sealed the deal, for his 1973 New Year's message – Cabral's last public address before his subsequent assassination – contained a tantalizing hint of what was to come:

> We must acknowledge that the enemy, possessing more planes and helicopters provided by its NATO allies, has significantly increased its bombing...we could and should have done more and better. This is what we must do and certainly will do in 1973, *especially as we are going to use more powerful arms and other weapons of war on all fronts* [emphasis added].[59]

Portuguese intelligence and security services had been aware of possible PAIGC *Strela* acquisition as early as 1972, but the missile's introduction in Guinea-Bissau the following year came as a surprise to both FAP aircrews and senior Portuguese commanders.[60] The PAIGC first used the missile on 20 March 1973, targeting, but missing, a two-ship flight of G.91s; multiple unsuccessful *Strela* firings against a Do-27 and a pair of G.91s followed three days later.[61] PAIGC *Strela* teams scored their first success on 25 March, downing a G.91 piloted by Lieutenant Miguel Pessoa. "It was a surprise," Pessoa afterward recounted. "Perhaps there was already information regarding the existence of the *Strela*, but we [pilots] had no knowledge." (See Illustration 12.2)[62] Five additional *Strelas* were launched against FAP aircraft later that day, but

58 Amílcar Cabral to Pedro Pires, 9 July 1972, Documentos Pedro Verona Pires, folder 07166.008, Fundação Mário Soares, Lisbon, 1; Luis Cabral, *Crónica da Libertação*, 433.

59 Amílcar Cabral, "A Sua Última Mensagem ao Povo da Guiné e Cabo Verde (Janeiro de 1973)," January 1973, Ronald H. Chilcote Collection, University of California-Riverside Library, 4–5.

60 Direcção Geral de Segurança, Delegação da Guiné, "Actividades do 'PAIGC,'" 3 April 1973, PIDE/DGS Collection, PC 6992/6993, folder 5, Arquivo Nacional Torre de Tombo, Lisbon, 2; Dalila Cabrita Mateus, *A PIDE/DGS na Guerra Colonial (1961–1974)* (Lisbon: Terramar, 2004), 307–309; Fraga, 111.

61 Direcção-Geral de Segurança, Delegação em Moçambique, "Relatório: Armas anti-aéreas inimigas," 10 April 1973, PIDE/DGS Collection, PC 332-CR (2), NT 6992–6993, folder 5, Arquivo Nacional Torre de Tombo, 1; Direcção Geral de Segurança, Delegação da Guiné, "Actividades do 'PAIGC,'" 1; Lisbon; Lieutenant Colonel José Manuel Pinto Ferreia, Portuguese Air Force, "A introdução do 'strela' na guiné," *Luís Graça e Camaradas da Guiné* (blog), 9 February 2009, https://blogueforanadaevaotres.blogspot.com/2009/02/guine-6374-p3859-fap-6-introducao-do.html.

62 Miguel Pessoa, "Um 'Fiat' abatido," in *A guerra de África (1961–1974)*, vol. 1, ed. José Freire Antunes (Lisbon: Temas e Debates, 1996), 987.

ILLUSTRATION 12.1 The Strela-2M shoulder-fired missile.
CREDIT: DEFENSE VISUAL INFORMATION SERVICE, WWW.DEFENSEIMAGERY.MIL; IN THE PUBLIC DOMAIN.

without further success.[63] Within two weeks, however, four more Portuguese aircraft – accounting for a total of some 10% of the FAP's operational inventory in Guinea-Bissau at the time – had fallen to the new, deadly, and frustratingly mysterious *Strela* missiles. (See Illustration 12.1)[64]

63 Direcção Geral de Segurança, Delegação da Guiné, "Actividades do 'PAIGC,'" 1.

64 Estado Maior do Exército, 2a Repartição, "Relatório periódico sobre a situacão no Ultramar No 04/73 (01 a 30 Abr 73)," May 1973, Annex A, report index 1827/NI, Annex A, file SGDN Cx 761, Arquivo da Defesa Nacional, Paço do Arcos, 1; Direcção-Geral de Segurança, Delegação em Moçambique, "Relatório: Armas anti-aéreas inimigas," 1; Fereira, 591–592.

One of these engagements, on 27 March 1973, claimed the life of Lieutenant Colonel Almeida Brito, the commander of all flying units at Bissalanca/BA12. Brito, with Miguel Pessoa as his wingman, had been targeted by *Strela*s during the missile's African debut on 20 March. Even after Pessoa's downing, however Brito continued to insist that the offending weapons must have been rocket-propelled grenades (RPGs) rather than MANPAD missiles. Manuel Catarino, *As Grandes Operações da Guerra Colonial, 1961–1974, XIV: O Terror does Mísseis – Guiné, 1973* (Lisbon: Correio da Manhã, 2009), 13–14; Pessoa, "Um 'Fiat' Abatido," 987; José Pinto Ferreira and Miguel Pessoa, "A Introdução do Míssil Russo SAM-7 Strela no CTIG," *Luis Graça & Camaradas da Guiné* (blog), 9 February 2009,

ILLUSTRATION 12.2 The rescue of Lieutenant Pessoa, 26 March 1973.
CREDIT: PHOTO COURTESY OF MIGUEL PESSOA; USED WITH PERMISSION.

The FAP's initial exposure to those weapons precipitated immediate and dramatic tactical changes mandated by higher headquarters;[65] additional guidance further limited the FAP's operational flexibility toward the end of

https://blogueforanadaevaotres.blogspot.com/2009/02/guine-6374-p3859-fap-6-introducao-do.html.

65 General Spínola's Directive 20/73 (May 1973), for example, ordered all aircraft operate in pairs at a minimum, instituted a general prohibition against flight operations between 200 and 6,000 feet, and directed compulsory maneuvering if aircraft did fly within those altitudes. In addition, the directive mandated spiral take-offs and descents for all aircraft and forbade repeated use of the same flight routes, "at least not within short periods of time." The directive further prohibited cargo drops by parachute from transport aircraft, cancelled all flights to several remote fields serving Army garrisons, significantly reduced the number of recovery and CASEVAC missions (as well as the locations where they could be conducted), and placed limitations on sortie durations. Comando-Chefe das Forças Armadas da Guiné, "Directiva N.º 20/73: Condições em que é executado o apoio aéreo às FA do TO face à evolução de situação," 29 May 1973, AHM/Div/2/4/228/2, Arquivo Histórico Militar, Lisbon, *passim*.

MAP 12.1 Confirmed *Strela* launches in Guinea: 1) 20 March 1973, 2). 22 March 1973, 3). 25 March 1973, 4). 28 March 1973, 5). 6 April 1973, 6). 31 January 1974.
CREDIT: UNITED NATIONS MAP 4063, REVISION 3, JUNE 2004; PUBLIC DOMAIN.

1973.[66] Consequently, Portuguese airmen, like their American counterparts in Vietnam the previous year, largely avoided the *Strela*'s engagement envelope and operated outside of the missile's effective altitude and range (See Map 12.1).[67] As Lieutenant António Graça de Abreu noted at the time,

> The pilots know the characteristics of the air-to-ground missiles, the Strella [sic] or SA-7, that they are effective between 200 and 2,000 meters

66 Comando-Chefe das Forças Armadas da Guiné, Quartel General, 3ª. Repartição, "Directiva para o apoio aéreo," 30 November 1973, file SGDN-1666, Arquivo da Defesa Nacional, Paço do Arcos.

67 See, for example, "Enemy View of Air: The Air Defense Buildup – Introduction of the SA-7" (briefing slides, Headquarters Pacific Air Forces, Hickham Air Force Base, HI, 26 May 1972), USAF Collection, Air Force Historical Research Agency, Maxwell AFB, AL; Charles A. Nicholson, "Project CHECO Southeast Asia Report – The USAF Response to the Spring 1972 NVN Offensive: Situation and Deployment" (Saigon: Headquarters Seventh Air Force, 1972), 22; Donald J. Mrozek, *Air Power and the Ground War in Vietnam* (McLean, VA: Pergamom-Brassey's, 1989), 92; and Major A. J. C. Lavalle, USAF, ed., *Airpower and the 1972 Spring Invasion* (Washington, DC: Office of Air Force History, 1985), 43–44.

> altitude, and they take the necessary precautions. The Dorniers and the helos fly very low, to *scrape*, close to the trees, swamps, and rivers, while the Nordatlas [sic] and DC-3 flies very high.[68]

Such counter-tactics and security measures, however, seriously compromised the FAP's effectiveness in providing support to Portuguese surface forces.[69] Aerial logistics, arguably the FAP's most critical contribution to the wider war effort, was drastically curtailed: 36 airstrips in Guinea were placed off-limits to Do-27 light transports, while medium cargo aircraft such as the C-47 and Noratlas were restricted to the use of just four airfields throughout the theater.[70]

Airborne reconnaissance likewise faced severe restrictions that curtailed its effectiveness. Before the *Strela* appeared, Do-27 aircraft – the FAP's primary aerial observation platform – typically flew hours-long sorties in low, slow orbits over familiar terrain. The frantic treetop passes Abreu described denied airborne observers the ability to assess the subtle changes characteristic of guerrilla activity, such as new footbridges, forest trails, or other evidence of increased movement or presence.[71] Such tactics also degraded the Do-27's long-established usefulness as an airborne command post (*Posto de Comando Volante*).[72]

Even more deleteriously affected were the FAP's fighter-bombers. The T-6, Portugal's ground-attack workhorse earlier in the war, was eliminated from its

68 Lieutenant António Graça de Abreu, Portuguese Army, *Diário da Guiné: Lama, Sangue e Água Pura* (Lisbon: Guerra e Paz, 2007), 124–125.

69 Comando-Chefe das Forças Armadas da Guiné, "Acta da reunião de commandos realizada em 15 de Maio de 1973," file AHM/DIV/2/4/314/2, Arquivo Histórico Militar, Lisbon, 3.

70 Comando-Chefe das Forças Armadas da Guiné, "Acta da reunião de commandos realizada em 15 de Maio de 1973," 4.

71 According to the FAP's published tactical procedures manual, promulgated just five days before the *Strela* made its theater combat debut, Do-27 aircraft conducting visual reconnaissance were charged with investigating "the existence of trails, changes [on the ground], beaten terrain under trees, deforestation, crops, canoes, and treading at the edge of the forest" – visual cues best observed in the course of low, slow flight profiles. Comando-Chefe das Forças Armadas da Guiné, Quartel General, Repartiçao de Operações, "NEP – Operações (2º volume) – Alteração Geral nº 7," 15 March 1973, F2/92/306/5, Arquivo da Defesa Nacional, Paço do Arcos, 6–9.

72 Comando-Chefe das Forças Armadas da Guiné, "Acta da reunião de commandos realizada em 15 de Maio de 1973," 4; Mario Canongia Lopes, "High-Winged Workhorses: Broussards and Dorniers in Portuguese Service," *Air Enthusiast* 73 (May–June 1998): 42–44; Riccardo Niccoli, "Portuguese Numerology: Serial Systems Used by the Aeronautica Militar and the Força Aérea Portuguesa," *Air Enthusiast* 75 (May–June 1998): 38.

onetime bread-and butter role.[73] The sole remaining alternative was the Fiat G.91 fleet, which was itself required to adopt single-pass, high-altitude delivery profiles using bombs only – rocket and machinegun firing profiles were deemed too dangerous in potential *Strela* areas of operation.[74] While such precautionary measures reduced risk, they also compelled the jets to operate where one observer deemed them to be "too high to be effective in bombing."[75] When FAP G.91s attempted extremely low-altitude ingress tactics, coupled with extreme egress maneuvers following weapons release, the results in terms of accuracy proved equally dismal.[76]

Operationally speaking, then, within two months the PAIGC's SA-7s had "seriously reduced the effectiveness of the Portuguese air force"[77] given, according to senior FAP commanders, "the explosive increase in the enemy's ability to neutralize our heretofore total freedom of aerial action."[78] The subsequent impact on the ground forces has been described as "devastating."[79] Senior Portuguese officials echoed this assessment, including Prime Minister Marcello Caetano, who noted that with the "unexpected" appearance of the *Strela* "our unchallenged air superiority, which had been our trump card and the basis of our entire military policy had suddenly evaporated."[80] "The situation deteriorated dramatically," Spínola agreed. "The introduction of the missiles opened, without any doubt, a new phase of the war."[81]

That "new phase" pivoted on the terrifying realization that Portugal's sole remaining military advantage vis-à-vis its insurgent opponents had been negated overnight. As another Portuguese officer in Bissau lamented, "Our planes no longer completely dominated the skies. That easy and constant cruising of the air space had suddenly become very dangerous."[82] Abreu concurred in his

73 Comando-Chefe das Forças Armadas da Guiné, "Acta da reunião de commandos realizada em 15 de Maio de 1973," 5.

74 Comando-Chefe das Forças Armadas da Guiné, "Acta da reunião de commandos realizada em 15 de Maio de 1973," 5.

75 Teletype, 100930Z APR 74, American Embassy Dakar to Secretary of State, 10 April 1974; João de Melo, *Os Anos da Guerra, 1961–1975: Os Portugueses em África; Crónica, Ficção e História* (Lisbon: Publicações Dom Quixote, 1988), 118.

76 Lopes, 68.

77 Chabal, *Amilcar Cabral*, 103.

78 Comando-Chefe das Forças Armadas da Guiné, "Acta da reunião de commandos realizada em 15 de Maio de 1973," A-1.

79 Chabal, *Amilcar Cabral*, 103.

80 Piero Gleijeses, "The First Ambassadors: Cuba's Contribution to Guinea-Bissau's War of Independence," *Journal of Latin American Studies* 29, No. 1 (February 1997), 78.

81 Cited in Gleijeses, 78.

82 Gleijeses, 78–79; Davidson, *The People's Cause*, 192.

diary entry of 29 April 1973, when he wrote "Aviation is almost paralyzed. Without aircraft, without helicopters, life for all of us [ground forces] has become complicated." Indeed, all FAP missions were curtailed to a greater or a lesser extent – air support, observation, aerial resupply, and – perhaps most critical from a soldier's perspective – casualty evacuation.[83] "For all this," Abreu said of the drastic reduction in FAP support, "the soldier walks frightened."[84]

This last line hinted at the tremendous psychological impact of the *Strela* on Portuguese forces in Guinea-Bissau. "The psychological shock suffered by our pilots was spectacular," according to a Portuguese officer in the provincial capital, Bissau.[85] By 6 April 1973, less than two weeks after its first successful engagement, the *Strela*'s success had "generated a climate of uncertainty among the pilots." The weapon's s not yet known parameters and demonstrated effectiveness consequently engendered aircrew mistrust in the survivability of their aircraft.[86] Abreu related the reaction of one Do-27 pilot four days after the first *Strela* kill: "[He] seemed worried to me, scared. We had lunch together, [but] he almost did not touch his food, he fixed his eyes on nothing." The pilot in question, a conscript aviator named Baltazar da Silva, was killed by a *Strela* eight days later.[87]

Determined leadership helped to slow the deterioration of aircrew morale, as notably demonstrated by Colonel Gualdino Moura Pinto, then-commander of the FAP contingent in Guinea. He addressed his command following the first spate of losses to allay their fears of the *Strela*, and succeeded in "fortifying" their morale by volunteering to fly a high-risk medical evacuation mission himself.[88] Nonetheless, the *Strela* had already dealt "a very serious blow to the morale of the pilots, accustomed until then to flying with minimal difficulty."[89]

The impact of the *Strela* on the ground forces' morale, though more indirect, met or exceeded the psychological dislocation experienced by

83 Miranda, 598; Dhada, *Warriors at Work*, 50; Abreu, 161. One Portuguese casualty, a Navy fusilier, suffered a particularly grim fate even after death: since his colleagues wished to return their colleague's body to Portugal, they refused to bury him and instead opted for a "field embalming" using quicklime and candle wax to preserve the corpse while they waited for the PAIGC's siege of the Guidage garrison to end, one way or the other. Francisco de Vasconcelos, "O 'Inferno' de Guidage," *Público*, 5 November 1995.

84 Abreu, *Diário da Guiné*, 96.

85 Cited in Gleijeses, 78–79.

86 Fraga, 112.

87 Abreu, Diário da Guiné, 91.

88 Fraga, 112. According to Lieutenant General Abecasis, Pinto was relieved of his command and placed in disciplinary detention for this action. Abecasis, 821.

89 Josep Sanchez Cervelló, "La inviabilidad de una victoria Portuguesa en la Guerra Colonial: el caso de Guinea-Bissau," *Hispania*, September–December 1989, 1039.

Portuguese airmen. As aforementioned, the FAP's defensive tactics against the *Strela* curtailed both the frequency and effectiveness of its ground support operations – support on which the Portuguese Army had grown increasingly, and dangerously, dependent.[90] As described by Portuguese Army Lieutenant Colonel Jorge Golias,

> The missiles completely changed the situation on the battlefield. One entered a situation, I do not say of total panic, but a certain fear, deep preoccupation with one's imminent defeat, of being walled in, blockade, immobility. And there came a time during which soldiers would not leave their garrisons.[91]

Lieutenant Abreu likewise lamented that, "If the PAIGC men down an aircraft or a helicopter, we [ground forces] are all screwed because air support is suspended yet again."[92]

The tactical, operational, and psychological effects of the *Strela* subsequently contributed to a string of PAIGC military victories, most notably against the garrison at Guiledje, the key to Portugal's defense of southern Guinea-Bissau (See Map 12.2).[93] Denied effective air support, Guiledje fell to PAIGC forces within a week, clearly demonstrating to the Portuguese that "without adequate air cover they could no longer resist the increased strength of the PAIGC army."[94] The insurgents repeated their success with an assault on another major fort at Copa in early 1974. The FAP attempted to support that besieged garrison on 31 January, only to lose another G.91 to a *Strela*.[95] As recounted by one senior PAIGC figure, Aristides Pereira, "With the loss of the colonial forces' air superiority, the possibility of a military collapse for them became likely.... [PAIGC units] began to attack forts by day, laying siege."[96] In Guiledje and Copa, they succeeded, and, although similar operations against Portuguese

90 As Portuguese military leaders observed in the spring of 1973, "From the military aspect, the main part of the fight belongs to the surface forces which cannot, however, guarantee success without continuous and effective air support." *Acta da reunião que teve lugar na Biblioteca do EMFA, pelas 09H30 de 24 de Abril de 1973 sobre a ameaça antiaérea no Ultramar,"* 24 *April 1973, file F3/17/37/64, Arquivo da Defesa Nacional,* Paço do Arcos, (3).

91 Cervelló, 1039.

92 Abreu, *Diário da Guiné*, 124–125.

93 Melo, 118.

94 Dhada *Warriors at Work*, 50; Chabal, *Amilcar Cabral*, 104. This "increased strength" included 122 mm rocket launchers, 130 mm cannons, 120 mm mortars, and armored vehicles from the Soviet bloc.

95 Chabal, *Amilcar Cabral*, 53.

96 Aristides Pereira, *O Meu Testemunho: Uma Luta, Um Partido, Dos Países* (Lisbon: Editorial Notícias, 2003), 205. Pereira was later the first President of independent Cape Verde (1975–1991).

MAP 12.2 Map showing 1). Guiledje, 2). Guidage, 3). Gadamael, and 4). Copa.
CREDIT: UNITED NATIONS MAP 4063, REVISION 3, JUNE 2004; IN THE PUBLIC DOMAIN.

forts at Guidage and Gadamael were ultimately unsuccessful, they inflicted significant Portuguese casualties.[97] As the Portuguese command had feared, their insurgent foes had made the anticipated but unwelcome transition from guerrilla harassment to conventional maneuver warfare, following Portugal's aerial edge being blunted.[98]

Following the fall of Copa on 12 February 1974, an outright Portuguese defeat in Guinea-Bissau appeared inevitable.[99] Portuguese commanders increasingly concurred that the loss of air supremacy increased the PAIGC's relative military might while worsening Portuguese morale – already tenuous after 11 years of fighting – to the point that outright defeat became an increasingly

97 Lieutenant Colonel João José Brandão, Portuguese Air Force, *Cronologia da Guerra Colonial: Angola-Guiné-Moçambique 1961–1974* (Lisbon: Prefacio, 2008), *passim*. Brandão's work includes a list of Portuguese military deaths (by disease, accident, or combat) in Africa at the end of each of his chapters, which are organized by year from 1961 to 1974.

98 Estado-Maior do Exército, Repartição do Gabinete, "Relatório da missão á Guiné e Cabo Verde do Chefe do Estado-Maior do Exército," 27 March 1974, report index Pº. 324.410–324.510-223.230 Nº. 4805, Arquivo da Defesa Nacional, Paço do Arcos, 7.

99 Chabal, *Amilcar Cabral*, 53.

likely outcome.[100] As Spínola reported in a dispatch to Lisbon during the "Year of the Missiles,"[101] "Nothing more is possible within the province but to wait calmly for the end. Briefly, we are drawing ever closer to the contingency of military collapse."[102] While one cannot quantify the precise extent to which the PAIGC's *Strela* use contributed to this conclusion, one can assert that the introduction of the shoulder-fired missiles was critical to the timing and rapidity of that collapse.

The realization of Spínola's warning would not be long in coming, though it developed in an unexpected manner.[103] The increasing climate of frustration and defeatism in Guinea has been accepted by most historians as one of the primary catalysts leading to the armed forces' *coup d'etat* of 25 April 1974, the so-called "Captain's Revolt" which overthrew the *Estado Novo* regime and engineered Portugal's rapid withdrawal from Africa.[104] As General Costa Gomes, former chief of the Portuguese armed forces and one of the coup's architects, famously remarked, "Our armed forces have reached the limits of neuro-psychological exhaustion."[105] While the PAIGC's *Strela* missiles were in no way exclusively responsible for that condition, those missiles' use at a critical moment in the war for Guinea persuaded Portuguese servicemen, to a sufficient degree, that their last remaining advantage had been weakened, if

100 Chabal, "National Liberation in Portuguese Guinea, 1956–1974," 75–99, 84; Chabal, *Amilcar Cabral*, 95, 104.

101 As christened by Manuel Catarino, 18.

102 Quoted in David Martello, *1974: Cessar-Fogo em África* (Mem Martins, Portugal: Publicações Europa-América, 2001), 24.

103 One prolific commentator on Portugal's war in Guinea noted that "When at last it arrived, eighteen years after six men had founded the party, eleven years after the party had launched armed struggle, the final success came easily, even gaily, above all surprisingly." Basil Davison, "Guinea-Bissau and the Cape Verde Islands: The Transition from War to Independence," *Africa Today* 21, No. 4 (Autumn 1974): 5.

104 See Henry Giniger, "War in Africa Eroded Stability," *The New York Times*, 26 April 1974; Kenneth Maxwell, "The United States and the Portuguese Decolonization (1974–1976)," working paper, Instituto Portugûes de Relaçãos Internacionais, Universidade Nova de Lisboa, 2002, 11; John Woollacott, "A luta pela libertação nacional na Guiné-Bissau e a revolução em Portugal," *Analise Social* XIX, Nos. 77–78–79 (1983): 1131, 1139–1140; and Davison, 8–9. Woollacott explicitly contends that the "Armed Forces Movement" responsible for the 1974 coup in Portugal "originated in the inhospitable marshes and savannahs of Guinea." He further states that "It was in *Guinea* that this revolt was conceived, inspired and organized." For his part, Davison observed that "The kernel of the Armed Forces Movement… took shape first in Guinea-Bissau, and was organized here with a political acumen which developed its structures fight down through the ranks, something that failed to develop when the AFM duly appeared in Portugal."

105 Al J. Venter, *The Chopper Boys: Helicopter Warfare in Africa* (London: Greenhill Books, 1994), 74.

not completely neutralized. The destruction of a half-dozen aircraft led to this perception, and a lightweight, user-friendly, shoulder-fired SAM led to those losses. As Costa Gomes later recollected, with the effective negation of Portuguese air superiority "the war was lost."[106]

In retrospect, the disproportionate impact of the FAP's losses in 1973–1974 strongly suggested an inappropriately excessive reliance on airpower. The Portuguese command in Guinea-Bissau had evidently persuaded itself that its meager Air Force contingent could compensate for growing inferiority vis-à-vis the PAIGC on the ground; in the words of a former FAP general officer, "They thought it was possible to accomplish everything with air support."[107] Airpower had been viewed as a relatively low-risk, low-cost alternative to the manpower commitments required to combat a growing insurgency.[108] Once that air cover was threatened to the point of neutralization, however, the entire strategy collapsed, and arguably brought the Portuguese African empire – and regime – down with the *Strela*'s six victims.

Yet the FAP's record in Guinea-Bissau demonstrated that, when destruction was the primary aim, even a small complement of airpower can accrue disproportionate advantages. It was certainly true that the support functions of airpower proved critical to the aggregate Portuguese effort in the territory from 1963–1974. Yet, the Portuguese learned – as American air arms discovered in Vietnam – that airpower practitioners always faced the risk that unforeseen novelties, whether technological or tactical, could upset the delicate balance between air superiority and aerial denial.[109] This lesson rang especially true in

106 João Paulo Guerra, *Memória das Guerras Coloniais*, 2nd ed. (Porto, Portugal: Edições Afrontamento, 1994), 221.

107 Fereira, 592.

108 Despite liberal employment of the "airpower expedient," more than a million Portuguese served in the *Ultramar* wars from 1961 to 1973. In fact, Portugal's military manpower commitments amounted to six percent of the labor force (see Pedro Lains, "An Account of the Portuguese African Empire, 1885–1975," *Revista de Historia Económica* 15, No. 1 (January 1998): 258). From another perspective, "On the eve of the regime change, Portugal's population of eight million was fielding a full 282,000 troops. Since the end of the Second World War, only Israel and the Vietnams have had such a large percentage of their populations under arms." Nancy Bermeo, "War and Democratization: Lessons from the Portuguese Experience" (*Democratization* 14, No. 3 [June 2007], 388–406), 391. This would equate to a US military force of over 10.6 million in 2010, based on that year's census data.

109 Prof. John F. Guilmartin, Jr., of The Ohio State University notes that "a modest increment of technological advantage applied to the right operational niche can make a difference tactically and, at least potentially, strategically...When the low end niche advantage is high tech, the results can be even more telling, at least over the short term, even against a high tech, high end opponent, particularly when applied by surprise." John F. Guilmartin, Jr., "Technology and Asymmetrics in Modern Warfare," in *Challenging the United States*

those instances where air superiority was initially won by default, and where airpower practitioners consequently dismissed the prospects of an effective defensive riposte.

Finally, and perhaps most importantly, the war for "Portuguese" Guinea reinforced the hard-won lesson that counterinsurgency was never just about mere destruction or military activity writ large. Portugal's campaign vividly demonstrated the dangers of over-emphasizing military power to achieve COIN ends – and the corresponding folly of relying on one component of that power. Granted, Spínola attempted a respectable range of social and political reforms, but these might justly be described as a case of "too little, too late." Despite his efforts, the regime's policy remained mulishly committed to efforts to destroy the PAIGC via purely martial means in order to maintain full metropolitan sovereignty over Guinea-Bissau.[110] With the abrupt and unexpected loss of Portugal's final "trump card," however, Guinea became the first figurative domino to fall, knocking the rest of the Portuguese empire – along with a regime that had ruled Portugal for nearly six decades – down with it.[111] As one Portuguese general officer aptly concluded, many years later, Lisbon's COIN campaign in Guinea-Bissau ultimately saw Portugal lurching "from military victory to military victory, until the final political defeat."[112] That defeat was inflicted by Portugal's own forces who, after 13 years of fighting, had lost their last hope for any degree of honorable return on their sacrifice.

---

*Symmetrically and Asymmetrically: Can America Be Defeated?* ed. Lloyd J. Matthews (Carlisle Barracks, PA: US Army War College – Strategic Studies Institute, July 1998), 37.

110 In fact, Portuguese Prime Minister Caetano ordered Spínola to suspend his diplomatic efforts to seek a political solution in 1972. See MacQueen, "Portugal's First Domino, 223.

111 General Carlos Fabião, Portuguese Army, "A descolonização na Guiné-Bissau – Spínola: a figura marcante da guerra da Guiné" (Lisbon: Associação 25 de Abril, 1984), 12. Prime Minister Caetano had also used the term "trump card" when referring to airpower in the previously cited Gleijeses essay, "The First Ambassadors," 78.

112 Miranda, 599.

CHAPTER 13

# The Vicissitudes Of Violence: Fear, Physiology, and Behavior under Fire

*Edward Coss*

I had the privilege of discussing with Joe Guilmartin, for hours beyond count, the human variable in the technology-and-violence equation. He helped me come to the conclusion that the one constant in the organized hostility of war, regardless of the weapon systems employed, was the unpredictable nature of the human response to the potential for imminent demise. The battlefield is a place of uncertainty. The vicissitudes of combat, the waves of cascading events, linear and nonlinear, place every soldier at the crossroads of mortality and lethality. There, lead and steel threaten flesh in such an overwhelming way as to appear for the combatant to be the impending *"ragnarök"* (the Norse conception of the apocalypse).[1] In battle, a soldier's reactions can be as singular as the causal events. Those reactions are internal, perception-based responses and very much idiosyncratic; the related range of brain chemistry reactions, acquired coping skills, and shaping behavioral norms are so diverse that the concept of a "universal soldier" response to combat loses all meaning.[2]

The extreme stress of combat, or any endeavor wherein death or grievous bodily harm is a potential outcome, initiates certain physical and psychological responses. Yet, how a soldier processes the unsettling and sometimes overwhelming events of a firefight or other combat engagements is highly individualized. The brain chemistry related to the human capacity to identify and interpret threats is an extraordinarily complex phenomenon. No two reactions to danger, even by soldiers undergoing the same event, are alike because each individual's brain deciphers the incoming stimuli differently and sets in

---

1 In old Norse, the "final destiny" or the "doom" of the gods. Accessed 11 January 2016. <http://www.pantheon.org/articles/r/ragnarok.html>. I am indebted to Majors John Spencer, Bobby Sidell, Ryan Canady, and Lieutenant Colonels Bill Staffa and Mike Wiser for acting as catalysts for exploring how and why men behave under stress. Their expertise, their network of combat veterans/friends, and their enthusiasm to understand the human dynamics of battle helped lay the groundwork for much of this paper.

2 This supports John Lynn's mantra that historians should "come to bury the universal soldier, not to praise him." John Lynn, Preface, *Battle: A History of Combat and Culture* (Boulder: Westview Press, 2003), xv.

 | DOI:10.1163/9789004393301_015

motion subtlety distinct biochemical responses, many of which have accompanying symptoms in the form of autonomic reactions. In the seminal World War II survey work of combat soldiers, Samuel Stouffer et al. ascertained that Pacific Theater veterans exhibited a wide range of responses (Table 13.1):[3]

A brief summary of the bio-physiology of fear reveals its intricate nature and the dynamic whereby perception and coping mechanisms put a soldier

TABLE 13.1 Biological Responses to Combat, WWII Survey

| Symptoms In percentages | *Division A* South Pacific (2,095 men) | *Division B* South Pacific (1,983 men) | *Division C* Central Pacific (1,299 men) | *Division D* Central Pacific (643 men) |
|---|---|---|---|---|
| 1. Violent pounding of the heart | 84 | 78 | 74 | 68 |
| 2. Sinking feeling in the stomach | 69 | 66 | 74 | 68 |
| 3. Shaking | 61 | 54 | 53 | 39 |
| 4. Sick at the stomach | 55 | 50 | 46 | 39 |
| 5. Cold sweat | 56 | 45 | 43 | 39 |
| 6. Feeling weakness or faint | 49 | 46 | 36 | 34 |
| 7. Feeling stiffness | 45 | 44 | 43 | 31 |
| 8. Vomiting | 27 | 21 | 18 | 8 |
| 9. Losing control of bowels | 21 | 12 | 9 | 4 |
| 10. Urinating in pants | 10 | 9 | 6 | 3 |

3 Samuel Stouffer et al., *The American Soldier: Combat and Its Aftermath*. Vol. 2 of *Studies in Social Psychology in World War II*. 3 vols. (Princeton: Princeton University Press, 1949–50), 201.

somewhere on the fight-flight-posture-submit continuum.[4] Everything starts with the limbic and cerebral cortexes, with the former providing raw data to the latter, which processes it into visual, auditory, somatosensory (tactile), and olfactory information. The cerebral cortex plays the primary role in perceptual awareness, conscious thought, and problem solving, among other higher-order functions.[5] The sights, sounds, feels, and smells of the battlefield, however fragmentary in nature, are organized into a recognizable, conceptual, three-dimensional environment, with the body as the central point.[6] Sifted, select information is then routed through the frontal lobe, which coordinates related actions and movement; constant feedback regarding spatial, tactile, and motor signals is relayed to the cerebral cortex though the thalamus, the relay station between the midbrain and the cortex. Depending on the data processed, the time frame, and the context in which the information is evaluated, the soldier creates a world view that is unique to his or her own circumstances and temporal interpretation.

This on-going stream of signals is dispatched via fibers to the sympathetic nervous system (SNS), a part of the autonomic nervous system located in the vertebral column; SNS receptors connect to almost every organ in the human body.[7] The SNS automatically activates the distress reaction, or the sympatho-adrenal response, when an individual believes his or her life is in jeopardy. Nerve impulses controlled by the hypothalamus cause the adrenal medulla, the innermost portion of the adrenal glands (that sit just above the kidneys), to convert amino acids (tyrosine and phenylalanine) into the stress hormones epinephrine (adrenaline) and norepinephrine.[8] Further, the hypothalamus

4 David Grossman, *On Killing: The Psychological Cost of Learning to Kill in War and Society* (New York: Hachette Book Group, 1996), 5–7. Grossman argues that the "fight or flight dichotomy applies to a set of choices for any creature faced with danger *other* than that which comes from its own species." Italics are in the original.

5 Patrick McCaffrey, "Cerebral Lobes, Cerebral Cortex, and Brodmann's Areas," *The Neuroscience on the Web Series,* Accessed 11 January 2016. <http://www.csuchico.edu/~pmccaffrey/syllabi/CMSD%20320/362unit4.html>

6 This collation of data is done by the parietal, temporal, and occipital lobes.

7 The SNS has a primary and ancillary function: to marshal the body's resources during stress (triggering what is often called the "flight or fight," response) and to constantly monitor and help maintain the internal environment of the body (homeostasis). "Sympathetic Nervous System." *Science Daily*. Accessed 12 January 2016. <http://www.sciencedaily.com/articles/s/sympathetic_nervous_system.htm>.

8 These stress hormones are called catecholamines. Dopamine is also released; this neurotransmitter works to inhibit the pituitary from releasing prolcatin, which is associated with lactation in females. "What is Dopamine," New Medical. Accessed 23 September 2011. <http://www.news-medical.net/health/What-is-Dopamine.aspx>. For an in-depth analysis, see Kim A Neve, ed. *Dopamine Receptors,* New York: Humana Press 2009.

causes the pituitary secretion of the adrenocorticotropic hormone (ACTH). ACTH, in turn, increases the release of corticosteroids (steroid hormones), one of the most important being cortisol.

Epinephrine increases blood pressure, heart rate, airway branch dilation, the constriction of skin and gastrointestinal tract blood vessels (slowing digestion), and decreases metabolism.[9] Norepinephrine also raises heart rate, amplifies the release of glucose, restricts blood flow (vasoconstriction), and heightens attention by affecting the amygdala, which assists in the storing and processing of emotional memory. In short, blood is directed away from the skin and the stomach and toward core muscles, diminishing the chances of bruising and bleeding. At the same time, the liver is converting glycogen to glucose (the type of sugar that can be utilized by the body immediately). Additionally, respiration increases, getting more oxygen to the lungs and to the extremities; pupils become dilated for better vision; and endogenous opiates are released to diminish pain reception, which in turn increases the body's ability to act or flee despite injury.[10] American radioman Bruce Bradley, who participated in the D-Day assault in 1944, captured the effect of such a neuro-chemical surge: "I had been tired, and all of a sudden I was not tired. I was alert."[11]

Simultaneously, the amino acid sequence of Neuropeptide Y (NPY), a neurotransmitter released by the brain in response to stress, slows down the activity in the amygdala, which controls autonomic responses associated with fear and regulates emotional behaviors and feelings.[12] This neurotransmitter generates the feeling of slowed time during moments of duress. There is evidence that generating higher levels of NPY during stressful situations, such as combat, enables soldiers to remain calmer, to think clearly, and to perform effectively.[13] NPY also works to block the transmission of pain signals to the brain, again assisting an individual to fight or run.

---

9 One-hundred and eight consultants, *Moseby's Dictionary of Medicine, Nursing, & Health Professionals* (China, 2009) 48, 656. Also see "Epinephrine," Accessed 1 December 2011. <http://eilat.sci.brooklyn.cuny.edu/newnyc/DRUGS/EPINEPHR.HTM>.

10 J. Douglas Bremer, *Does Stress Damage the Brain? Understanding Trauma-Related Disorders from a Mind-Body Perspective* (New York: W.W. Norton & Co., Inc, 2002), 7.

11 Christopher Hamner, *Enduring Battle: American Soldiers in Three Wars, 1776–1945*, Lawrence: Kansas University Press, 2011, 72.

12 Neuropeptide Y also helps regulate feeding behaviors that include intake, carbohydrate preference, metabolism, and the formation of fat. *Science News Digest for Physicians and Scientists*, "Neuropetide Y and Body Weight," Accessed 9 January 2012. <http://www.bioscience.org/news/scientis/obesity2.htm>.

13 Stephen Rosen, *War and Human Nature* (Princeton: Princeton University Press, 2005), 101. Deane Aikins, a professor of psychiatry at Yale University, is currently exploring the relationship between this hormone and the ability to function in high stress situations.

In conjunction with these biochemical triggers, ACTH, through the release of cortisol, elevates blood sugar, aids in metabolism, and suppresses immune response; it also acts as a diuretic.[14] The latter phenomenon is, on one end of the spectrum, common among athletes, soldiers, and anyone experiencing anxiety before a game, an exercise, or an interview. This response to non-lethal stress is manifested by the need to urinate. During moments of overwhelming alarm, though, the end result can be uncontrolled urination; as previously shown, Stouffer, et al. discovered that in one Pacific theater division, nearly one man in ten urinated in his pants as a result of fear.[15]

On the positive side, cortisol works in conjunction with epinephrine and norepinephrine to generate vivid, long-lasting memories of high emotional value related to the triggering event.

The neuro-chemical responses to extended exposure to a combat environment, however, have a significant and negative effect on a combatant's capacity to function. As Bruce Siddle has observed:

> Functional magnetic resonance imaging (fMRI) scans have demonstrated that as stress increases, the efficiency of the prefrontal cortex breaks down and we become more instinctive or primal in action. A breakdown of analytical processes results in a cascade of deteriorating performance. Perception, analysis, evaluation, precision motor skills, the processing of time that's so critical…[and] memory of past experiences (successful and unsuccessful) are all lost.[16]

The neuroplasticity of the brain, which Elizabeth Stanley describes as its capacity to "generate new synapses and neurons and discard old ones in

---

He has worked with soldiers undergoing Survival, Evasion, Resistance and Escape (SERE) training and female combat veterans. Aalok Mehta, "What makes a 'supersoldier'?" Dana Foundation, 17 February 2009, accessed 30 September 2016. <https://danablog.org/2009/02/17/aaas-what-makes-a-supersoldier/> and n.a., "Researcher Shows Success with PTSD Treatment for Female Veterans," Wayne State University, 10 February, 2015, accessed 30 September 2016, <https://wayne.edu/newsroom/release/2015/02/10/researcher-shows-success-with-ptsd-treatment-for-female-veterans-5530>. Aikins has not published, formally, on either topic.

14 "Cortisol," *Moseby's Dictionary of Medicine,* St. Louis (MO: Elsevier, 2016), 464.

15 Stouffer et al., *The American Soldier: Combat and Its Aftermath*, 201.

16 Bruce Siddle, "The Stress Paradox: Safe and Rational Protocols for Pre-hospital Tourniquet Use," *War on Trauma: Lessons Learned from a Decade of Conflict, Journal of Emergency Medical Services,* Supplement, October, 2008: 28–32, accessed 4 October 2011, <http://www.lawofficer.com/article/training/stress-paradox> Also see Stouffer et al., *The American Soldier,* 285.

response to sensory input from the environment," is a phenomenon with both positive and deleterious effects. These neurochemical changes are the process by which humans learn and adapt. Elevated levels of cortisol over protracted periods, though, also engender permanent changes in brain structure.[17] In his study of the effects of stress on the brain, J. Douglas Bremer described these alterations as "insidious and detrimental."[18] Too much cortisol can skew perception, can impair cognitive performance, and can lead to the following: insomnia, lower immunity, thinning of bones, diabetes, asthma, ulcers, higher blood pressure, weight gain, anxiety, and depression.[19] High levels of cortisol can damage the hippocampus, which affects memory and the ability to learn.[20] Current research on soldiers' stress has revealed that combatants who had been deployed in Iraq exhibited cognitive deficits in terms of memory and focus for more than two months after returning home. In addition, when compared to soldiers who had not deployed, the 654 men and women in question scored lower on cognitive tests, but outperformed their counterparts in reaction times, a brain function related to in-country survival.[21]

Soldiers often suffer from spikes in their cortisol levels immediately after combat and even months later when they recall and relive battlefield experiences.[22] Elevated levels of this hormone may contribute to a cycle of sleep

17 Elizabeth Stanley, "Neuroplasticity, Mind Fitness, and Military Effectiveness," *Bio-Inspired Innovation and National Security* (Washington DC: National Defense University Press, 2010), 258.

18 Bremer, *Does Stress Damage the Brain?*, 7.

19 Stuart Yudofsky and Robert E. Hales (2007). *The American Psychiatric Publishing Textbook of Neuropsychiatry and Behavioral Neurosciences* (New York, 2007) 814–815. This is the 5th edition. Also, *Moseby's Dictionary of Medicine, Nursing, and Health Professions*, 464 and 489–490. These symptoms are also described as part of Cushing's disease and syndrome. Rosen provides supporting evidence. Rosen, *War and Human Nature*, 100, 105, 113. Also see Bremer, *Does Stress Damage the Brain?*, 7–9. Bremer noted that prisoners of war in WWII "were found to be seven times more likely to have had a stroke at some times in their lives than were non-POWS. The original study on stroke was conducted by L. Brass and W. Page, "Stroke in former Prisoners of War (POW)," *Journal of Stroke and Cerebrovascular Disease*, no. 6 (1996): 72–78.

20 Stanley, "Neuroplasticity, Mind Fitness, and Military Effectiveness," 259. Also see Bremer, *Does Stress Damage the Brain?*, 108–110.

21 Jennifer Vasterling, et al., "Neuropsychological Outcomes of Army Personnel following Deployments to the Iraq War," *Journal of the American Medical Association* 296, no. 5 (2006): 519–529.

22 The Charles Morgan, Kevin Colwell, and Gary Hazlett study of soldiers attending the Survive, Evade, Resist, and Escape (SERE) course showed that Special Forces officers produced high levels of NPY during stressful events resulting in calmer responses and more effective stress reactions. After 24 hours the NPY levels returned to baseline readings. Charles A. Morgan, Kevin Colwell, and Gary A. Hazlett, "Efficacy of Forensic Statement

loss, followed by elevated cortisol levels (the following day), and feelings of anxiety that lead to more sleep loss, thus robbing the soldier of sleep's regenerative powers. Sleep deprivation brings about a litany of woes: apprehension, fatigue, lethargy, and depression, in addition to the more obvious cognitive and psychomotor deficiencies, which are considerable.[23] All of these factors contribute to the development of Post-Traumatic Stress Disorder (PTSD).[24] As

---

Analysis in Distinguishing Truthful from Deceptive Eyewitness Accounts of Highly Stressful Events," *Journal of Forensic Sciences*, Vol 56, issue 4 (2011): 1227–1234.

23 June Pilcher and Allen Huffcutt, "Effects of Sleep Deprivation on Performance: a Meta-analysis," *Journal of Sleep Research & Sleep Medicine*, Vol 19, issue 4 (1996): 318–326. In a recent study, Army Rangers and Navy Seals who purposely were sleep deprived during a training exercise performed as poorly on cognitive tests as if they had been sedated or legally intoxicated (.08 percent blood alcohol concentration). Elizabeth Stanley and Amishi Jha, "Mind Fitness: Improving Operational Effectiveness and Building Warrior Resilience," *Joint Forces Quarterly*, no. 55 4th quarter (2009) 145. Also see A. Williamson and Anne-Marie Feyer, "Moderate Sleep Deprivation Produces Impairments in Cognitive and Motor Performance Equivalent to Legally Prescribed Levels of Alcohol Intoxication," Occupational and Environmental Medicineoem.bmj.com *Occup Environ Med*, 57 (2000): 649–655.

24 Post-Traumatic Stress Syndrome (PTSD), though labelled differently across the decades of the 20th-century (shell shock, battle fatigue, etc.), was not identified as a psychiatric diagnosis until 1980, when it was included in the American Psychiatric Association's *Diagnostic and Statistical Manual of Mental Disorders*. Jim Reardon, *Upside: The New Sciences of Post-Traumatic Growth* (New York: Touchstone, 2015) 26. Research on soldiers in the United States military reveals that, not surprisingly, nearly 30 percent of soldiers with extensive combat experience likely have PTSD. Additionally, and significantly, Army suicide rates climbed throughout the conflicts in Iraq and Afghanistan, reaching the highest total in July 2011, the worst month since suicide data was first collected in 1980. The final number of suicides in 2011 suicides fell below those of 2010, but the incidents of domestic violence and sex crimes have jumped. Surveys of combatants, according to Stanley and Dr. Amishi Jha, confirm the existence of long-term stress problems: nearly 50 percent of returning National Guard and reservists, 38 percent of regular Army, and 32 percent of Marines experienced mental health issues. Stanley and Jha, "Mind Fitness," 145. Their summation is built on a fairly extensive survey of the literature linking deployments and mental health issues. Bremer observes that about half of the American civilian population will "experience a traumatic stress at some time in their lives. Of these, about 15% will develop chronic symptoms of post traumatic stress disorder." Bremer, *Does Stress Damage the Brain?*, 19. Perhaps as many as 470,000 service members may be affected. Pauline Jelinek, "Army Reports Decline in Suicides but Rise in Other Destructive Acts," *Washington Post*, 20 January 2012. About half of the 470,000 are in the Army. Mark Thompson, "Army Suicide Rate Hits New High," *Time Magazine, Battleland* (online), 12 August 2011, accessed 3 January 2012, <http://battleland.blogs.time.com/2011/08/12/army-suicide-rate-hits-new-high./> Rates for Army and National Guard nearly doubled from 2009 to 2010, while those on active duty declined. Bob Brewin, "Army Reserve and Guard Suicide Rates more than Doubled Last Year," *Government Executive*, 19 January 2011. Accessed 3 January 2012 <http://www.govexec.com/dailyfed/0111/011911bb2.htm>. According

Bremer observed, "excessive or repetitive stress, [caused] some individuals... [to] develop long-term changes in these same brain systems that mediate memory and stress response...and can no longer properly adapt to new stressors."[25] Prolonged exposure to "war-zone stress" or singularly overwhelming moments increased the probability of a PTSD diagnosis.[26] This situation is immensely

---

to an article in *Army Times* by Michael Hoffman, "Of the 112 guardsmen who killed themselves in 2010, more than half had not deployed. Officials also said unemployment was not a common theme among Guardsmen and Reservists who committed suicide. About 85 percent of the guardsmen and more than half of the reservists who killed themselves had jobs, said Maj. Gen. Raymond Carpenter, the Army National Guard's acting director." Michael Hoffman, "Guard, Reserve Suicide Rate sees Big Spike," 19 January 2011. Accessed 9 January 2012, <http://www.armytimes.com/news/2011/01/army-guard-reserve-suicide-rate-sees-big-spike-011911w/>. Jelinek, "Army Reports Decline in Suicides but Rise in Other Destructive Acts."

25 Bremer, *Does Stress Damage the Brain?*, 6. Also see "Stress: Constant Stress puts your Health at risk." Accessed 4 January 2012, <http://www.mayoclinic.com/health/stress/SR00001>. Veteran Affairs Secretary and retired four-star general Eric Shinseki gave a short speech in 2009 in which he stated that, "Veterans for some unfortunate reason lead the country in homelessness, mental health problems, depression, substance abuse, and suicides." He also observed that Veteran Affairs must assist these men and women to "have a better outcome than they currently have." Jennifer Jenkins, "Charges against Veteran are Infuriating Mental Health Advocates," *The Washington Post*, 21 February 2012. Retsina Meyer and colleagues note that "Humans with PTSD and other anxiety disorders exhibit hyperactivity in the amygdala and amygdala-related processes, such as fear learning to novel stimuli in disorders in laboratory settings, are enhance in humans with PTSD." Retsina Meyer et al, "A Ghrelin-Growth Hormone Axis Drives Stress-Induced Vulnerability to enhanced Fear," *Mol Psychiatry*, no, 19 (12): 1284–1294, 1285.

26 According to researchers Amy Wagner and Matthew Jakupcak, in a study from 1991 of National Vietnam Veterans' Readjustment Study (NVVRS) data revealed that in "all psychiatric diagnosis *except* PTSD [italics in original]," 63 percent of male veterans with high "war zone stress" experiences "had at least one lifetime psychiatric diagnosis," compared to 45 percent with "low to moderate stress exposure; and 30 percent had a current diagnosis, compared to 13 percent of those with low to moderate stress exposure..." The original work was William E. Schlenger et al, "The Prevalence of Post-Traumatic Stress Disorder in the Vietnam Generation: A Multimethod, Multisource Assessment of Psychiatric Disorder," *Trauma Stress*, no. 5 (1992): 333–363. In a related study examining PTSD, it was noted that 36 percent of men exposed to high stress combat "met the criteria for current PTSD compared to eight percent with low to moderate stress." Amy Wagner and Matthew Jakupcak, "Combat-Related Stress Reactions among U. S. Veterans of Wartime Service," *The Oxford Handbook of Military Psychology*, Janice Laurence and Michael Matthews, ed. (Oxford: Oxford University Press, 2012), 17. From 2000 through 2014, there were approximately 128,000 cases of PTSD across all services. Hannah Fischer, "A Guide to U.S. Military Casualty Statistics: Operation Inherent Resolve, Operation New Dawn, Operation Iraqi Freedom, and Operation Enduring Freedom," Congressional Research Service (Washington DC, 2014) 2. The role of Traumatic Brain Injury (TBI) on "cognitive and social functioning" is just beginning to emerge. It is estimated that since 2000 through 2014,

important, as about 61 percent of American casualties in Iraq and Afghanistan have been related to improvised explosive devices (IEDs), a decidedly unpleasant technology that troops on the ground justifiably feel somewhat powerless to neutralize.[27] It makes sense that such an environment sows the seeds of doubt, raises anxiety, and starts the release of cortisol, even with soldiers at home, contemplating a deployment.

How a soldier deciphered the events of a firefight or other combat events, both in the immediacy of combat and afterwards when he or she intellectualized his or her experiences, was dependent upon a plethora of factors: observations during the firefight; evolving circumstances; past experiences; group unity; training; leadership; incident intensity; fatigue; and weapons effectiveness. These variables, alone and collectively, shaped the soldier's appraisal of on-going events in ways that increased, or decreased, his or her effectiveness in combat. When variables changed on the battlefield and within the primary group, the viewpoints of individual combatants shifted, and so did physiological reactions. These variables include things such as the loss of an officer or esteemed comrade, arrival of a replacement NCO, a weapon-system malfunction, bad news from home disconcerting to a key squad member, shortages of supplies or ammunition, and assignment of a dangerous task to a unit just before the end of a difficult deployment. The behavior of individual combatants during times of danger varied considerably based on these kinds of variables.

Major Mike Taylor, recipient of the Army's 2012 MacArthur Leadership Award, reflected on this phenomenon and its related challenges as they pertained to his time as a rifle company commander in Afghanistan: "[Soldier] reactions are different every time…I'd like to honestly say that I can predict how I or anyone else will act in combat, but I can't. It's different in every instance…you can't predict bravery over time under different circumstances."[28] He observed that on one mission he watched one of his SAW (squad automatic weapon) M249 gunners "fearlessly employing his weapon and advancing through a farm field under fire," but that "six hours later, in the same intensity of conflict, not 200 meters away, he was not employing his weapon [at all]; he's hunkered down in survival mode."[29] Taylor also added that one of his

there were over 300,000 cases of TBI across all services, diagnosed from "unclassifiable" to "severe or penetrating." Fischer, "A Guide to U.S. Military Casualty Statistics," 4.

27 Tom Vanden Brook, "IED casualties dropped 50% in Afghanistan in 2012," *USA Today*, 18 January 2013.

28 Captain Michael Taylor, interviewed by author, Command and General Staff College, Ft. Belvoir, 9 March 2012.

29 Ibid. This response might have been due to several factions, most likely related to fatigue, to perceptual awareness, or even to vasoconstriction, driven by the rush of

men went back under heavy fire to rescue a comrade, "acting with complete valor."[30] But, two weeks later, the "valor and fearlessness just wasn't [sic] there. He didn't act like that all the time. Turns out, he was a strong infantryman, dependable more often than not, but not all the time."[31] Taylor concluded that a leader "can't judge by snapshots," because men act differently depending on the situation. Taylor commented that he believed that a soldier's preparations (mental, physical, and psychological) and expectations greatly affected how a man reacted under fire. In the first part of the mission mentioned above, the SAW gunner participated in a proactive assault; in the second, the unit traveled to its forward operating base where it was attacked by nearly 100 men, a number over six times what the platoon was told to expect by the company's intelligence support team.[32]

Personal interpretations of events drove the biochemical responses of these soldiers' bodies. Using various studies testing the physiological reactions of men under great duress, Stephen Rosen found that perception had a profound effect on the body's reaction to uncertainty: "If the subjects feel they are in control, the presence of stressors in the environment creates high levels of adrenaline…and is associated with the subjective feelings of excitement and arousal, not anxiety." Perceived helplessness, on the other hand, resulted in elevated levels of cortisol over extended periods, which caused distress and depression. This distress-induced melancholy, Rosen argued, was "cognitively, behaviorally, and biochemically similar to, if not indistinguishable from, the disease known as depression."[33]

In a shifting and potentially deadly setting, the normative standards set by leadership and the small group fused with training, experience, and weaponry to guide the efforts of the soldiers during battle.[34] Leadership and discipline

norepinephrine, which caused blood to drain from the forebrain. These symptoms have led to humans "freezing" during times of elevated threat. Lt. Colonel David Grossman in an article by Adam Linehan, "This is your Brain on War," *Task and Purpose*, <http://taskandpurpose.com/this-is-your-brain-on-war> (Accessed on 21 June 2016). / This concept works in conjunction with a lack of belief in one's actions, which causes a prolonged release of cortisol.

30 Taylor interview, 9 March 2012.

31 Ibid.

32 Taylor interview, 9 March 2012. Both events took place in October 2010. The soldier in question was awarded a Silver Star.

33 Rosen, *War and Human Nature,* 100, 105, 113.

34 Because of their Byzantine nature, the interconnections between these variables cannot be definitively parsed and individually weighed. Carl von Clausewitz noted the difficulty in quantifying the morale aspects of cohesion, writing that forces contributing to morale "will not yield to academic wisdom. They cannot be classified or counted. They have to be

play a crucial role in combat effectiveness, for soldiers under stress revert to the nearly-automatic actions learned through repetitive training.[35] In his time with the soldiers of the 2nd Battalion, 503rd Infantry Regiment in a remote outpost in Afghanistan, journalist Sebastian Junger discovered that men such as Private James Monroe were capable of "methodically [taking] apart a SAW [squad automatic weapon] while rounds are slapping the rocks all around them."[36] When asked about the role training played in the ability of a soldier to withstand and function in combat, Major John Spencer, a seasoned officer with the 173rd Airborne Combat Brigade, confirmed that, without question, soldiers reverted to the "mechanized muscle movements" they learned through the repetitions of training.[37] This "muscle memory," as soldiers referred to the phenomenon, routinely was cited by combatants as the key default mechanism that kept men/women functioning under fire.

Junger also came to realize that for these seasoned, well trained, and well led men—at that time and place—"combat jammed so much adrenaline" into their systems that they were usually able to sublimate fear.[38] This combination of physiology and rhythmic responses, what soldiers refer to as "muscle memory," helps humans deal proactively with a threat; the action subconsciously elevates a belief in self efficacy, allowing soldiers to continue responding, in varying degrees, during times of duress.

Combat experience also increased situational awareness and provided cognitive data points that allowed decisions to be made more rapidly and, often, with greater appropriateness.[39] This procedure, which has been termed, "intuitive decision-making," permits an individual to progress from "controlled processing," during which the soldier takes in novel situations and slowly works through various solutions, to a more "automatic process" in which high-order tasks can be performed in a fluid and swift manner. It results from the creation of schema that psychologist Michael Matthews defines as "the development

---

seen or felt." Carl von Clausewitz, *On War*, ed. and trans. Michael Howard and Peter Paret (Princeton: Princeton University Press, 1976), 184.

35 Leadership and group cohesion appear to be two factors that act as both sustaining and combat motivations. Also see Charles Moskos' commentary on the role of exterior factors affect unit effectiveness. Charles Moskos, *The American Enlisted Man; the Rank and File in Today's Military* (New York: Russell Sage Foundation, 1970), 147.

36 Sebastian Junger, *War* (Hachette Book Group: New York, 2010), 74.

37 Major John Spencer, interviewed by author, Command and General Staff College, Ft. Belvoir, 30–31 August 2012.

38 Ibid., 73–74.

39 Michael Matthews, "Cognitive and Non-Cognitive Factors in Soldier Performance," *The Oxford Handbook of Military Psychology*, 203–208.

and storage of, and ready access to, complicated knowledge structures...that can be matched to a given situation."[40]

Moreover, the mental and mechanical components of training, coupled with the trust and support of the primary group and a belief in leadership, allows a type of compartmentalization to occur. This set of interactions works to regulate individual brain chemistry in a way that permits each man and woman, in varying degrees, to perform in the extremes of a combat environment. A soldier's mental construct of his or her situation can keep the harrowing "what ifs" of the battlefield at bay. When asked about his time in the Korengal Valley (Afghanistan) and how he handled the stress of that deployment, Private Brendon O'Bryne, also of the 2nd Battalion, 503rd Infantry Regiment, replied that he was "Numb...Wasn't scared, wasn't happy, just f***ing numb. Kept to myself, did what I had to do. It was a weird, detached feeling those few months." As for the fear of dying, he captured the kind of mental dynamics that allowed soldiers to function: "No, I was too numb. I never let my brain go there. There were boundaries in my brain, and I just never let myself go to that spot."[41]

Each of the four officers interviewed for this paper exhibited an ability to minimize thoughts pertaining to their own perilous situations while maximizing situational awareness regarding obstacles to mission success and threats to subordinate soldiers. In each case, the officer prioritized incoming stimulus into a "mission, men, and me" hierarchy of needs and related responsibilities.[42]

These leaders described this capacity to override their survival instincts as a function of forced cognitive processing and repetitive training. When queried the first time about their most daunting combat situations, all four men took several minutes to recall the events and were somewhat surprised that their decisions were driven by certainty in the moment. Upon reflection, each officer assessed that no other options were acceptable, regardless of the personal danger involved.

Lieutenant Colonel (ret.) William Staffa—a helicopter pilot in Vietnam with more than 2,400 combat flying hours, two Silver Stars and three Distinguished Flying Crosses—reminisced on this connection between perceived helplessness and fear. He shared during an interview that he only remembered being really "frightened on one single occasion. What triggered the remembrance was the list of symptoms [I read] in your paper ... I felt stiff ... I thought the

40 Matthews, "Cognitive and Non-Cognitive Factors in Soldier Performance," 206–207.

41 Junger, *War*, 64.

42 This description is gender specific because, at the time, all the soldiers involved were men.

aircraft was having a hydraulic issue. Turned out to be me. It was a situation where I felt I had no control over anything; [it was a] strange circumstance."[43]

Staffa used the term "auto-get 'em" when describing how he channeled his thoughts once the enemy was engaged.[44] Staffa believed that as a combatant you were also "less apt to be afraid if you're on the offense."[45] He mentioned that, mentally focusing on certain tasks while purposely minimizing negative thoughts (fear) became standard for him each time he flew. In subsequent interviews, Staffa described this mental "locking of a door" as nothing special and the only way to take care of business when in flight. It is interesting how technologies become focal points of certainty and comfort in dangerous situations.

Lieutenant Colonel Michael Wiser, a decorated Airborne Ranger, shared several instances of how he sublimated fear in combat during the invasion of Iraq in 2003. In each, he wrote of being afraid as he considered the "what might happen" stuff.[46] For him, the anxiety melted away as he put mission requirements into play. He called it "mission mode," whereby he went into "offensive, not receiving mode." In an attack by 30 to 40 insurgents on his forward operating base (FOB) in July 2003, he recalled the compelling need to be proactive. Due to a compound wall, he could not see "the same angles" as his men on the rooftops; consequently, he was hampered in his ability to direct fire. Concerned about the danger of fratricide, the ongoing threat to his men, and "the fear of failing as a leader," Wiser transformed his distress into a kind of driving anger. Feeling a surge of self-belief and recognizing that the men were following training, Wiser took actions that exposed him to serious danger but enhanced his unit's ability immediately to eliminate the three dozen enemy combatants.[47] His need to "understand the threat" initiated a physiological response, in this case triggered by his pet "get 'em" phrase, which was "You son of a bitch!" Wiser hustled out of the main gate into the fire zone wearing "body armor and maybe a helmet, I'm not sure."[48] Wiser remembered bullets whizzing

43 Lt. Col. William Staffa, email interview by author, Command and General Staff College, Ft. Belvoir, 5 March 2012.

44 Ibid.

45 Ibid.

46 Lt. Colonel Michael Wiser, interviewed by author, Command and General Staff College, Ft. Belvoir, 13 September 2013 and 14 February 2014.

47 Ibid. Wiser noted, though, that "This was *not* like training; you don't get shot at during training."

48 Wiser laughed during the interview when he realized that he almost always used that phrase when taking the initiative against the enemy. Wiser Interview, 13 September 2013 and 14 February 2014.

by and the sound of rocket propelled grenades screaming overhead; yet, his focus was on identifying pockets of enemy fighters so he could guide the fire from his men. As he stood, exposed, calling out targets, he remembered thinking that "failing is forever"; he had only a fleeting recognition that he might be killed at any moment. His unit was successful in repelling the attack with no American casualties.

In remembering that day, Wiser realized that he and his men may have been experiencing what he called "reciprocating [shared] courage," a concept he had never considered before the interview.[49] He also laughed when admitting that his profane "You son of a bitch!" expression was likely his bio-physiological prompt, one that initiated an epinephrine, norepinephrine, and NPY surge. In some ways, this term acted as shield against fear; verbalizing it during moments of distress lent the perception of control in an inherently chaotic environment. In his mind, this enhanced his belief that he could shape the outcome of engagements. The key was his perception that his actions and those of his men, once unleashed, had efficacy. This synchronized his physiological response with his need to confront and eliminate the dangers that threatened his men and, as a corollary, his own life.

Major Spencer experienced similar feelings in July of 2008 when he took over a broken company of the 1st Battalion, 68th Armor Regiment after it survived a near-catastrophic situation.[50] Alpha Company was previously tasked to provide security for an AEROSTAT (a high powered camera mounted on a helium balloon) at Joint Security Station (JSS) Ur, just west of Sadr City, a suburban district in Baghdad. The JSS was, according to Spencer, "a patch of sand fenced in by 20 feet high Alaska [concrete] barriers." Undermanned and living in the shade of their vehicles in 100-degree heat, the company manned concrete towers, pulling rotating shifts of five hours. On the spot leadership options were limited due to troop shortages and mission requirements; thus the officer and his men had to "embrace the suck" and to maintain their position in what was obviously a hazardous outpost. Covered in fine powdered sand, what they called "moon dust," and exhausted from day and night shifts, the men of Alpha company were hit by Improvised Rocket Assisted Munitions (IRAM) hidden in the large semi-trucks that insurgents had parked right next to the JSS, which was itself adjacent "to the largest food distribution storage area in Baghdad"—and therefore always frequented by trucks. The enemy parked a truck adjacent to the outer wall and lobbed nine IRAM, each filled with 50 to

49 Ibid.

50 Major John Spencer, interviewed by author, Command and General Staff College, Ft. Belvoir, 30–31 August 2012.

200 pounds of explosives, over the barricade and into the compound. The result was waist-deep craters, burning trucks, shattered concrete barriers, and the near-complete devastation of the living quarters. The generators burned for hours. When sharing the videos of attack and the aftermath within the destroyed compound, Spencer still was amazed that only three soldiers and one civilian had been wounded.[51]

The next day, Spencer was "dropped off at JSS Ur and unceremoniously assumed command." Spencer quickly realized that his company (Alpha) had discipline and morale issues. Spencer knew that the unit was obviously traumatized by the events of the day before. He also noted deficiencies in standards and processes, organization, and discipline, which in his estimation contributed to the previous day's terrible events. He called it a "Lord of the Flies" situation, in which the men had reverted to extreme survival mentalities, the unit ceasing to function as a collective. Spencer went about reinstating standards and, working with the small pockets of cohesion remaining in the company, began to rebuild it. Following the "mission, men, and me" methodology, Spencer attempted to reconstitute the company even while it was tasked to continue executing combat missions, including complicated combined-arms house raids.[52]

Reflecting back on this period, Spencer recalled that constant leadership demands pushed fear out of his mind. By reducing his sleep time to three or four hours a day, Spencer dedicated the acquired time to constantly checking on all the platoons. Feeling unending pressure to lead and make the unit mission ready, Spencer remembered feeling exhausted and overwhelmed, but rarely afraid, even in combat. He believed that his image of himself was as a leader and the seemingly endless tasks and responsibilities prohibited any time for reflection for the recognition of danger. Like Wiser, Spencer, in recalling his various times under fire, realized that he, too, had a phrase that acted as his biochemical "on switch." Spencer's expression, one that he muttered out loud to himself rather than to others, was "Oh no, you didn't!" Smiling at the recollection, he observed that his belief in the American soldier and the awesome arsenal of firepower at his command, gave substance to this battle mantra.[53] This understated war cry, one that initiated the chain reaction of biochemicals that saturated his system, permitted Spencer to take proactive steps in deadly situations where hesitation could bring disastrous consequences; it allowed him to be the leader that he envisioned in his mind's eye.

51 Major John Spencer, interviewed author, Command and General Staff College, Ft. Belvoir, 30–31 August 2012.

52 Ibid.

53 Ibid.

Spencer conceded that, "I'll admit I've been afraid in combat, afraid of being killed or wounded, afraid of failing my fellow soldiers as their leader. But my worst nightmare of all was being separated from my unit without a weapon. The confidence that most soldiers display in combat comes from rigorous training that, hopefully, triggers instinctive responses that become part of their muscle memory. Yet it also comes from the feel of your weapon in your hands, from the radio you can use to call for other soldiers with weapons."[54] He argued that the idea that "I have more [fire] power than you" gave him what he called "self-efficacy shaped by combat dominance." This perceptual interpretation of battlefield challenges aligned well with the neurophysiology requirements necessary to keep Spencer believing and preemptively acting on those beliefs; his faith in his weapons and his men helped him visualize a direct connection between his actions and combat outcomes, even if this conceptual framework was perceived as illusionary by an outside observer. Spencer's association of his rifle, his radio, and his men's response with what he called "overwhelming American firepower" made sense in tactile, cognitive, and biological terms. Spencer described the role of being "conditioned [to] the sounds, the smells, the sights" after multiple "exposures to danger." He reminisced about "harnessing the power of the .50 cal…feeling the power made me feel better." He stated that such clout "acts as a barrier to negative thoughts, blocking out the chaos and fear."[55]

Spencer described a night ambush on an Iraqi defensive position in late 2003, a difficult task under most conditions. On the way to the objective, Spencer's lead vehicles were attacked on the highway. Uttering his "Oh no you didn't…but I'm glad you did" mantra, Spencer unleased the firepower of his entire platoon. Initially, he said that he felt, "no fear, just all excitement." The growing uncertainty of the situation, however, and his platoon sergeant yelling that he "saw lots of guys popping up behind berms," made Spencer feel like he was losing control. He was slammed with a rush of doubt and a related cortisol dump. His breathing became labored and he could sense the sudden change between the "positive rush and the negative body reaction." He felt an impending "fear of total failure." His descent into the neurophysiological hell of perceived helplessness was mitigated by his training and the men's adherence to "tactics, techniques, and procedures."[56]

Following established tactical methodologies, Spencer directed his group to bound forward, clearing the berms and the objective. Before much destruction

54 Spencer interview, 30–31 August 2012.
55 Ibid.
56 Ibid.

took place, Spencer was informed that, "It's the local Iraqi police…and the police chief," who admitted shooting at Spencer's men first, thinking they were highway robbers. Sorting things out "seemed like a lifetime" to Spencer who was relieved to discover that none of his men were wounded and only five of the policemen were injured. The two nearby police vehicles, however, "were smoked." "We never got the real story," Spencer recalled. Apparently, the police had taken out one of the highway teams who were "stopping vehicles to steal money and kidnap." There was a pick-up truck nearby with two dead civilians, who apparently were "bad cops out of uniform," as well as several AK47s, and "AK rounds all around."[57]

During the many hours of his interview, Spencer recalled the dichotomy of combat as it was experienced by enlisted soldiers and officers. The singular nature of leadership, wherein much of the bonding is stripped away, was, for him, replaced by an allegiance to the mission and the well-being of subordinates. The end result was a different kind of fear; this one based on mission failure and the ongoing worry that soldiers might die because of his choices and the speed with which he could decipher events. Spencer commented quietly that "Being a leader can be lonely." He argued that the enlisted men had each other to help mitigate the stress, but the officer stood apart and only rarely could tap into the small group for emotional support. Spencer returned to this "loneliness" component of battle several times, describing the void he felt when making decisions and the absence of the sustaining camaraderie he had experienced previously as an enlisted soldier. Enlisted soldiers could rely on one another to develop a common belief in what they should do to get through the moment.[58] These collective perceptual constructs helped them to make sense of the combat environment in ways that permitted some freedom of action.

Spencer remembered only occasionally fearing death or personal injury; instead, he noted the difference between proactive "power and excitement" and reactive "chaos, fear of failure, and total loss of control." During the interview, Spencer first mentioned, then reflected on, and finally agreed that his physiological reactions were based on the fluid nature of the encounter and his shifting perceptions of whether or not his actions could, in any way, influence

57 Spencer interview, 31 August 2012.

58 It is worth noting that in the Stouffer study, when enlisted men were asked the question, "Generally, on your combat experience, what was most important to you in making you want to keep going and do as well as you did?" they listed "prayer" and "solidarity with group" as their top motivating forces. Enlisted men identified "leadership and discipline" as the least important incentives. Officers, on the other hand, classified "leadership and discipline" as most significant, followed by "solidarity with group." Stouffer et al., *The American Soldier*, 109.

the situation.[59] Spencer noted that, "The central idea is control. It is my mind's conscious or subconscious belief that there is [such a thing] as control of a situation, control of whether I am safe or not, control of the outcome…In a firefight, I have a belief in control." He grinned when he recalled his fondness for the .50 caliber machine gun, which helped him believe he could exert his will in a given situation: "Just harnessing the power of the weapon, feeling the power, made me feel better." Spencer also commented that "conditioning, the exposure to danger, being around the sounds, smells, etc. helps a soldier from being overwhelmed."[60]

As a soldier who had worked himself up through the ranks, Spencer recalled nostalgically his times in combat as "a Joe" in a Ranger unit. The guidance and encouragement of the soldiers in this assemblage allowed him to grow and to function under difficult circumstances. More than once, he stated that the group shaped him for the next 20 years of his life in the Army.[61] He agreed, without reservation, that the emotional and psychological validation provided by the small group played a particularly important role in helping soldiers cope with combat. Allegiance to this significant social collective, with whom the combatant shared hardship and peril, helped the soldier believe that his or her actions had efficacy. The seemingly equal distribution of danger worked to reduce the apprehension of individual soldiers and allowed each man or woman to function, to varying degrees, with a higher degree of effectiveness than would be the case when facing the danger alone.[62]

In the quintessential work on men in combat, Stouffer et al. discovered that, "The group in its informal character, with its close personal ties, served two principal functions in combat motivation: it *set and enforced group standards* and it *supported and sustained the individual* in stresses he would otherwise not have been able to withstand."[63] This communal sense, which has always

59 Spencer Interview, 30–31 August 2012.

60 Ibid.

61 Ibid.

62 Roger Little, in his analysis of group relations and men under fire in the Korean War, concluded that "the primary basis for solidarity…was the recognition of mutual risk" and "that relationships were established to enhance effectiveness and survival." Roger Little, "Buddy Relations and Combat Performance," Morris Janowitz, ed., *The New Military* (New York, 1964) 218. Some recent work on group processes reinforces Little's contention. A study by Roger Nibler and Karen Harris revealed that "cohesive effects were more likely to emerge in the process of decision making rather than in outcome." Roger Nibler and Karen Harris, "The Effects of Culture and Cohesiveness on Intragroup Conflict and Effectiveness," *The Journal of Social Psychology* 143 (2003): 627. Also see Chapter 6 "Banded Brothers," Edward J. Coss, *All for the King's Shilling: The British Soldier under Wellington* (Norman: University of Okalahoma Press, 2010), 191–210 for an overview of combat dynamics.

63 Stouffer et al., *The American Soldier*, 130–131. Emphasis in the original.

been associated with combat, is linked directly to research on the relationship between belief (self-confidence), motivation, and an individual's capacity to take action. When asked, "When the going got tough, what kept you going?," the most common responses by thousands of combat soldiers in WWII were, "Prayer" and "Thinking that you couldn't let the other man down."[64] This collective spirit lay at the heart of how and why men sublimated their fear and self-preservation instincts to act in ways that safeguarded the group.

An unknown private in the Napoleonic-era British infantry (the 71st Regiment of Line) captured the effect of group resolve on his vacillating morale just before a British charge, timed to unhinge a French attack in column:

> In our first charge I felt my mind waver; a breathless sensation came over me. The silence was appalling. I looked alongst the line. It was enough to assure me. The steady, determined scowl of my companions assured my heart and gave me determination. How unlike the noisy advance of the French.[65]

The significance of this loyalty-based motivation was expressed by a number of American soldiers fighting in Iraq during 2007.[66] Specialist (SPC) Sterling Jones, a member of the 173rd Airborne Brigade, left little doubt about the mechanism he believed guided his behavior under fire:

> I'm not doing this for recognition from my country... Truthfully, I could give a shit what anybody thinks. Except for those guys to my left or my right because that's what it's about... In the end you don't have anything else. Your mama ain't there. Your wife ain't there. Your sisters, your brothers, your cousins, none of those people are there. Those people can't help you."[67]

Another enlisted man in the same conflict expressed his combat motivation as follows: "Everybody did what they had to do. It was just looking out for one another. We weren't fighting for anyone else but ourselves. We weren't fighting for some higher-up who is somebody; we were fighting for each other." His comrade explained why this was the case: "We eat, drink, [and go to the bathroom]

64 Ibid., 177–181.

65 Anonymous, *A Soldier of the Seventy-First*, ed. Christopher Hibbert (London: Leo Cooper, Ltd., 1975) 18.

66 Leonard Wong, et al., *Why They Fight: Combat Motivation in the Iraq War* (Carlisle, PA: U.S. Army War College, 2003) 12–13.

67 Specialist Sterling Jones on being a soldier. Interview. Accessed on 12 March 2011. <http://restrepothemovie.com/video/>.

together—everything—together. I think that it should be like that…I really consider these guys my own family, because we fight together, we have fun together…We are to the point that we even call the squad leader 'dad.'"[68]

This channeling of will helps to protect the group by causing the release of hormones that make further and perhaps life-saving actions possible. Yet, the amorphous nature of behavior under fire, predicated as it is on the fluidity of hurried, snap-shot observations made under often harrowing conditions, makes predicting a soldier's conduct less than precise. As noted in *All for the King's Shilling*,

> [B]iochemical responses help soldiers take actions that preserve each other. The group cohesion that results is a complex phenomenon. Rather than an absolute, all-or-nothing concept, group unity is a fluid and emotionally-sensitive construct predicated on perception. Hierarchical adjustments occur almost continually; individual and collective experiences and observations—and the evolving awareness of individual members regarding threat level, group status, norms, leadership, group effectiveness, and the value of membership—continually cause fluctuations in unity and performance. Thus, as Marshall concluded over forty years ago, "morale is not a steady current…but an oscillating wave."[69]

This brings us back to the cerebral cortex, wherein sensory stimuli are processed and conscious thought initiated. Biochemical responses are all based on the soldier's *perception* of a firefight and the belief in the value of his or her own responses. The forces that affect how a soldier discerns and synthesizes the torrent of sensory input during combat, and the related actions he or she then initiates, must be considered within the context of that unit, at a precise time and in a specific locale. Sensitivity to incidents and the related physiological reactions are dependent upon a discrete set of circumstances.

Thus, as Major Taylor argued, the ever-changing circumstances of combat and all the related violent outcomes make it nearly impossible to predict the behavior of a soldier in battle with any certitude beyond "more likely than not." Events, anticipated and otherwise, break like waves over combatants. Individual interpretations of situational stressors combine with an entire spectrum of belief in self-efficacy—that actions might or might not affect the combat environment—resulting in alterations in brain physiology. Training, leadership, group cohesion, and the mental ability to compartmentalize affect this neurochemistry in ways that allow soldiers to cope, in varying degrees, with

68 Wong, et al., *Why They Fight*, 13.

69 Coss, *All for the King's Shilling*, 194.

the onslaught of stimuli and the very real threat of existential oblivion. Historians attempting to unravel the conduct of men and women in combat across historical eras must acknowledge that there are no personal, social, cultural, or neurological absolutes.

As Major Taylor noted, and Dr. Guilmartin and I discussed over many a fine afternoon, the ever-changing circumstances of combat and all the related violent outcomes make it nearly impossible to predict the behavior of a soldier in battle with any certitude beyond "more likely than not" or "most of the time." We shared a fascination with the often enigmatic human response to the stressors of conflict. Working with "Dr. G." to explore the vicissitudes of combat, the unpredictability of human behavior, and the role of technology as it applies to battlefield perceptions and self-belief, was a privilege. He helped me to see that great care is needed when historians assess the conduct of men and women in battle, as the urge to generalize is omnipresent. His insights as an airman-scholar and his gentle, guiding hand allowed me, as John Keegan phrased it, to make a "personal attempt to catch a glimpse of the face of battle."[70] For that profound gift, and a thousand other kindnesses, I am forever grateful to Joe Guilmartin.

---

70 John Keegan, The Face of Battle (New York: Viking Press, 1976) 78.

CHAPTER 14

# *Des Fusils Aux Idée*: Technological Skepticism and Maculinity in the French Army, 1954–1962

*William Waddell*

The opposite of manliness isn't cowardice; it's technology.
NASSIM NICHOLAS TALEB[1]

*Président de la République* Charles de Gaulle announced a debatably overdue triumph in his New Years' address to the army in 1962. Henceforth, French arms would be dedicated to "modernization," to a return of its units from the winding-down Algerian War, and to a reduction in effectives. Specifically, he had in view the creation of a smaller, more technologically sophisticated force, built to defend the *métropole* against Soviet aggression and backed by the deterrent power of France's relatively new *force de frappe*. It was unwelcome news to many who had spent their careers fighting in the "dirty wars" for France's shrinking overseas empire, particularly the paratroops and the members of the Foreign Legion. Given the failure of the April 1961 putsch against de Gaulle, further resistance on the part of the professional troops *d'outre-mer* (overseas) was out of the question, though, as *Le Monde* journalist Jean Planchais noted at the time, their tempers remained "heated."[2]

Most of this animus came, for these men, from de Gaulle's perceived abandonment of the empire; but, it was also rooted quite strongly in his turn to technology and to Europe. In many ways, and especially in the minds of many of the long-serving colonial professionals, these were closely related phenomena. De Gaulle had quite nearly said as much several years earlier at an address to *l'École Militaire*. In touting the as-yet-to-be-realized *force de frappe* in 1959, de Gaulle spoke broadly and gave a "mission planétaire" to this coming

1 Nassim Nicholas Taleb, *The Bed of Procrustes: Philosophical and Practical Aphorisms* (New York: Random House, 2010), 13.

2 Jean Planchais, "Une nouvelle étape vers une armée modernisée et regroupée en Europe," *Le Monde*, January 1, 1962.

 | DOI:10.1163/9789004393301_016

arm of French power. Nuclear weapons would restore prestige and flexibility to French diplomacy, redraw in brighter colors the dismal failure of the colonial venture, and, eventually, allow France's rupture with NATO. To do these things, however, as observers then recognized, was expensive, and required a fundamental reorientation of France away from endless combat in Algeria. Building de Gaulle's new, modern Army in Europe meant *ipso facto* abandoning the old war in North Africa.[3] This was almost certainly to de Gaulle's liking; he had no more mind to fight the war in Africa, and the cost of making France "modern" provided more ground from which to make that case.

Abandoning that war was more difficult, however, than recognizing new political realities for many of the forces involved. For a small but important minority within the French military elite, these wars (Algeria and Indochina) were almost all that they knew. These were defining events; ones that had inscribed their lessons in blood across two decades, on two continents, and against two tenacious opponents. Though, we often think of Western armies as particularly keen to embrace technology, things were somewhat different for a vocal subset of the French professional soldier during the wars of decolonization.[4] Embracing technology and abandoning their struggle in this context meant betraying who they believed they were as soldiers and as men.

To fight these wars had required the cultivation of specific characteristics, a certain *virtus*, among France's fighting class. Chief among them was a commitment to the *la parole donnée*, the word of honor. Indochina and Algeria were arenas where responsibility mattered; this was a brand of visceral leadership, the "close communion in hardship, danger and death" that Jean Lartéguy put into the mouth of his character, Colonel Raspéguy, the fictional avatar for the famed *para* commander, Marcel Bigeard.[5] Physical courage, trust, and fighting prowess on a near intimate level were the things that mattered in the myriad, bloody encounters that made up these wars. Writing long after the wars, and his own rebellion against the State at the head of a regiment of legionnaires, Hélie Denoix de Saint-Marc recalled that "an army is not a mechanical body…

3 Jean Planchais, "La France doit avoir sa force de frappe atomique déclare le général de Gaulle," *Le Monde,* November 6, 1959.

4 Geoffrey Parker, "The Western Way of War," in Geoffrey Parker, ed., *The Cambridge History of Warfare* (New York: Cambridge University Press, 2005), 1–2. Among the five "foundations" upon which Parker argues Western warfare rests, number one is "heavy reliance on superior technology." Parker goes so far as to argue that the West has been "preternaturally receptive to new technology…"

5 Jean Lartéguy, *The Centurions*, trans., Xan Fielding (New York: Penguin Books, 2015), 295.

it is a living corps. Responsibility and obedience must be accompanied by an elevated conscience, always awake, irreducible."[6]

The nuclear clique, in particular, seemed to require the exact opposite. General Pierre Gallois, *père de la bombe* as he was known, along with de Gaulle were the chief advocates of a nuclear-armed France, who saw the logic of warfare as inverted by the coming of the atomic age. No longer would it be necessary to stockpile arms, men, and munitions.[7]

Even the possession of a tiny nuclear arsenal rendered war all but impossible; the risks were too great. Whereas, in ages past, the actual prosecution of warfare could be "an intelligent operation," the prospect of nuclear destruction made war ridiculous.[8] In fact, Gallois saw the day coming and soon in which the world might be forced to "give up war altogether" as a meaningful activity.[9] He explicitly rejected any notion of nuclear escalation, arguing that "starting from any stage of destruction, means to risk climbing one by one the steps which lead to the irremediable;" that, in effect, nuclear arms presented war as a "staircase toward absurdity."[10]

This evolved context was war as *irresponsibility,* and Gallois saw no place for it. In fact, he saw no place for war or warriors much at all. He argued that new technological realities would "sweep away the old structure of conventional confrontations."[11] The only question remaining for French defense policy was to ensure sufficient resources were put into technological development so as to maintain a credible threat, and therefore the status quo, as "obligatory, firm, and immutable."[12]

While there may have been a kind of resistless logic to Gallois' points in regards to the danger posed by Great Powers, to many of those engaged on the ground in France's war in Algeria the coming of the bomb promised less resources for the fight, and no increase in useful capability. To them Mao's principle of revolutionary war was, at once effective, lethal, and spreading; moreover the practitioners of subversive war did not give a fig for French nuclear deterrence.

Another *Armée de l'Air* (Air Force) officer, like Gallois, emerged from the first of France's post-WWII colonial failures with decidedly different lessons.

6 Hélie de Saint Marc, *L'aventure et L'espérance* (Paris: Editions des Arènes, 2010), 217.

7 Pierre Gallois, *The Balance of Terror: Strategy for the Nuclear Age*, trans., Richard Howard (Boston: Houghton Mifflin Company, 1961), 2.

8 *Ibid.*, 7.

9 *bid.*, 57.

10 Gallois, 62.

11 *Ibid.*, 233.

12 *Ibid.*, 13.

Writing some months after the fall of Dien Bien Phu, Lionel Max Chassin cared little for the growing technical capabilities of atomic weapons. The Americans had been unwilling to use them to save the French position, and, more importantly, he was unsure if French possession of one would have at all affected the outcome of the Indochina conflict.[13] For him, while the existence of the bomb might impair the ability of nations to fight one another, it made more likely what he termed "local wars of the jungle." Chassin warned that, despite her technical superiority, the West had no adequate answer for its opponents' "fanaticism, contempt for life, endurance, frugality, discipline, auto-criticism, and grip on the masses."[14]

For Chassin, technology was rarely decisive in war, especially in "guerres idéologique."[15] In these types of struggles – i.e. the kind he saw as now predominant on the globe – technology was no guarantee of success; in fact, technology was not even a good guarantor of power. Chassin prophesied an "immense international civil war" between the "spiritualism" of the West and the crass materialism of Communism.[16] At base Chassin, despite his *armée de l'Air* heritage, saw a basic paradox in the West's reliance on high-technology forces. Despite the massive expense and therefore apparent importance of these weapons, it was not possible to change "ideas by cannon fire, and even less so by atomic bombs."[17] In the face of such an enemy, motivated by unflinching ideology, Chassin found that the West's "technical superiority showed itself to be powerless."[18]

Though he was one of the first to highlight this apparent disconnect between technical sophistication and strategic bankruptcy, Chassin was hardly the last, nor a solitary voice. Colonel Nemo, another old Indochina hand writing as the war in Algeria picked up pace, noted the dangerous technical fascination of Western warfare to the detriment of more important factors, at least in his thinking. Writing for the *Revue de Défense Nationale* in 1956, Nemo related a story of an unnamed general officer reviewing a motor-park of tanks,

13 L.M. Chassin, "Réflexions Stratégiques sur la guerre d'Indochine," *Revue de Défense Nationale*, (1954): 509–512.

14 Chassin: 513.

15 L.M. Chassin, "Esquisse d'une Stratégie mondiale," *Revue de Défense Nationale* 9, (November 1949): 407–408.

16 *Ibid.*, 412.

17 *Ibid.*, 411.

18 L.M. Chassin, *Bélisaire, Généralissime Byzantin, 504–565* (Paris: Payot, 1957), 7. In the preface to this historical work into the life and career of the Byzantine general Belisarius, Chassin states that the work was inspired by the war in Indochina and how he saw similarities with the situation of the 6th century. This quote comes from the preface to the book where he draws explicit connections to the French troubles in Indochina.

artillery, and engineer equipment waiting to load for shipment back to France from Indochina. "How, with all of this force intact, have we come to this point?" Nemo's answer was that the French and her local allies had invested too much in equipment, and had ignored the "political and social factors" of war.[19]

War, in Nemo's estimation, was no longer chiefly a contest of arms, and to see it as such was damaging the French effort. He was particularly adamant that a fixation on atomic war made these proclivities even more dangerous. Nemo chastised his readers for an untoward obsession with "robot destructors" and for forgetting "that it is men who make war, with their passions, with their moments of heroism, with their flaws, and with their fear. The final challenge is men, and the opinion of men."[20]

Others were equally concerned. Lieutenant-Colonel (later Col.) Gabriel Bonnet, writing in the closing days of the Indochina War, lamented that "the machine" in French military thinking had "overridden the man."[21] Some of this was appropriate, he argued. Tanks and planes had made war faster paced and harder hitting. But, reliance on them had overshadowed deeper realities concerning the "laws of war," not subject to the "rhythm" of "scientific and technical evolution."[22] While he was not dismissive of the immense material and technical needs involved in "guerre moderne," he argued forcefully that these were not all that made up the "force of a country."[23] Certainly, one could argue that here was necessity retooled as virtue by an agent of a pauperized French state, but he believed it nonetheless. "Modern war" was, according to Bonnet, more than anything else a "war of souls, a test of morale."[24] He did have seemingly incontrovertible evidence on this point. With no tanks or airplanes, the Viet Minh had nevertheless triumphed.[25]

The Viet Minh certainly did not have "the bomb" either, nor did this seem to matter to the final outcome. Lieutenant-Colonel F.O. Mitsche, also writing in the French defense review, dismissed nuclear strategy as "mutual suicide." The threat of complete annihilation was simply not credible to his thinking, and so the threat of nuclear exchange was the functional "negation of all strategic thought."[26] Nuclear weapons were not capable of dissuading a revolutionary

19 Nemo, "La guerre dans le Milieu Social," *Revue de Défense Nationale*, (May 1956): 610.

20 *Ibid.*, 606.

21 Gabriel Bonnet, "Guerre et Technique," *Revue de Défense Nationale* 18, (May 1954): 549.

22 *Ibid.*, 547.

23 *ibid.*,: 557.

24 Bonnet: 556.

25 Gabriel Bonnet, "L'importance du facteur 'Moral' dans le complexe Armée-Nation," *Revue de Défense Nationale*, (November 1955): 436.

26 F.O. Mitsche, "Désarmement ou réorganisation?," *Revue de Défense Nationale* 24, (1957): 240.

opponent; and more importantly, they were not capable of combating him. As Mitsche said of them: "They can do nothing but destroy; they cannot practice infiltration nor occupy terrain."[27]

This lack was a significant problem for France's colonial soldiers. In the waning years of the 1950s, Mitsche argued that the flourishing "esprit atomique" in defense circles (and arguably the associated hobbling of the overseas military to make it possible) was more dangerous than the "Maginot spirit of France in 1939."[28] It was certainly understandable, as far as Mitsche was concerned. It was nearly a "natural phenomenon" to believe (perhaps hope) that "all problems are, above all, technical problems" to be solved by technical means. But, as Mitsche concluded "other factors intervene in battles."[29] The rapid development of technology in arms obscured this fact for the time being and at the expense of proven methods. Colonel Villiers de L'Isle-adam bemoaned this fact: "Technical developments prevent, for the moment, the employment of the classical means of conquest."[30]

These authors' various plaints did not come back void. Commandant Georges Robin, one of the legionnaires who participated in the *putsch des généraux* in April 1961, related in his memoirs how he read Chassin's articles and was affected by the aviator's stark distinctions.[31] He came to share the misgivings concerning what he termed the French "obsession" with nuclear arms. Like others he saw in them the offer of superficial force coupled with actual "military and diplomatic powerlessness."[32] For Robin, this "obsession" with nuclear weapons, with destructive power, was born of the Western world's "[sensitivity] to that which is quantifiable…"[33] The events of his wars, however, were not readily counted or measured, nor were they more easily ignored or avoided altogether. Of course, this would not do for Robin and men like him. The empire and the wards of French civilization required something else, something better. "We had to pass from terrain to men," he argued, "des fusils aux idées…" (from rifles to ideas).[34]

Much of this sentiment crystallized in and around the growing doctrine of "guerre révolutionnaire," viz. the French adaption of Maoist technique

27 F.O. Mitsche, "Désarmement ou réorganisation?," 240.

28 *Ibid.*, 241.

29 *Ibid.*, 241.

30 Villiers de L'Isle-adam, "Cette Guere de Notre Siècle," *Revue de Défense Nationale*, (2e trimestre 1956): 884.

31 Georges Robin, *Commandant Rebelle: Algérie, 1958 – de l'obéissance à la Revolte* (Saint-Amand-Montrond: Editions JC Lattès, 1998), 85.

32 *Ibid.*, 80.

33 *Ibid.*, 92.

34 *Ibid.*, 93.

and attitude for its own somewhat vague Western ends in contrast to the brutal simplicity of the Marxist dialectic.[35] In this respect, one can view the anti-technological bent of a certain faction of the French military as a natural outgrowth of sustained interaction with revolutionary movements. But, it was not only this development.

Colonel Nemo's remarks, and others like it, reflected a growing and nearly atavistic sense that the course of modern warfare was headed in directions incommensurate with what it meant to be a soldier and a man. Nemo was worried that French arms might become hopelessly trapped in its "techniques and traditions" at the expense of the deeper, older prerogatives of the French warrior.[36]

In these ways, technology threatened the very essence of war, an enterprise of men. One contributor to the *Revue de Défense Nationale* in 1957 urged his readers to remember that "war is the department of moral forces" and that, at base, it is "led by *men,* beings of flesh and nerve."[37] War was an issue of relationship and responsibility. Material means were of little value without the *chef* and men dedicated to his service.[38] What mattered most was that this *chef* worked toward "at every occasion, *the mastery of self*…"[39] Unlike what we must presume this observer saw taking hold, he argued for an older, and yet newer, conception of leadership which would involve "above all a human conception of the *chef*."[40]

These human qualities mattered "above all," because the war fought in the colonies was a war of men, of infantry. The grand master and chief prophet of *guerre révolutionnaire*, Jacques Hogard, expounded at the same time as the case made above that the army suffered from a "grave crisis of morale" which could only be ameliorated by, among other things, a "restoration of real infantry, numerous, sober, rustic, capable of marching on foot as well as being transported by plane or helicopter…" This infantry, as of old, needed a reinvigoration after years of neglect, stemming from, as he saw it, "an erroneous

35 Even now the best single explanation of French "guerre révolutionnaire" is Peter Paret's slim volume on the subject. See Peter Paret, *French Revolutionary Warfare from Indochina to Algeria: The Analysis of a Political and Military Doctrine* (New York: Frederick A. Praeger, Publishers, 1964).

36 Nemo, 623. Interestingly, elsewhere in that same article Nemo argues that a "war of religion" is the only "real war." See page 611.

37 François Kuntz, "Le Moral au Combat," *Revue de Défense Nationale* 25 (July – December 1957): 1083–1084.

38 *Ibid.*, 1095.

39 *Ibid.*, 1091.

40 *Ibid.*, 1095.

interpretation of atomic perspectives."[41] Given the wars they were fighting, infantry was or needed to be the "queen of the battlefield" once again, as Hogard maintained.

The real danger was the stalking horse of communist infiltration in the third world, not direct confrontation in Europe. In the empire, and hidden behind nationalism, communism threatened the "encirclement of the West."[42] How could a thermonuclear weapon address this problem? General de la Chapelle quoted Napoleon to make the point. "There are only two forces on earth, the saber and the spirit; and it is the spirit that always has the last word."[43] War was an issue of soul. "An army fights," he argued, "with its soul just as much as with its engines of war."[44] Their opponents, according to men like Chassin, waged a "struggle based on subversion" not brute force. To oppose the communist "faith," as Chassin believed, France in its role as guarantor of Western civilization, had to deploy another faith. To counter "an insidious and powerful ideology" necessitated a "superior ideology capable of winning of the hearts of men."[45] More investment in technology was not the issue. Commitment was the issue and they saw it as a spiritual one.

The "faith" Chassin required, the "force of ideas" that Hogard wished to deploy against communism, however, had no real content.[46] Hogard believed that winning this battle, or at least holding the line, required "two pillars: a sufficiently dense military implantation and the will to see it through."[47] Both of these were lacking in 1957 as he wrote the words.

The first requirement was perhaps in the offing, given the Fourth Republic's massive and growing commitment of forces to Algeria; but where to find the second?[48] The conscript forces holding down static posts had little personal stake in the conflict and suspect morale, especially in the opinion of the regulars. Writing under the pseudonym "X," one contributor to the *Revue de Défense Nationale* in 1956, presumably himself a regular, noted that the "rappelé" had

41 Jacques Hogard, "L'armée française devant la guerre révolutionnaire," *Revue de Défense Nationale* 24 (January 1957): 85.

42 L.M. Chassin, "Vers un encerclement de l'Occident," *Revue de Défense Nationale* 22 (1956 (1e trimestre)): 532.

43 Napoleon quoted in Chapelle, "Les aspects particuliers de la guerre possible," *Revue de Défense Nationale* 22 (1956 (1e trimestre)): 156.

44 Chapelle, 151.

45 Chassin, "Vers un encerclement de l'Occident," 548.

46 Jacques Hogard, "Le Soldat dans la guerre révolutionnaire," *Revue de Défense Nationale*, (February 1957): 217.

47 Hogard, "Le Soldat dans la guerre révolutionnaire," 218.

48 Benjamin Stora, *Appelés en guerre d'Algérie* (Evreux: Gallimard, 1997), 24. By April 1956 there were approximately 400,000 French soldiers stationed in Algeria.

no idea why he was recalled, or why he was fighting.[49] What could animate the will to follow the war through to the end and prevent disinterest and technocratic escapism from taking hold?

Against the unifying principles of Marxism, these men struggled to find a riposte, despite the acknowledgement that one was necessary. Colonel René de Metz recognized the need for some kind of "sacred fire" to animate the troops, but was confounded about its source. At one time, Christianity might have provided that force, he thought, but no longer in a thoroughly secular French state.[50] The traditional Republican values of "Liberté, Egalité, Fraternité" were equally troublesome, because communism had already co-opted so much of their metaphysical appeal.[51]

Many French officers (and opinionated observers) felt the search for a new élan was seriously encumbered by the influence of their senior partner and arch technophile, the United States. NATO (read America), as French commentators believed, had little interest in France's grinding struggle in North Africa. It threatened to push the teetering *tiers monde* into the open arms of the Soviets. More directly, it put considerable strain on a perilously overstretched military to fight the war in Algeria and to meet its obligations to NATO. America, of course, wanted up-to-date conventional forces poised on the Red frontier ready to stop the onrushing communist horde. Algeria required those same units to dismount, forsake their heavy equipment in favor of the dirty work of pacification. To many, it seemed that French formations could do one of those things well, not both. Various proposals filled the pages of the *Revue Défense Nationale* at the time calling for "grandes unités" mechanized, armored (and perhaps equipped with tactical atomic weapons) to protect continental France.[52] These would be explicitly modeled according to American preference.

The prolix, and ever truculent, Combaux murmured darkly that Americans were considering moving atomic weapons down to the faddish "divisions pentomiques" and that France would be supremely foolish to follow suit. In the first place, it would put France right into the "tragic dilemma" awaiting Western arms, viz. to be forced, when faced with conflict, to "give way or unleash a general conflict."[53] Commentators described American war-making proclivities

49 X, "Le Prix du Sang," *Revue de Défense Nationale* (1er trimestre 1956): 184.

50 René de Metz, "Une idée-force : laquelle?," *Revue de Défense Nationale*, (1er trimestre 1957): 354.

51 Metz, 355.

52 Claude Delmas, "La France et Sa Défense Nationale," *Revue de Défense Nationale* (1957): 1444.

53 These discussions obviously bear more than a passing resemblance to the same strategic arguments concerning "flexible response" following Eisenhower's administration.

as disturbingly "totale" and marked by an unwavering commitment to the "absolute defeat" of its enemies which hamstrung diplomacy and laid up foreign policy as a hostage to "la technique militaire."[54] General Charles Ailleret, who interestingly had achieved a measure of fame by opposing the *putsch des généraux* in 1961, offered up the preeminence of the Strategic Air Command within the US Air Force as proof positive of Americans' technologically-driven tendency toward irresponsibility.[55]

As in later American strategic debates, Combaux and those likeminded saw in nuclear weapons the "arm of terror and demented panic," in other words "of chaos."[56] They preferred a doctrine of "limited war" in which nuclear arms played a deterrent role, but which reserved a large scope to conventional forces of an other-than-American architecture. These formations (the American ones) were, he assured his readers, "technologically over-equipped" and far too expensive for Europeans to replicate.[57] Such luxurious units also gave the Soviets exactly what they wanted – a Western world pouring money and resources into hyper-sophisticated formations based in Europe and destined to do nothing, while the Communist menace outflanked them through subversive war in Africa. Algeria was not a peripheral issue (perhaps even altogether superfluous in some estimations); on the contrary, it was the heart of the matter and combating the nibbling cancer of communist insurrection was not to be done the American way, with money and gadgetry. "[T]his war," argued Combaux, "[would] be won less with machines than in stirring up the human masses."[58]

Seeing a direct invasion of Western Europe as suicidally inconceivable, the partisans of "guerre psychologique" doubled down on inoculating their people against subversion, against this "invisible invasion of ideas."[59] The proposals for how to combat this danger flirted openly with radical concepts. More than a few of France's champions were convinced that in this battle of ideas France was decidedly unarmed. After all, according to men like Colonel Nemo, "there are no true wars except religious wars."[60] But lacking an adequate "sacred fire"

---

E. Combaux, "Armes Atomiques et Non Atomiques dans la Défense de l'Eurafrique," *Revue de Défense Nationale* 26 (1958): 65, 71.

54 It's crucial to note that *technique* in French carries the sense of "technique" qua method, and also the idea of the "technical" or "technological."

55 Charles Ailleret, "Guerre Nucléaire Limitée ou 'Drole de guerre,'" *Revue de Défense Nationale* 26 (1958): 418.

56 Combaux, "Au Dela de Clausewitz," *Revue de Défense Nationale* 24 (1957): 519.

57 Combaux, 64.

58 Combaux, 71.

59 James Achard, "Possibilités et Volonté," *Revue de Défense Nationale* (1957): 1691.

60 Nemo, 611.

meant France needed to fabricate one by aping the communist organization, and purging civil society of its debilitating weakness.[61] Quite explicitly, this meant returning France to a pre-modern setting where leadership and manly virtue might once again take hold.

Given the inescapable reality of partisan war, these theorists foresaw a society rent by political division and propaganda. A "technically perfected" military, engaged in fighting far afield and comfortable in an apolitical professional bubble, was most unsuited to confronting the threat of subversion, particularly once it metastasized and spread to continental France herself, as many implied it already had.[62]

France needed a "territorial" military organization centered on the people featuring independent and self-sustaining militias ready to engage in bitter and prolonged struggle against a shadowy ideological opponent. It was no accident that Colonel Nemo talked about a return to a kind of "féodalisation" of war and society; a rejection of the modern and materialist Fourth Republic for the sterner martial, masculine values of ancient France.[63] Happily for France, this reversion was not to be.

No matter the Republic (the IV or de Gaulle's V) involved, preserving mainland France eventually came down to beating out the fire for a manly, but medieval France. Raymond Aron, the most perspicacious and even-handed of France's political commentators, could see the writing on the wall. In late 1959, a few weeks after de Gaulle declared his intent to keep work going on France's "force de frappe atomique," Aron argued that France's desire to assure its own deterrent capacity was entirely "reasonable." He reminded his readers (quoting Nietszche) in *Le Figaro* that the state is, after all, a "cold monster" properly committed to its own security in nearly as mercenary a sense as is conceivable, and implicitly shorn any romantic notion of empire. This security, an increasing number of observers believed, required France to possess its own nuclear armament. Aron himself believed this acquisition might necessitate France foregoing any modernization towards the rest of the Army (de Gaulle's desire), even if the war in Algeria could be brought to a speedy close.[64] In the end, it hardly mattered.

French strategic thinkers long had recognized that her various security problems were pulling the military establishment apart. The need to "balance"

61 Metz, 353.

62 Nemo, "Combat de Mélée et Défense Nationale," *Revue de Défense Nationale* (1957): 1293.

63 *Ibid.*, 1295–1300.

64 Raymond Aron, *La Coexistence: Mai 1955 à Février 1965*, ed. Georges-Henri Soutou, 3 vols., Raymond Aron: Les Articles de politique Internationale dans Le Figaro de 1947 à 1977, vol. II (Paris: Editions de Fallois, 1993), 586–587.

between its "capacité de dissuasion," the imperative to maintain intervention forces à la NATO, and to counter simultaneously the internal discord in Algeria was not only costly; it also represented three nearly mutually exclusive operational requirements. It would be wrong to suggest that de Gaulle ultimately withdrew from Algeria for organizational reasons alone – bucking up against the perceived Anglo-American cabal within NATO was also a significant concern – but departing the war did succeed in cutting this strategic Gordian knot. However, it did not do so without the former general famously incurring the so-named "anger of the legions."[65] Theirs was a France alive and fighting, or dead, but with honor intact.

In his novel detailing the heady and dangerous closing days of the Fourth Republic in Algeria, Jean Lartéguy relates a series of interesting observations through the voice of a female journalist and love interest to one of the chief protagonists. Lartéguy based his characters on the real life men involved in these actions; while they are perhaps painted in starker relief than their real life analogues, the differences are not all that exaggerated. In the later scenes of this work, *Les Prétoriens*, this journalist comments perplexedly on the strange "hodgepodge of commonplaces" that animated these soldiers – part Christianity, part some vague notion of the "West" and France's unique place within it. Furthermore, they lived in a "simplified and brutal universe," quite unlike her own, in which there was no place for women.[66]

In essence and at base, an amorphous, but nonetheless, powerful notion of masculinity worked to give purpose to those committed to the defense of the French empire. There was no coherent, codified system of values that one can speak of during this time. It was instead a potent mélange of more visceral notions that permeated the thinking and writing of those within this world. They, of course, spoke about the eternal quality of valor as "courage for war" and the need to cultivate and to inculcate "virtus" and "force d'âme" (fortitude, spiritual strength) among leaders.[67]

Others put forward the importance of decisiveness in "l'homme d'action" (the man of action). In a time full of uncertainty, so the argument went, it would be a mistake to retreat into scientific thinking, into the desire to know fully before moving; that would be the recipe for doing nothing.

General Bouscat recalled the life and times of Herbert Lyautey and commended him to the army of his own day. Lyautey, the progenitor of the colonial soldier as a species, was a "man of action" who made himself anew in the

65 Lartéguy, frontispiece.

66 Jean Lartéguy, *Les Prétoriens* (Paris: Presses de la Cité, 2013), 411–413.

67 Kuntz, 1088–1089.

empire. Bouscat, quoting Lyautey, deplored the "sterile" garrison life he found in the *métropole*, marking out that the "real life of France, the fertile life" was to be found overseas.[68] The French soldier had to be bold, and impart the germ of French civilization to a passive, but fecund outer world. In keeping with this sentiment, the army had to formulate a "doctrine of action" in order to root out the "formalism" which stood ready to "emasculate" it.[69]

If the problem – increasingly cast almost as a problem of meta- or grand-strategy – was this opaque notion of France as "dévirilisé" (emasculated), then the remedy had to be some kind of *virilisation*, equally vague and far reaching. Gabriel Bonnet, colonel and instructor at the *Centre de Préparation à l'École Supérieure de Guerre*, stated this case plainly: the officer as "chef du jeunesse" (leader of youth) must "teach virility in all its forms in order to make a man."[70] He recommended, among other things, a sports regimen to bring the men together, political and psychological training, and, tying it all together, advocated for a close connection with his men in order to communicate his "faith, his élan" in order to establish an "instinctive accord" among his charges.[71] Bonnet's ideal officer would, at all times, "embody the dominant spirit of male vigor which gives to the *chef* physical prestige joined with a moral and intellectual superiority."[72]

On the surface here was an apparently tactical set of concerns – i.e. how to build competent infantrymen – that immediately gained a strategic hearing. The diagnosis of what was lacking for an infantry company could and was extended to the level of the nation at large. General André Zeller, one of the chief conspirators of 1961, maintained that the "army was not an instrument, not a means. It [was] an expression; one of the aspects of the Nation."[73] The unity and "male vigor" needed at the level of the company was equally needed across the national landscape because communist subversion operated across that entire scope. Jacques Hogard, much like Nemo, envisioned and advocated for a vastly expanded system of local reserves, essentially militias composed of "politically reliable" men to compensate for the mass of unreliables that he perceived perched throughout French society. These troops could be called upon to backfill regular units in times of crisis.[74] Though he did not mention it in this way, it was clear that he was propounding for France the organization

68 R. Bouscat, "Lyautey et L'histoire," *Revue de Défense Nationale*, (November 1954): 385.

69 P. Daure, "De la Décision," *Revue de Défense Nationale*, (2e trimestre 1957): 682–685.

70 Gabriel Bonnet, "L'officier – 'Chef de Jeunesse,'" *Revue de Défense Nationale*, (1957): 768.

71 Bonnet, "L'officier – 'Chef de Jeunesse,'" 771.

72 Bonnet, "L'officier – 'Chef de Jeunesse,'" 771.

73 André Zeller, "Armée et politique," *Revue de Défense Nationale*, (1er trimestre 1957): 517.

74 Hogard, "L'armée française devant la guerre révolutionnaire," 84.

of the Viet Minh military into regular, regional, and local troop castes. And, just as the military needed the accompanying unity of effort and action, this overt militarization of society would require the same. It was not a prospect he thought possible until the "interior enemy [was] reduced to silence," an ominous portent of things to come for French civil-military relations.[75]

Perhaps, it is appropriate to speak of the doctrine(s) of *guerre révolutionnaire* as representing a crisis in some small stratum of French masculinity. The series of successive failures from the ruin of 1940, to Indochina, and on to the unrolling catastrophe of Algeria posed significant identity challenges to the professional soldiers of the French empire. Technology offered no hope, and, in many ways, aggravated the situation. It dangled the bauble of a deceptive technical rationale for abandoning the empire before weary eyes, while promising to neuter the elite troops and what, in their minds, it meant to be French.

But, *guerre révolutionnaire qua virilisation* offers no clear-cut answer to the subversive methodology of France's opponents. This is probably why it did not work, and ends itself in a paroxysm of near pointless violence. Yet, there is an almost sexual notion of how it was supposed to work, one suspects. In Jean Lartéguy's early (and better known) novel, *The Centurions,* the cast of intrepid paras passes into communist prison camps after the fall of Dien Bien Phu. One of the main characters, an officer named Philippe Esclavier, falls seriously ill and is attended by a Viet Minh nurse of impeccable communist credentials and suppressed natural beauty. Despite her commitment to cause and country, and hatred for the "colonial mercenary," she finds herself overcome by the image of a naked Esclavier and, despite herself, daydreams of her "strong and vigorous" captive/patient.[76] Eventually, fully smitten, she disavows communism and her own people for the sake of her ill-fated love for Esclavier. If only there were more men like him, Lartéguy seems to say, the empire would remain French out of sheer passion. But there were not. Later in the story, a local asks another of these French supermen why he is fighting so hard to keep Algeria in France. His answer: "What I'm fighting against in Algeria is this mechanization of man."[77]

The relationship between technology and war is fraught with paradox. In the first place technology when it appears is often quite disruptive. It challenges methods, practices and assumptions. It also challenges the warrior's notion of self. Technology is a threat to the heroic. In the late 17th century a controversy

75 Hogard, "L'armée française devant la guerre révolutionnaire," 88.

76 Lartéguy, *The Centurions*, 167.

77 Lartéguy, *The Centurions*, 472.

opened up in France between the "Ancients" and the "Moderns," two sides of a cultural clash pitting those who revered the ancients against those devoted to novelty, particularly science and technology. Though primarily a literary debate, a chief argument deployed by the "moderns" was that technology – for example the telescope – demonstrated that "modern" man was better than his predecessors.[78] Opponents of the "moderns" argued that, by departing from traditional practice in their storytelling, the new novelists were taking ancient, heroic figures and making them weak, feminine.[79]

As technology passes into more common use its role moves from disruption to a source of dependence. Warriors can be defined by their technology. When we talk about types of soldiers we mention artillerymen, pilots, lancers, tankers, or submariners. Even the infantryman – the one perhaps least defined by his weapons – is usually called hoplite, rifleman, sharpshooter, grenadier, paratroop, and so on. It is an inescapable tension. It helps explain how the warrior class can be the most enthusiastic about technology *and* the most skeptical. An attitude too heavily focused on the instrumental quality of technology can lose perspective on its equally profound definitional power.

78 Joan DeJean, *Ancients against Moderns: Culture Wars and the Making of a Fin de Siècle* (Chicago: University of Chicago Press, 1997), 43.

79 DeJean, 54–55.

CHAPTER 15

# Cyber: War?[1]

*Michael Pavelec*

Cyber power is the combination of computer technology, the internet, and the interconnectedness it provides to exert influence and power over opponents. It has opened a new chapter in the study of war and warfare. Some consider cyber power and cyber attacks revolutions that will change the very nature of war. This essay presents an argument for the reconsideration of what cyber power is, what it can do, and what its effect on society, information, and war will be. We are potentially in the midst of a revolution that will change the definition of war and how future wars are conducted and won.

An academic debate is raging about whether cyber attacks constitute a method of war[2] (or warfare), as the U.S. and the world struggles to define the future of war. In the wake of Stuxnet, the Sony attack, and the recent U.S. Department of Defense Office of Personnel Management (OPM) breach, it can be argued that state-sponsored and individual cyber attacks are a new form of war. In this chapter, I will attempt to clarify the debate by first outlining a personal definition of war, then discussing the idea of cyber attacks (actions, breaches, hacks) as both war and not war. In the end, I hope to spark a debate on the definition of cyber attacks as well as a novel approach to the definition of war.

## 1 What Is War?

When considering a definition of war, strategic analysts and philosophers often turn to the classical theorists of war. From the ancient to the modern,

1 This paper was written for the conference and eventual publication for the Dr. John (Joe) F. Guilmartin, Jr. memorial Festschrift, in memory of my Ph.D. advisor at The Ohio State University (1999–2004). I would like to especially thank Dr. Rob Ehlers and Dr. Sarah Douglas for putting together the Festschrift as well as the conference (Columbus, Ohio, 23–24 Sept 2016). The thoughts and ideas presented in this paper are my own, and in no way reflect the position or opinions of the U.S. Air Force or the U.S. Department of Defense.

2 See specifically Lawrence Cavaiola, David Gompert, and Martin Libicki, "Cyber House Rules: On War, Retaliation, and Escalation," in *Survival* 57, no. 1, (February–March, 2015): 81–104.

 | DOI:10.1163/9789004393301_017

theorists have categorized war was "acts of violence" aimed at achieving political purpose; see specifically Thucydides, Sun Tzu, and Clausewitz.[3] Of these, Clausewitz is most often cited by military professionals and remains a favorite within U.S. Joint Professional Military Education (JPME), where I teach. Clausewitz's dictum that "War is thus an act of *force* to compel our enemy to do our will"[4] is the touchstone for the modern warrior and defense professional. But the concept of a cyber attack does not fit nicely into pre-21st century military philosophy. Thus, I will attempt to redefine war for the modern era. Based on extensive study of a variety of disciplines (philosophy, physics, sociology, psychology),[5] I have come to the conclusion that modern war is based less on the actions of belligerents and more on the *perception* of agency and action. Today, war is less about forceful action, and more connected to the perception of warlike actions.

Traditionally, war was about the actual violence to compel an enemy to submit to the will of the powerful. Thucydides stated that the "… strong do what they have the power to do and the weak accept what they have to accept."[6] Clausewitz emphasized that it was about the force, focused at the strength of the enemy, in order to coerce an enemy to submit.[7] Further, he argued that once other instruments of power failed to resolve a conflict of interest, military force had to be applied against military force in an epic struggle of direct actions of armies on armies. Once an enemy's army was destroyed or driven from the battlefield, the winner could impose his will on a vanquished enemy. This continuation of politics by other means hinged the idea of a direct approach, and defeating an enemy by degrading his strength. During the Napoleonic Era – which was the focus of Clausewitz' theories – multiple examples confirmed his ideas. Sun Tzu offered the most insight – even though his writing is the oldest – when he argued that attacking an enemy's strategy should be the primary focus,[8] and that to win without fighting is the acme of skill.[9] Sun Tzu clarified the indirect approach, attacking an enemy's plans and strategy

3 Thucydides, *History of the Peloponnesian War* (New York, NY: Penguin, 1974), x; Sun Tzu, *The Art of War* (London, UK: Oxford University Press, 1963), 39–44; Carl von Clausewitz, *On War* (Princeton, NJ: Princeton University Press, 1976), 83.

4 Clausewitz, 83. Emphasis added.

5 Including the excellent insight gained from M Mitchell Waldrop, *Complexity: The Emerging Science at the Edge of Order and Chaos* (New York, NY: Simon and Schuster, 1992) and James Gleick, *Chaos: Making a New Science* (New York, NY: Viking Press, 1987). Both of these books offer valuable insight into the idea of war as a complex (chaotic) system.

6 Thucydides, 402 (5:89).

7 Clausewitz, 83.

8 Sun Tzu, 77 (III:4).

9 Ibid, 77 (III:3); "…to subdue the enemy without fighting is the acme of skill."

in order to avoid the physical violence of war. Furthermore, he argued that it was important to attack an enemy's weakness with strength if it came to fighting, rather than focusing on an enemy's strength as outlined in Clausewitz. After World War I, where Clausewitz' theories were put into question with epic battles like the Somme and Verdun, theorists reinvigorated the idea of the indirect approach; see specifically Basil Liddell Hart[10] and the airpower theorists Douhet,[11] Trenchard,[12] and Mitchell.[13] During World War I, theorists, military thinkers and planners, and academics realized that strength-on-strength warfare in the modern industrial era was bankrupt, especially when war moved at the speed of foot, horse, and train. The stalemate that emerged after the initial battles in 1914 forced militaries to reconsider ways to recapture maneuver and mobility lost on the Western Front. When million-man armies faced off against each other, there were no flanks, and movement on the battlefield ceased.

Technology, as Dr. John F. Guilmartin, Jr., defined it, provided an answer.[14] Historically, the side with the better technology has an advantage, if that technology is used correctly for strategic decision. Although technology does not always overcome poor strategy and planning, it has the potential to provide advantages to the more technologically advanced side. It is not a panacea, but is an advantage. In his classes, and in direct relation to the underlying purpose of this project, Dr. Guilmartin discussed the role of technology in war, and the potentialities it offered. He opined that technology and war were intimately interrelated, and that both technological evolution and revolution moved faster in war than in peace, as each side sought to outperform the other, especially in protracted war. The interconnectedness of war and technology is one of the lasting themes I have carried with me throughout my academic career, including the idea that technology and war are inextricably interconnected in the martial context.

10 Basil Liddell Hart, *Strategy* (New York, NY: Praeger, 1967).

11 Guilio Douhet, *The Command of the Air* (Tuscaloosa, AL: University of Alabama Press, 2010).

12 Hugh Trenchard, *Air Power* (London, UK: Directorate of Staff Studies, Air Ministry, 1946).

13 William Mitchell, *Winged Defense: the Development and Possibilities of Modern Air Power – Economic and Military* (Tuscaloosa, AL: University of Alabama Press, 2009).

14 Dr. Guilmartin's working definition is as follows: "technology, n., (1): the application of knowledge to achieve a physical effect by means of an artifact, object or thing (distinct from art, which involves the application of knowledge to achieve aesthetic effect). (2): the artifact itself; the class of artifacts to which it belongs. (3): the knowledge needed to design, manufacture, operate, sustain and logistically support the artifact or thing and its user(s). Author's working definition." From John F. Guilmartin, Jr., "Technology and Strategy: What are the Limits?" Public Presentation, Strategic Studies Institute, Carlisle, PA. 1994.

From the very beginnings of recorded military history, land and sea warfare included technological evolution for advantageous results. Armies that had better equipment could outperform enemies with greater numbers. On the seas and waterways, navies that could move more quickly for positional advantage could overwhelm other navies and threaten distant shores. The paradigm did not change significantly until the advent of gunpowder, which arguably changed everything from military effectiveness to social structure.[15] From there, war and warfare remained trapped in the existing model with few significant changes, until World War I. Then, a literal and figurative explosion of technology changed the warfighting paradigm from the ground up. Although it was not the first use of many "new" technologies, their ubiquity and purpose changed warfare fundamentally. Submarines, airplanes, improved high explosives, chemical warfare, machine guns, tanks, fast-firing artillery, and wireless radio, among others, combined to change warfare and fighting. By the end of the war, technology became an integral component of the ways in which militaries fought and died. The test of war proved the value of technology at the start of the twentieth century. Similar, but improved weapons dominated the battlefields of World War II. But one technology again revolutionized warfare: the atomic bomb. At the very end of the war, the U.S. attacked Japan with two atomic bombs. The scale of war changed. One plane with one bomb destroyed an entire city. Technology changed war as well as society.

Importantly, nearly all of the above-mentioned technologies emerged from high-cost government programs. The average citizen could buy a truck or firearm, but only governments with highly advanced scientific and technical programs funded by taxpayers had the ability to procure submarines, military aircraft, or atomic bombs. The entry level for high-technology military weaponry was the state; and only the very wealthy could afford the highest technology. Today, only nine countries have nuclear programs, only a one has a truly aircraft carrier based navy – although at least three others are trying – and only a few dozen can muster an effective air force. High-technology military hardware rests in the hands of a privileged few.

However, technology is not stagnant, it is dynamic. By the end of the millennium, computer technology (hardware and software) and the internet opened up new potentialities for everything from information sharing, banking and economics, and communications, to cat videos. Nearly all of the internet, computers, and other elements comprising cyber power are benign and helpful, but

15 Geoffrey Parker, *The Military Revolution: Military Innovation and the Rise of the West 1500–1800*, 2nd edition (Cambridge: Cambridge University Press, 1996); Clifford J. Rogers, *The Military Revolution Debate* (Boulder, CO: Westview Press, 1995).

they are also a potential threat to the very structure of society. First and foremost, unlike high-cost military procurement programs, cyber power is ubiquitous and cheap. Any individual with a few hundred dollars and an internet connection can enter the cyber domain. With the right skill set, they can manipulate the domain for good or ill, with little accountability or legal oversight. The cyber domain is accessible by all: Governments and militaries compete with super-empowered individuals for control within and on the periphery of the domain. Furthermore, accountability is a major issue in the cyber domain; individuals as well as governments and militaries are able to cover their tracks to create plausible deniability for their actions. This makes tracing nefarious action to specific individuals or even organizations difficult. Further, whereas military power is usually wielded by established and recognized governments, or at least non-state actors or insurgents with recognizable organizations, actors in the cyber domain may be state-sanctioned government employees, corporate cyber "warriors," or even individuals, working outside the traditional state construct, frequently without any government's knowledge. Thus, in this argument, cyber power, and activity within the cyber domain, has the potential to change the future of war and warfare at its most fundamental level. Just as gunpowder, airplanes, and atomic bombs changed war in the past, this new technology has the potential to revolutionize the military, war, and society as we move further into the Twenty-First Century.

## 2 Cyber Power

Cyber offensives do not represent an overt show of force; the only things being destroyed or altered are digital data on the receiving end. There may be some hardware destruction – more on this later – but for the most part a cyber attack is about disrupting or destroying enemy digital information, or denying access to it at critical times, rather than inflicting the traditional death and destruction associated with military force.

The significant concept for consideration is the idea of perception. It is the perceived threat of violence or attack, rather than an actual attack, which provides a new basis for the definition of war. In the end, it is the perception and response that constitutes the definition of war for the 21st century. Thus, war can be defined as "The perception of threat, and the response to that threat, in efforts to change the political calculus of a nation or organization." If a nation (organization, state, belligerent) does not perceive a threat and act on it, it is not war. There are multiple examples of the use of actual physical force where "war" did not emerge, likewise there are multiple examples where there was no actual violence, but the perceived threat elevated the environment of international relations to declared "war." Using examples, I will outline why cyber action can be perceived as both "war" and "not war" depending on the perceptions of the belligerents, and thus support my contention for a new definition of "War."

## 3 Cyber as Not War

The easiest case to make is that cyber attacks are not war. Cyber attacks do not fulfill Clausewitz's requirement of violence and force. In the 19th century, war revolved around massive armies (or navies) facing off against one another and the act and outcome of physical violence. During the Napoleonic Wars, the subject of Clausewitz's analysis, when armies and navies were destroyed, nation-states submitted to the will of the victor with very few exceptions. That is the main reason why Napoleon was able to subjugate the Austrians and Prussians, among others, and the reason why military historians label the era the Napoleonic Wars. The Russians did not fight fair, according to prevailing European rules of warfare, and even though their military was degraded and their cities occupied, they refused to submit. But the act of violence and the force component were very much present in the 1812 campaign. When Napoleon was finally defeated, it was because his army was destroyed, in 1814 and again in 1815, and he was forced to bend to the will of the victorious coalition facing him.

In recent history, there have been a number of spectacular cyber attacks that are not characterized as war. In 2010 the Stuxnet computer virus was unleashed against Iran's nascent nuclear enrichment program.[16] The computer virus, an actual cyber attack, was launched against the computers and hardware that enriched Iran's uranium processing industry, to disrupt and destroy their ability to manufacture weapons-grade uranium. The virus worked like a charm. The manufactured glitch manipulated the centrifuges according to the code designers' instructions, degrading some and destroying others. It was a sublime use of computer technology to disrupt Iran's ability to enrich uranium. The virus was eventually discovered, but not before delaying Iran's nuclear ambitions for years. But besides the destruction of hardware, there was no violence or force involved. Not a single person died, and Iran's nuclear program

16 As chronicled in Kim Zetter, *Countdown to Zero Day: Stuxnet and the Launch of the World's First Digital Weapon* (New York, NY: Crown Publishers, 2014).

was not halted outright. The coercion was accomplished through subterfuge. It was not war.

Another example in support of the argument is the recent Sony hack. In the fourth quarter of 2014, the movie mogul Sony was getting ready to release a marginal film that poked fun at North Korea and Kim Jong-un.[17] The Sony hack was industrial espionage against a corporation; the movie was a convenient scapegoat to point the finger at North Korean hackers, both independent and state-sponsored. Other than exploiting Sony user data and corporate emails, nothing was destroyed or even degraded.[18] However, the hack did achieve the political objective of pushing the release date of the film from the intended Christmas release into the New Year (2015). Sony seemingly acquiesced to veiled North Korean threats based on the content of the film. Thus, the cyber attack did not use force, and was not war.

The final example for "not war" is the recent attack on the U.S. Department of Defense's Office of Personnel Management. The DoD estimates that there are as many as 10 Million cyber attacks on its servers each day.[19] In June 2015, hackers attacked the OPM system and harvested data from approximately 22 Million personnel files.[20] The breach did not do any violence, there was no force involved, and no one died as a result. Thus, even though the likely culprit is Chinese hackers – either state-sponsored or independent – the U.S. government and military did not perceive or define it as an act of war.

Espionage and spying are an integral part of international relations. Through cyber technology, advanced states (and non-state actors) can spy from anywhere on the globe, find secrets, and potentially disrupt computer networks and capabilities. But to this point – disregarding some of the fantastical Hollywood plotlines – computer viruses and cyber attacks have not included the violence and force components of the classical military theorists. Few, if any, have died due to cyber attacks, governments have not toppled, and massive damage to state infrastructure has not occurred. There is speculation that the capacity to produce these kinds of effects may eventually evolve from the current state of computer and cyber technology, but thus far, cyber attacks cannot

17 *The Interview*, http://www.imdb.com/title/tt2788710/?ref_=fn_al_tt_1 (accessed 22 July 2015).

18 See specifically the coverage by *Wired* magazine from October 2014 through February 2015 on the first and subsequent Sony hacks, attributed to North Korea. http://www.wired.com/tag/sony-hack/ (accessed 22 July 2015).

19 http://csis.org/event/cybersecurity-discussion-general-keith-b-alexander-director-national-security-agency (video: accessed 22 July 2015).

20 http://abcnews.go.com/US/exclusive-25-million-affected-opm-hack-sources/story?id=32332731 (accessed 22 July 2015).

be classified as war based on the perception-focused definition of it given earlier.

## 4 Cyber as War

The more difficult argument is that cyber attacks can be classified as war. Cyber attacks are a new form of modern warfare, and can be regarded as such. They are simply another way to use the military or state-sponsored forces, or even individual actors, to project power to achieve political and thus policy objectives. Cyber attacks are simply one more tool comprising a complex set of increasing capabilities residing within the larger instruments of power. In other words, they are part of the kit bag for international relations that countries and people use as they interact with each other in the modern, globalized world to achieve advantages over competitors and adversaries. Consequently, using the same three examples above, there is a coherent argument to be made that cyber attacks are in fact acts of war.

Stuxnet damaged and destroyed Iranian Uranium enrichment facilities with a computer virus. During World War II, the Allies tried to damage and destroy German and Japanese productive capacity. In 1981 Israeli Air Force jets bombed the Iraqi nuclear facility at Osirak, south of Baghdad, and in 2015 (?) destroyed a nuclear reactor in Syria. The difference was that in the latter three cases, the attacks were physical rather than virtual, were attributable, and had obvious kinetic effect with consequent damage and deaths. Interestingly, the bombing campaigns during World War II occurred during a declared war and were largely within the existing rules of international conduct.[21] Paradoxically, the Israeli air strikes on nuclear reactors did not happen during a declared war, nor did they lead to war between Israel and Iraq or Israel and Syria. Although these were overt, kinetic attacks causing death and destruction, the preventative Israeli acts of war did not lead to further or formal hostilities. Arguably, the Israeli strikes were a gamble to use a widely recognized act of war to prevent future or further conflict that would likely have placed Israel in a less advantageous position. In 1981 and 2015 (?), the gamble paid off for the Israelis. If the Stuxnet virus was a preventative act of war to forestall Iranian force projection, then it may have succeeded; it is too soon to tell definitively.

21 Although this is still being debated. See specifically A. C. Grayling, *Among the Dead Cities: the History and Moral Legacy of the WWII Bombing of Civilians in Germany and Japan* (New York, NY: Walker and Company, 2006).

It would be a stretch to declare the Sony hack an act of war. However, in the 21st century, corporations contribute in major ways to state power in the globalized world and indeed represent major economic powers in their own right. From the declaration of "corporate personhood" in the U.S.[22] to their transnational character, multinational corporations are a natural evolution within the capitalist system. Corporations still rely on the governments in their home countries for physical protection, but recognize modern cyber threats and have subsequently bolstered internal cyber protection.[23] Corporations spend millions of dollars on internal cyber defense capabilities; it is not as huge leap to speculate that they will add offensive capabilities in order to retaliate against attacks, as well as a tool for corporate espionage. When corporations fight each other in the virtual world, what is the role of the state in the current (but declining) Westphalian system?

The recent Chinese hack of the OPM system was an act of war. The Chinese gained illegal access to a secure U.S. government computer system to exploit its workers and information.[24] The non-destructive cyber attack compromised government personnel records for the primary objective of forcing political change and thereby gaining advantage over the U.S. In this case the intended outcome was exposing U.S. secure information, which would in turn lead to changes in government (people fired), sensitive information leakage (covert operatives abroad exposed, individuals' information disclosed), and further domestic mistrust of the government and cyber networks.[25] The Chinese hack was an act of war with potential long-term political outcomes, and thus falls within the definition of "War" proposed above. Cyber attacks may or may not cause actual physical damage, but are definitely acts of war in a new era of warfare.

---

22 *When did Companies Become People? Excavating the Legal Evolution*, National Public Radio (NPR), 28 July 2014, http://www.npr.org/templates/transcript/transcript.php?storyId=335288388 (accessed 22 July 2015).

23 Any casual internet search will reveal the massive web content on "corporate cybersecurity." One article for consideration is the Bob Moritz and David Burg article "How Corporate America can Fight Cybersecurity Threats," in *Fortune* (online, 17 February, 2015), http://fortune.com/2015/02/17/how-corporate-america-can-fight-cybersecurity-threats/ (accessed 22 July 2015).

24 Including, but not limited to, exposing U.S. covert operatives overseas. See Peter Roff,, "Hacker's Paradise," from *U.S.News and World Report*, 1 July 2015, http://www.usnews.com/opinion/blogs/peter-roff/2015/07/01/chinas-opm-hack-just-the-beginning-of-us-cybersecurity-problems (accessed 22 July 2015).

25 Pundits are fear-mongering, but the information breach is indeed a cause for concern. See specifically Jose Pagliery, "OPM Hack's Unprecedented Haul: 1.1 Million Fingerprints," from *CNN Money*, 10 July 2015, http://money.cnn.com/2015/07/10/technology/opm-hack-fingerprints/ (accessed 22 July 2015).

## 5 Cyber as *both* War and Not War

In his timeless thought experiment, physicist Erwin Schrödinger proposed an idea about a cat in a box. The cat was nestled in the closed box with a vial of radioactive material, which would decay and at a random time expose the innocent cat to a lethal dose of radiation. Without any way of observing the cat, Schrödinger proposed that the cat was both alive and dead, because it was impossible to determine the reality until it was observed. Schrödinger's idea built on Werner Heisenberg's brilliant "Uncertainty Principle"[26] on quantum mechanics. In the cyber realm, there are constant attacks from abroad, but they are not perceived until they are observed, at which time they become reality. An unobserved cyber attack cannot be perceived unless it creates conditions that have actual effects and the affected party figures out that they are real. It is an excellent example of subterfuge; an unobserved attack is both happening and not happening at the same time. This leads to one of this work's later recommendations regarding the need for vigilance. We know what is in the box; we just need to assume that attacks are occurring and look for anomalies in the system.

Furthermore, cyber attacks are inherently hard to attribute. Until they are claimed, it is very hard to figure out who did it. It took months after the initial Stuxnet infection for the Iranians to figure out that they were suffering from a computer virus that was crippling their Uranium enrichment industry. They speculated that it was either the U.S. or Israel (or both), but neither has ever claimed responsibility. Journalists from around the world have argued that the U.S. and Israel were both involved, but the event remains shrouded in secrecy.[27] While cyber attacks may be war, they may be un-attributable, thus, not war. Put another way, can there be a war if there is no one to blame or against whom we can respond?

Finally, cyber attacks fall in the realm of both war and not war simultaneously because they are outside the existing paradigm of war.[28] They seem to be something else. They may be attacks, but without actual damage they remain unclassifiable as war. Even if there is actual damage, and an attack can be attributed, the offended party may stop short of declaring it an act of war or

26 Werner Heisenberg, "Über den Anschaulichen Inhalt der Quantentheoretischen Kinematic und Mechanik," in *Zeitschrift für Phisik*, no 43, 1927, 172–198.

27 Tucker Reals, "Stuxnet Worm a U.S. Cyber-Attack on Iran Nukes?" from *CBS News*, 24 Sept., 2010, http://www.cbsnews.com/news/stuxnet-worm-a-us-cyber-attack-on-iran-nukes/ (accessed 22 July 2015).

28 In a Kuhnian sense, see specifically the definition of a paradigm in Thomas Kuhn, *The Structure of Scientific Revolutions* (Chicago, IL: University of Chicago Press, 1970).

even using armed force short of declared war. This brings us full circle to the concept of "perception": If the attacked does not perceive the attack as an act of war, is it in fact war?

## 6 Cyber as *neither* War nor Not War

Thus cyber attacks are neither War nor Not War. We have entered a new era of War *and* warfare, where cyber attacks are simply one more tool for a sophisticated belligerent to use against an opponent. With the ubiquity of interconnectedness provided by the internet, the entire concept of War has changed. Humanity has progressed, through technology, to a post-war society. Obviously there will be conflict. Obviously there will be killing in the name of ideology and politics. Obviously there will be War in the traditional sense. But cyber presents a paradigm shift for humanity. The future of conflict will be less about political change within the established liberal democratic capitalist model (nearly all of the civilized world has subscribed to this idea), and more about economic manipulation of the established system in the struggle between the declining state system and the rising power of corporate globalism. The future of War will be less about power and more about wealth. The difference that cyber brings to the fight is that it has the potential to completely supplant all other forms of military (or corporate) force. In the 14th century, gunpowder changed warfare but was so expensive that only emerging states could afford the technology, which led in turn to the rise of nation-states, their bureaucratization, formal taxation, professional militaries, and a host of other massive societal changes.[29] In the late 19th and early 20th centuries, the internal combustion engine changed warfare and society again, especially with the introduction of airpower by World War I.[30] At the end of 1945, atomic power undermined many of the traditional concepts of warfare but has yet to change society from petroleum-based consumption to nuclear power.[31] But these Revolutions were incredibly expensive and could only be obtained by wealthy, sophisticated, and technology-oriented governments. In the case of nuclear power (atomic and thermonuclear), only nine states today possess the hardware; only seven

29 See specifically the Michael Roberts "Revolution in Military Affairs" argument, first presented as lectures in 1955, then in print in 1965. The entire argument is encapsulated by Clifford Rogers, *The Military Revolution Debate* (Boulder, CO: Westview Press, 1995).

30 Among the best books that discuss the influence of aviation on military and society is Peter Fritzsche, *A Nation of Fliers* (Cambridge, MA: Harvard University Press, 1992).

31 See for example Richard Rhodes, *Nuclear Renewal: Common Sense About Energy* (New York, NY: Viking, 1993).

actually produce nuclear weapons. And, neither airpower nor nuclear power changed the existing War paradigm. War up to the end of the 20th century was based on the traditional concept of power and force to coerce an enemy. The destructive potential of these two weapon systems created the conditions that all war since 1945 has been, by definition, limited. This is a good thing: The U.S. and USSR never launched ICBMs at each other, even though both watched the other nervously throughout the Cold War. There have been a number of limited conflicts in the interim, between states, involving non-state actors, and everything in between, but all of the conflicts since 1945 have been limited.

Cyber power has the potential to shift the paradigm.[32] The technology is cheap, ubiquitous, and unlimited. Cyber warriors can be state sponsored, but can also be individuals, anywhere in the world, launching probes and attacks from centralized or remote locations. They can destroy or disrupt government or corporate systems with anonymity and impunity. They are at large, and they are active. The cheap technology allows anyone with the will and skills to act against anyone else in the virtual realm for political or economic purpose. This is no longer War; it is something else.

## 7 A New Definition of War

War remains. Groups will use weapons, force, and power to attempt political change. Cyber will be a tool within the accepted War paradigm. But Cyber power also opens the door for a new definition of War. Used on its own, cyber power has the potential to change the very nature of War. Without the traditional violence and destruction of weapons, cyber can influence political change without creating massive damage or casualties. In the 21st century, cyber will change the perception of what war is and is not. Cyber power will force defense intellectuals and governments – to say nothing of corporations and individuals – to rethink their vulnerabilities and weaknesses as we evolve into the predominantly digital age. Countries and corporations that are highly dependent on the cyber domain will be especially vulnerable until they evolve their capacity for virtual protection and defense. This defense will no longer be limited to governments alone; from individuals to corporations to states, all will have to focus on protecting themselves in the cyber domain.

Countries with massive conventional militaries that lack cutting-edge cyber capabilities will likely become obsolete. The U.S. government still

32 Kuhn, *op cit.*

considers overwhelming military hardware superiority to be the panacea for all contingencies. The U.S. still builds multi-million and billion-dollar-plus systems like submarines, ICBMs, aircraft carriers, and air-superiority fighter planes, while at the same time maintaining a massive ground force. Only a small portion of the U.S. military budget is allocated to the analysis and exploitation of the cyber domain.[33] For a government and country so dependent on the internet and the cyber domain, this is a dangerous oversight.

The new method of war is still balanced in favor of the offensive. It is easier and cheaper to launch a cyber attack than it is to defend against one. That may change, but today offense is more powerful and prevalent. Cyber security remains largely reactionary. The U.S. government, military, and corporations are having an increasingly hard time adapting to the skills and abilities of cyber attacks (some label it cyber terrorism depending upon the attackers' motives and the defenders' perceptions). The bad guys are winning; the defenders will need funding, experience, skills, and viable counterattacks to become more successful.

In the end, cyber power will redefine the nature of War. First, cyber power will be able to "attack" an enemy without the overt damage of a nuclear strike or a bombing raid. A cyber attack may not even harm any people or destroy any mechanical systems. It may only be a tool for espionage or theft. Second, cyber power may change the political outlook of entire countries without causing any damage whatsoever; the ability of a cyber attack to destabilize an entire society is within the realm of possibility of the technology. Third, a cyber attack will only be described as War if the offending party perceives it as such. It will take an act of political will to label a cyber attack as an act of war to determine if it falls within the traditional definition. Until then, it will not be War, it will be something else. Unfortunately, the U.S. is not in a good position to respond or defend, and will continue to be harassed by cyber attacks until our government comes up with clear policy and guidelines for activities within the cyber domain, then fund the proper organizations to deal with the enormous technological implications now and into the future. Even so, there is no panacea – the attacks will be constant, and vigilance plus capability will be required to stop at least some or most of them.

---

33 Although that budget is expanding, as the U.S. military realizes the importance of the cyber domain, see Brian Fung, "Cyber Command's Exploding Budget, in 1 Chart," *The Washington Post*, 15 January 2015 (accessed 22 July 2015). The budget was increased from just over $200 million to over $400 million from FY 13 to FY 14, but today only represents .06% of the defense budget. The U.S. military's new fighter jet, the JSF-35 received $10.6 billion for FY 14.

As I refine my syllabus for this academic year, I will rely on the latest information on cyber capabilities from esteemed colleagues in the U.S. and around the world. I am confident that there are a number of defense intellectuals and academics taking the threat seriously and putting effort, time, and resources into understanding the cyber domain. I hope to build on the work of Paul Springer, Peter Singer, Martin Libicki, Jeffrey Carr, and Derek Reveron as well as others, and their excellent scholarship on cyber capabilities and threats.[34] But at the same time it is important to return to the classical military theorists to re-think the concept of War, rewrite the definition of War for the 21st century, and consider how Joe Guilmartin would have approached this vitally important topic and its implications for our country's security and prosperity.

34 Paul Springer, *Cyber Warfare* (Santa Barbara, CA: ABC-Clio, 2015), Peter Singer, *Ghost Fleet* (Boston, MA: Houghton Mifflin Harcourt, 2015), Martin Libicki, *Cyberdeterrence and Cyberwar* (Santa Monica, CA: RAND, 2009) as well as his new article with David Gompert, "Waging Cyberwar the American Way," in *Global Politics and Strategy*, (57, no. 4, 7–28), Jeffrey Carr, *Inside Cyber Warfare* (Sebastopol, CA: O'Reilly, 2012), and Derek Reveron, (ed.), *Cyberspace and National Security* (Washington, DC: Georgetown University Press, 2012).

CHAPTER 16

# The Precision-Information Revolution in Military Affairs and The Limits of Technology

*Peter R. Mansoor*

On September 9, 1943, the Italian battleship *Roma* was making its way south towards Malta and the safety of Allied air cover. While traversing the Strait of Bonifacio between Sardinia and Corsica, the ship and its escorts came under attack from a specially trained German air squadron equipped with a new weapon: the Fritz X guided bomb. Six Dornier Do 217 aircraft, each carrying two bombs, flew over the Italian flotilla at an altitude of 5,500 meters. Shortly after 3:30 pm, the Germans attacked. The *Roma* and its escorts fired back and took evasive action, but to no avail. German bombardiers guided bombs onto the *Roma* and *Italia*, causing them to lose speed and drop out of formation. Shortly before 4 pm another Fritz X penetrated the *Roma* and detonated inside the forward engine room. The forward magazine exploded, causing catastrophic damage to the ship. In a matter of minutes the *Roma* capsized and sank, taking 1,253 of her crew with her to a watery grave.[1]

Two days later on the morning of September 11, 1943, the *Brooklyn*-class light cruiser USS *Savannah* was positioned in the Bay of Salerno, supporting U.S. ground troops ashore with naval gunfire. The ship was being tracked from 5,700 meters overhead by a Dornier Do 217 aircraft, which released a Fritz X bomb towards it. The bombardier guided the bomb onto the roof of the Savannah's No. 3 gun turret. The Fritz X passed through three decks before exploding in the lower ammunition-handling room, causing secondary explosions that nearly doomed the ship. Heroic damage control efforts by *Savannah*'s crew saved the vessel, which was forced to sail to Malta for repairs.[2]

1 Brendan McNally, "The Sinking of the Battleship Roma and the Dawn of the Age of Precision Guided Munitions," Defense Media Network, June 1, 2014, http://www.defensemedianetwork.com/stories/the-sinking-of-the-battleship-roma-and-the-dawn-of-the-age-of-precision-guided-munitions/.

2 "USS Savannah (CL-42), 1938–1960 – Damaged by German Bomb, 11 September 1943," Naval Historical Center, https://www.ibiblio.org/hyperwar/OnlineLibrary/photos/sh-usn/usnsh-s/cl42-l.htm; "USS Savannah (CL-42)," Naval Warfare, March 10, 2009, http://navalwarfare.blogspot.com/2009/03/uss-savannah-cl-42.html.

 | DOI:10.1163/9789004393301_018

ILLUSTRATION 16.1 German "Fritz-X" Guided Bomb in the Air Power Gallery at the National Museum of the U.S. Air Force. (By U.S. Air Force [Public domain], via Wikimedia Commons - https://upload.wikimedia.org/wikipedia/commons/5/57/Bomba_Fritz_X_en_el_Museo_de_la_Fuerza_Aerea_de_USA.jpg).

For several days German bombardiers showered the Allied fleet at Salerno with guided bombs, hitting the cruiser USS *Philadelphia*, the light cruiser HMS *Uganda*, the battleship HMS *Warspite*, and several merchant ships. All of the ships were temporarily put out of action and had to retire for repairs.

The Fritz-X was a gravity bomb that could have its course adjusted while in flight (See Illustration 16.1). It consisted of an 11-foot-long casing containing 320 kilograms of explosive. A gyroscope ensured the boxlike tail section remained pointed upwards so that it could receive signals from the controlling aircraft overhead. Adjustments were made in flight by a bombardier who lined up a flare in the bomb's tail with the target aim point using a joystick connected to a radio controller. The joystick controlled oscillating spoilers in the tail section that provided pitch and yaw control. Provided the bomb was initially released to land within 500 meters of its target, a skilled operator could hit within 30 meters of the aim point 90 percent of the time.[3]

The German onslaught sent the Allies searching for countermeasures. They were fairly certain the bombs were radio-controlled, but did not know the frequency spectrum used. The controlling aircraft, however, had to fly straight

3 McNally, "The Sinking of the Battleship Roma;" "Guided German air to ground weapons in WW2," http://www.1jma.dk/articles/1jmaluftwaffegroundweapons.htm.

and level to keep connectivity with the bomb while in flight, which made them vulnerable to fighter aircraft and anti-aircraft fire. Two U.S. destroyer escorts, the USS *Frederick C. Davis* (DE 136) and the USS *Herbert C. Jones* (DE 137), specially outfitted with radio search receivers and jammers, were hastily dispatched to the Mediterranean. In an attack on a convoy off the Algerian coast on November 6, the operators on the ships determined the frequency of the German transmissions and the ship's jammers were tuned to cover that particular part of the spectrum. Improved jammers were soon rushed across the Atlantic and proved effective (but not foolproof) in impeding German use of radio controlled bombs at Anzio, during the invasion of Normandy (where more than forty ships were equipped with jammers), and during the invasion of southern France (where more than sixty ships were so equipped). The Germans launched twenty radio-guided bombs against the invasion forces off the coast of southern France, recording only one hit against a LST (landing ship, tank).[4]

One could argue that the Japanese, in the form of *kamikaze* aircraft, were the most successful at fielding guided weapons during World War II. The fact that these aircraft were controlled by human pilots rather than technological guidance systems is irrelevant; their impact on the battlefield was the same. The Japanese launched Operation *Kikusui* ("Floating Chrysanthemum") during the invasion of Okinawa between April 6 and June 21, 1945, during which 10 major waves (roughly 1,400 aircraft altogether) of *kamikaze* targeted the Allied naval flotilla offshore, sinking 36 ships, damaging 368 others, and killing more than 5,000 American sailors. It was the largest loss of ships and life sustained by the U.S. Navy during the war.[5]

The U.S. military also experimented with limited numbers of guided weapons during World War II, among them acoustic homing torpedoes, radio-guided glide bombs (AZON), and a radar-homing glide bomb (BAT) which contained an internal, autonomous radar seeker. These weapons scored a number of ship and submarine kills before the war's end.[6] A U.S. Navy program to develop surface-to-air missiles (Project Bumblebee) to defend against platforms carrying guided weapons, made even more imperative by ship losses to kamikazes

4 Chief of Naval Operations, "How Radio-Controlled Bombs Were Jammed," *C.I.C.* (*Combat Information Center*), December 1945, reprinted at Lone Sentry blog, September 19, 2010, http://www.lonesentry.com/blog/how-radio-controlled-bombs-were-jammed.html.

5 C. Peter Chen, "Okinawa Campaign, 1 April 1945 – 21 June 1945," World War II Database, http://ww2db.com/battle_spec.php?battle_id=15.

6 David R. Mets, "The *Force* in US Air Force," *Aerospace Power Journal*, Fall 2000: 57–77, 61–62, http://www.au.af.mil/au/afri/aspj/airchronicles/apj/apj00/fal00/mets.pdf.

during the battle for Okinawa, eventually resulted in the first operational SAMs: the medium-range Terrier (1955) and the long-range Talos (1959).[7]

The age of guided munitions had begun, along with the battle between measures and countermeasures that continues to this day.[8]

The end of the war, cutbacks in defense spending, and the advent of atomic weapons combined to slow significantly the development of guided weapons after World War II. The Korean War saw more testing of radio-controlled guided bombs, but no new developments. By 1960, the U.S. Navy had developed a wire-guided torpedo, the Mark 45, which carried a nuclear warhead (making it less than precise in its effects). Surface-to-air missile development proceeded apace, with the U.S. Army developing the Nike for continental defense. The Soviets demonstrated their capabilities with the downing of a U-2 spy plane flown by Gary Francis Powers on May 1, 1960. The Soviets used a salvo of fourteen SA-2 Guideline surface-to-air missiles to hit the U-2 at an altitude of more than 60,000 feet while it traversed Soviet airspace.[9] The Soviets also developed the first ship-to-ship missile to sink a vessel in combat. On October 21, 1967, Egyptian missile boats launched four Soviet-made SS-N-2 Styx anti-ship missiles that sank the Israeli destroyer *Eilat* off the Sinai coast.[10] But the most successful guided weapon developed in the first two decades after World War II was the air-to-air AIM-9 Sidewinder missile, which uses infrared homing for autonomous guidance and has a range of 22 miles. Sixty years and more than a dozen variants later, the Sidewinder is still in use by air forces around the world.[11]

In the 1960s and early 1970s the U.S. military made the most significant gains in guided weapons and their countermeasures, due in large measure to the Vietnam War. The SA-2 surface-to-air missile (SAM) threat over North Vietnam during Operation Rolling Thunder led to the development of "Wild Weasel" aircraft specifically outfitted to detect the radar emissions associated with

7 George Galdorisi, "U.S. Navy Missile Defense: Getting Surface-to-Air Missile Development Started," Defense Media Network, October 18, 2012, http://www.defensemedianetwork.com/stories/u-s-navy-missile-defense-getting-surface-to-air-missile-development-started/.

8 The Fritz X was one of two German precision guided munitions used in large numbers during the war. The other, the Henschel Hs 293, was a radio-controlled, rocket-propelled glider with stubby wings that enjoyed limited success against unarmored targets.

9 Office of the Historian, U.S. Department of State, "U-2 Overflights and the Capture of Francis Gary Powers, 1960," https://history.state.gov/milestones/1953-1960/u2-incident.

10 John C. Shulte, "An analysis of the historical effectiveness of anti-ship cruise missiles in littoral warfare," Masters Thesis, Naval Postgraduate School, 1994, 3. http://calhoun.nps.edu/bitstream/handle/10945/27962/analysisofhistorooschu.pdf?sequence=1.

11 Carlo Kopp, "The Sidewinder Story: The Evolution of the AIM-9 Missile," *Australian Aviation*, April 1994, http://www.ausairpower.net/TE-Sidewinder-94.html.

surface-to-air missile batteries, and equipped with radiation-seeking missiles capable of targeting and destroying SAM radars. Jamming pods also interfered with the SA-2 tracking mechanism, continuing the battle between measures and countermeasures begun during World War II.[12]

But it was the offensive potential of guided munitions that came to the fore during the latter stages of the Vietnam War. By 1967 research and design teams from Texas Instruments and Air Force personnel at Eglin Air Force Base conceived of an innovative idea to retrofit gravity bombs with kits that enabled them to hone in on laser light. Once the pilot or weapons systems officer released the bomb, he would only have to keep the target "painted" with light from a laser designator; no longer would he have to steer the bomb into the target. The result was a radical improvement in bombing accuracy (the laser guided bombs were engineered to hit within 25 meters of the target 80 percent of the time) at a cost of $16,000 per bomb.[13]

During this same period, the Air Force launched a program to create an air-to-ground guided missile to replace the outdated AGM-12 Bullpup, first fielded in 1959. Developed and produced by Hughes Missile Systems Division, the AGM-65 Maverick missile entered the Air Force inventory in 1972 (See Illustration 16.2). Since its introduction, various models of the Maverick have used television, laser, and infra-red guidance systems. According to the manufacturer, the Maverick has become "the most widely used precision-guided missile in the free world."[14]

Laser-guided bombs were used extensively during Operations Linebacker I and Linebacker II in 1972. To stem the Easter Offensive that began on March 30, aircraft equipped with laser guided bombs destroyed key bridges along the North Vietnamese lines-of-communication. Tasks that in World War II often required dozens of aircraft, hundreds of bombs, and multiple sorties could now be performed by a single aircraft in one strike. A RAND study calculated the impact of this newfound accuracy:

> Of the more than 10,500 LGBs delivered throughout Southeast Asia from February 1972 through February 1973, 5,107 (48.1 percent) were assessed as direct hits; and another 4,000 achieved a circular error probable (CEP) of 25 feet. Compared to the 500-foot CEP of F-105s using manual

12 Barry D. Watts, *Six Decades of Guided Munitions and Battle Networks: Progress and Prospects* (Washington: Center for Strategic and Budgetary Assessments, 2007), 6.

13 Peter deLeon, *The Laser-Guided Bomb: Case History of a Development* (Santa Monica, CA: RAND, 1974), 22–23.

14 Raytheon, "Maverick AGM-65," https://web.archive.org/web/20131104071216/http://www.poweredmodelairplanes.com/pma/a10/maverick04.pdf.

ILLUSTRATION 16.2 AGM-65 Maverick missile moments before impacting on an M-48 tank. (By USAF [Public domain], via Wikimedia Commons – https://upload.wikimedia.org/wikipedia/commons/2/20/AGM-65_M-48_pre_impact.jpg)

dive-bombing to deliver unguided bombs against heavily defended targets in North Vietnam during 1965–1968, this was a remarkable improvement. Laser-guided bombs achieved an estimated CEP of 10–15 feet. Thus, although LGBs required clear air and were mostly limited to daytime operations in Vietnam, their accuracy was 33–50 times better than unguided bombs.[15]

RAND analysts quickly divined the impact of laser guided bombs and other emerging "smart" weapons on the future of warfare. Simply put, they were the future. Air Force brass, however, refused to embrace guided munitions as the centerpiece of air-to-ground warfare. Two more decades and another war would pass before Air Force culture adjusted to the changing technology and

15 Watts, *Six Decades of Guided Munitions and Battle Networks*, 9.

ILLUSTRATION 16.3 An RGM-84 Harpoon missile, immediately after launch from the cruiser USS *Leahy* (CG-16), near the Pacific Missile Test Center, California. (Public domain via Wikimedia Commons - https://upload.wikimedia.org/wikipedia/commons/2/20/AGM-84_Harpoon_launched_from_USS_Leahy_%28CG-16%29.jpg).

manifest potential of guided munitions.[16] When it finally embraced guided weapons, however, the Air Force went all in.

Like the U.S. Air Force, the U.S. Navy belatedly awoke to the potential of anti-ship guided missiles. The Navy launched studies to produce an anti-ship missile in 1965, an effort energized by the sinking of the *Eilat* two years later. The weapon was initially designed to target submarines on the surface of the water; its code name "Harpoon" was patterned after the weapons used to hunt whales (See Illustration 16.3). Despite being given a boost by Admiral Elmo Zumwalt upon his assumption of duties as chief of naval operations in 1970, the Harpoon did not enter service until 1977, a full decade after its development began. The versatile weapon, "a high-subsonic, seaskimming missile with active radar terminal guidance," could be launched by aircraft, or from ships or

16 Watts, *Six Decades of Guided Munitions and Battle Networks*,".

submarines.[17] For the United States Navy, however, the striking power of the fleet still largely resided in the aircraft wings housed on carrier flight decks.

As if American naval officers needed any further prodding, the sinking of several vessels of the small Pakistani Navy by three Indian missile boats equipped with Styx missiles in two separate attacks during the 1971 Indo-Pakistan War again reinforced the importance of these new weapons and the need for the development of countermeasures against them.[18] The first combat between missile-equipped ships took place less than two years later during the 1973 Yom Kippur War, with the Israeli Navy beating its Syrian and Egyptian counterparts. The Israeli boats, whose Gabriel anti-ship missiles were outranged by their opponents' Styx missiles, were equipped with jammers and chaff, which successfully deflected incoming missiles and allowed the Israelis to close within killing distance.[19] Nineteen of fifty Gabriel missiles hit their targets, while none of the forty-seven Styx missiles fired by the Syrian and Egyptian crews found their marks.[20]

The next decade brought further proof of the efficacy of anti-ship guided missiles. During the 1982 Falklands War, Argentinian Super Etendard aircraft firing French-made Exocet anti-ship missiles sank the destroyer HMS *Sheffield* and the container ship SS *Atlantic Conveyor*. On May 17, 1987, during the Iran-Iraq conflict, an Iraqi F-1 Mirage launched two Exocet missiles that hit the guided missile frigate USS *Stark*, killing 37 sailors and injuring 21 others. Defensive systems (such as the Close In Weapon System and chaff dispensers) were not activated, making the ship a vulnerable target. Valiant efforts by the crew, however, saved the vessel from sinking.[21] Exocet missiles proved effective in hitting and sinking or damaging undefended targets; from 1980 to 1984, 52 of 53 Exocets hit their targets, causing havoc among the oil tanker fleet in the Gulf.[22]

The U.S. Army was also late to embrace guided weapons, and some of its branches, such as the armor community, never have. But given the manifest

17 Watts, *Six Decades of Guided Munitions and Battle Networks*, 151–152.

18 Pakistani losses included two destroyers, a minesweeper, three cargo ships, and an oiler as well as the damage or destruction of fuel storage tanks and naval installations ashore. Admiral S.M. Nanda, *The Man Who Bombed Karachi* (Noida, India: HarperCollins India, 2015).

19 "Yom Kippur War: The Battle of Latakia," Jewish Virtual Library, http://www.jewishvirtuallibrary.org/jsource/History/latakia.html.

20 Shulte, "Anti-ship cruise missiles in littoral warfare," 8.

21 Commander, Cruiser-Destroyer Group 2, "Formal Investigation into the Circumstances Surrounding the Attack on the USS Stark (FFG 31) on 17 May 1987," June 12, 1987, http://web.archive.org/web/20120323222902/http://www.dod.gov/pubs/foi/operation_and_plans/USS_Liberty_Pueblo_Stark/65rev.pdf.

22 Shulte, "Anti-ship cruise missiles in littoral warfare," 8.

effectiveness of laser-guided munitions in the latter stages of the Vietnam War and the futility of unguided artillery munitions hitting armored vehicles without massive expenditure of ammunition, the Army in 1971 began research and development into guided munitions that could be fired from helicopters, cannons, and howitzers. By 1982 the Army was ready to field its new creation, the M172 Cannon-Launched Guided Projectile, or Copperhead: the world's first cannon-launched, fin-stabilized, terminally guided munition designed to hit point-targets with the assistance of a laser designator in the hands of a forward observer.[23] The defense research and development community was ecstatic about the possibilities, going so far as to claim that the Copperhead would make armored warfare obsolete: "It [the Copperhead munition] will deny the Russians the opportunity of massing tanks because we're going to tear the hell out of them with these kind of systems," boasted Dr. William Perry, undersecretary of defense for research and engineering (and a future secretary of defense). "[Precision guided munitions] have a greater potential for revolutionizing tactical warfare than the introduction of radar in World War II."[24] The Copperhead failed to live up to its billing, but its much more successful counterpart, the AGM-114 Hellfire missile, has been wildly successful. The laser-guided missile, designed for use on the Army's new AH-64 Apache attack helicopter, reached full operational capability in 1985 and is still the weapon of choice for use on helicopters and armed unmanned aerial vehicles. More than 60,000 Hellfire missiles of all variants have been produced to date.[25]

In addition to the advent of air-to-ground guided missiles, U.S. and Soviet development of anti-tank guided missiles (ATGM) with potent shaped-charge warheads, such as the AT-3 Saggar and the TOW (tube-launched, optically tracked, wire-guided anti-tank missile), made survival difficult for armored vehicles. The 1973 Yom Kippur War showed the vulnerability of armored vehicles to these weapons; in just three weeks the Israeli Defense Force and Syrian and Egyptian armies combined lost 2,650 tanks, more than existed in the entire arsenal of the U.S. Army in Europe at the time.[26] Although the development

23 Picatinny Arsenal, "Smart Munitions," http://www.pica.army.mil/Picatinny/products_services/products01.aspx.

24 William Currie, "New 'smart' weapons make 'Star Wars' appear tame," *Chicago Tribune*, January 21, 1979, http://archives.chicagotribune.com/1979/01/21/page/32/article/new-smart-weapons-make-star-wars-appear-tame.

25 "Boeing/Lockheed Martin (Rockwell/Martin Marietta) AGM-114 Hellfire," Directory of U.S. Rockets and Missiles, http://www.designation-systems.net/dusrm/m-114.html.

26 Richard W. Stewart, ed., *American Military History*, Vol. II (Washington: Center of Military History, 2005), 377; http://www.history.army.mil/books/amh-v2/amh%20v2/chapter12.htm.

of improved composite layered armor, active countermeasures, and combined arms tactics to suppress the ATGM threat meant that the tank was anything but dead, the loss rates in the Yom Kippur War called into question the ability of NATO's armored forces to halt a determined Soviet assault.[27]

What this plethora of guided munitions lacked was a concept that tied them together into a coherent operational doctrine. New ideas along these lines originated not with the services but with the Office of the Secretary of Defense and agencies such as the Defense Advanced Research Projects Agency (DARPA). As early as 1975 analysts "explored the prospective utility of 'near zero miss' conventional weapons to substitute for 'massive nuclear destruction,' emphasized the importance of developing all-weather guided munitions, and conceptualized a reconnaissance-strike capability using guided munitions and advanced sensors."[28] This idea was the genesis in the late-1970s of Assault Breaker, "a Department of Defense (DOD) concept using standoff weapons to attack moving, rear echelon armor massed deep behind enemy lines."[29] The concept was aimed at defeating the echeloned waves of the Red Army should it attack NATO forces in Western Europe using new capabilities to supplement U.S. Air Force air interdiction sorties. Airborne radars would locate enemy armored columns, which would then be attacked by air- or ground-launched missiles dispensing self-guided submunitions over the target.[30] If successful, Assault Breaker would preclude the need to use tactical nuclear weapons to prevent a Soviet takeover of Western Europe.

The intellectual ferment surrounding Assault Breaker and associated programs paid huge dividends. Efforts to develop a long-range targeting radar eventually resulted in the creation of the Joint Surveillance Target Attack Radar System, or JSTARS. The munitions program and associated delivery system morphed into the Army's new Army Tactical Missile System (ATACMS) and Multiple Launch Rocket System (MLRS) (See Illustration 16.4). Army doctrine writers meanwhile had rejected the linear, attrition-based doctrine of "active defense" in favor of a maneuver-based concept known as "AirLand

27 John Craddock, "The Tank is Dead – Long Live the Tank," USAWC Military Studies Program Paper, April 15, 1993, http://www.dtic.mil/dtic/tr/fulltext/u2/a264578.pdf; Tony Wilkins, "Is the Main Battle Tank Obsolete?" Defence of the Realm, March 18, 2016, https://defenceoftherealm.wordpress.com/2016/03/18/is-the-main-battle-tank-obsolete/.

28 Watts, *Six Decades of Guided Munitions and Battle Networks*, 11.

29 Comptroller General, "Report to the Congress: Decisions to Be Made in Charting the Future of DOD's Assault Breaker," February 28, 1981, i, http://www.gao.gov/assets/140/132235.pdf.

30 Ibid.

ILLUSTRATION 16.4
*U.S. Army Tactical Missile System* (ATACMS) (*Public domain via Wikimedia Commons - https://upload.wikimedia.org/wikipedia/commons/d/d6/ATACMSMay2006.jpg*).

Battle."[31] In November 1984 NATO adopted Follow-on Forces Attack (FOFA) as a new mission concept to improve alliance capabilities to counter Soviet conventional forces, and thereby delay the point at which the NATO commander would have to authorize the use of tactical nuclear weapons.[32] Together with AirLand Battle doctrine, FOFA gave reason and purpose to the growing arsenal of guided munitions:

> FOFA attack and AirLand Battle Doctrine focused military planning and weapons development on spatial and temporal dynamics: conceptualizing how to interdict Soviet follow-on-forces located 24, 48, and 72 hours forward from NATO defensive positions; coordinating Army and Air Force Deep Battle and Air Interdiction efforts out to 150 kilometers behind Soviet forward lines (roughly 72 hours from arriving at NATO

31 John L. Romjue, "The Evolution of the Airland Battle Concept," *Air University Review* 35, no. 4 (May-June 1984): 4–15, http://www.airpower.maxwell.af.mil/airchronicles/aureview/1984/may-jun/romjue.html.

32 Office of Technology Assessment, *Technologies for NATO's Follow-On Forces Attack Concept* (Washington: Government Printing Office, July 1986), 1.

> defensive positions based on Soviet Doctrine); planning Air Force deep strike missions 300 kilometers into Soviet rear areas; and developing technology and doctrine to enable commanders to quickly adapt fire missions and air strikes. FOFA was the first "systems-of-systems" architecture. Nearly 100 current and planned NATO systems were eventually integrated into FOFA attack planning, establishing many of the legacy capabilities in use today.[33]

Meant to counter the growing Soviet conventional arms superiority in Europe, the "systems-of-systems" surrounding guided munitions in time came to dominate the thinking of the U.S. military regarding warfare in general.

Ironically, the Soviets, against whom Assault Breaker was directed, were the first to realize the full implications of guided munitions and associated surveillance and reconnaissance systems. As improved sensors and reconnaissance systems and guided weapons emerged in the 1970s, Soviet military leaders theorized about their implications, classifying them as part of an emerging "military-technical revolution."[34] The Soviets were deeply concerned about the West's research into these new technological developments and the fielding of guided weapons, which would enable the realization of deep attacks against Red Army forces advancing to contact. Chief of the Soviet General Staff Marshal Nikolai Orgakov stated in an interview in May 1984 that developments in conventional weaponry, including terminally guided weapons, would "make it possible to sharply increase, by at least an order of magnitude, the destructive potential of conventional weapons, bringing them closer, so to speak, to weapons of mass destruction in terms of effectiveness."[35] Indeed, one military officer on special assignment to RAND contended, "The Soviets have listened to the claims being made by Western advocates of ET [emerging technologies] weaponry and have apparently accepted these claims with more enthusiasm than has been displayed by many in the United States and Western Europe."[36]

---

33 Robert Tomes, "The Cold War Offset Strategy: Assault Breaker and the Beginning of the RSTA Revolution," War on the Rocks, November 20, 2014, http://warontherocks.com/2014/11/the-cold-war-offset-strategy-assault-breaker-and-the-beginning-of-the-rsta-revolution/.

34 For an examination of Soviet strategic culture and its impact on Soviet theorizing concerning guided munitions, see Dima Adamsky, *The Culture of Military Innovation: The Impact of Cultural Factors on the Revolution in Military Affairs in Russia, the US, and Israel* (Palo Alto, CA: Stanford University Press, 2010).

35 Michal J. Sterling, "Soviet Reactions to NATO's Emerging Technologies for Deep Attack," RAND, August 1985, 8, http://www.rand.org/content/dam/rand/pubs/notes/2009/N2294.pdf.

36 Ibid., 23.

Comparing guided munitions to nuclear weapons was no minor assertion. The Soviets believed the U.S. military was on the verge of creating reconnaissance-strike and reconnaissance-fire complexes that could dominate the battlefield and neutralize the advantage in conventional weapons attained by the Soviets in Europe over the previous two decades. Such systems-of-systems would be comprised of three major elements: automated command and control mechanisms, sophisticated sensors to enable thorough battlefield coverage, and guided munitions to attack pinpoint targets.[37] While American military officers saw guided munitions as technical achievements that could improve tactical outcomes and enable operational maneuver, the Soviet military viewed them "as a yawning discontinuity that had significant ramifications in terms of the methods of employing corps and armies, and that shaped the nature of war in a revolutionary way."[38] Given that Soviet Marshal Mikhail Tukhachevsky pioneered the concept of deep operations in the 1930s, Soviet military leaders were well-prepared intellectually to conceptualize a new doctrine from the technological advances of the 1960s and 1970s.[39]

The Soviet Union, however, lacked the economic wherewithal and technological know-how to develop the information technology, weapons, and sensors needed to take advantage of this theorizing. In the 1980s only the United States possessed the means to design and build so-called "smart" weapons and the systems required to control and employ them. This situation became utterly clear during the first major post-Cold War conflict, the Gulf War of 1991.

The Gulf War was the first conflict in which guided munitions played a major and arguably decisive role in the outcome.[40] Hellfire-equipped AH-64 Apache helicopters led the way the first night of the air campaign, destroying an air defense radar installation and opening a path through which fixed wing aircraft penetrated to their targets inside Iraq.[41] Sea-launched Tomahawk land-attack cruise missiles, developed by McDonnell Douglas in the 1970s, were used in combat for the first time, as well as air-launched cruise missiles transported

37 Barry D. Watts, "The Evolution of Precision Strike," Center for Strategic and Budgetary Assessments, 2013, 5, http://www.voltairenet.org/IMG/pdf/The_Evolution_of_Precision_Strike.pdf.

38 Adamsky, *The Culture of Military Innovation*, 32.

39 Although Tukhachevsky was executed during Stalin's purges, his reputation and ideas were restored in the 1960s just as various nations began serious development of guided munitions. William J. McGranahan, "The Fall and Rise of Marshal Tukhachevsky," *Parameters* VIII, no. 4 (December 1978).

40 Although laser-guided munitions played a role in Linebacker I and II in 1972, they were not decisive to the outcome of the air campaign.

41 Clarence A. Robinson Jr., "Gulf War 20th: Apache Raid," Defense Media Network, February 2, 2011, http://www.defensemedianetwork.com/stories/gulf-war-20th-apache-raid/.

to their launch points aboard venerable B-52 bombers.[42] F-117 stealth fighters, operating under cover of darkness, carried laser-guided munitions to targets in Baghdad. Precision-guided munitions devastated the Iraqi Air Force and air defense network, destroyed key command and control and transportation nodes, and hit other key targets. After achieving air superiority and strategic effects inside Iraq, coalition forces turned their attention to destroying armored targets (euphemistically labeled "tank plinking") in the Kuwaiti Theater of Operations in an attempt to fulfil General Norman Schwarzkopf's goal of reducing Iraqi armored strength by 50 percent prior to the commencement of the ground attack.[43] The American public was enthralled by videos of guided munitions hitting specific aim points on buildings, bridges, oil terminals, and other targets.

Air attacks isolated Iraqi forces in Kuwait. Echoing the achievements of laser-guided munitions during Linebacker I and II, precision-guided munitions dropped by coalition aircraft "destroyed 41 of 54 key Iraqi bridges, as well as 31 pontoon bridges hastily constructed by the Iraqis in response to the anti-bridge strikes, in approximately four weeks."[44] JSTARS-equipped aircraft proved effective in identifying Iraqi vehicular columns moving to and from Kuwait.[45] Coalition aircraft (the vast majority U.S.) dropped 10,468 guided bombs, fired 2,151 anti-radiation missiles, and fired 5,508 guided missiles (principally Mavericks) at various targets, just less than 10 percent of the total air-delivered tonnage expended during the air campaign.[46] In addition, U.S. Army and Marine Corps helicopters fired 3,035 Hellfire missiles during the war.[47] But precision-guided munitions had an effect far greater than their raw numbers; they have been "credited with causing approximately 75 per cent of the serious damage inflicted upon Iraqi strategic and operational targets."[48] Laser-guided

42 "Submarine Armament: Cruise Missiles," National Museum of American History, http://www.americanhistory.si.edu/subs/weapons/armament/missiles/index.html. U.S. military services experimented with cruise missile technology in the 1950s, but shelved it in favor of more powerful and technologically mature ballistic missiles. Improvements in cruise missile technology and guidance systems convinced the services to resume research and development into cruise missiles in the 1970s.

43 Thomas A. Keaney and Eliot A. Cohen, *Gulf War Air Power Survey Summary Report* (Washington: Government Printing Office, 1993), 14–15, http://www.afhso.af.mil/shared/media/document/AFD-100927-061.pdf. Results achieved were between 25 and 40 percent, but varied by unit – see 105–107.

44 Richard P. Hallion, "Precision Guided Munitions and the New Era in Warfare," APSC Paper 53, Air Power Studies Centre, 1995, http://fas.org/man/dod-101/sys/smart/docs/paper53.htm.

45 Keaney and Cohen, *Gulf War Air Power Survey*, 109.

46 Ibid., Table 8, 203; Hallion, "Precision Guided Munitions."

47 Keaney and Cohen, *Gulf War Air Power Survey*, 204.

48 Hallion, "Precision Guided Munitions."

bombs were particularly lethal; due to the advent of stealth technology, aircraft could remain in a stable flight path while guiding munitions to their targets with little risk from enemy defenses. "Laser-guided bombs were particularly effective because their employment came as something of a surprise to the Iraqis," The Gulf War Air Power survey concluded. "Their reaction is understandable, because the LGB [laser-guided bomb] performance also surprised the United States."[49] When the Iraqi Air Force refused to rise to fight, coalition aircraft destroyed its hardened shelters one-by-one with laser-guided bombs. What was left of the Iraqi Air Force after the carnage was over fled to Iran, never to be seen again. The Gulf War Air Power Survey concluded, "Desert Storm reconfirmed that LGB's possessed a near single-bomb target-destruction capability, an unprecedented if not revolutionary development in aerial warfare."[50]

Shortly after the conclusion of the Gulf War, the DOD Office of Net Assessment under Andy Marshall delved into the concept of revolutions in military affairs (or, in Soviet parlance, military-technical revolutions) to determine whether advanced information systems enabled by the microprocessor, advanced sensors, and precision-guided munitions had fundamentally revolutionized warfare.[51] Marshall assigned the project to Andrew Krepenevich, a West Point-educated and Harvard-trained PhD who had served twenty-five years in the U.S. Army prior to becoming a defense analyst. His subsequent report shaped the debate over the future of U.S. defense policy for the next decade.[52]

---

49 Keaney and Cohen, *Gulf War Air Power Survey*, 226.

50 Quoted in Hallion, "Precision Guided Munitions."

51 Revolutions in military affairs are distinct from the better known historical concept of military revolutions. Military revolutions are fundamental changes in warfare brought on by broad, sweeping changes in political or societal organization, economic systems, or technology. Examples include the introduction of gunpowder artillery and artillery fortresses (*trace italien*) that both influenced and were influenced by state formation from the 14th to the 17th centuries, and the democratization of warfare engendered by the French Revolution. Revolutions in military affairs, on the other hand, are conceptual changes in warfare fostered by new military technology, doctrine, and organizations that create radical improvements in military effectiveness. Some examples of revolutions in military affairs include the development of mobile, combined arms warfare ("blitzkrieg"); strategic bombing; and carrier warfare. For an examination of military revolutions and revolutions in military affairs from a historical standpoint, see MacGregor Knox and Williamson Murray, eds., *The Dynamics of Military Revolution, 1300–2050* (Cambridge: Cambridge University Press, 2001), Chapter 1.

52 Andrew F. Krepinevich, Jr., "The Military-Technical Revolution: A Preliminary Assessment," Office of Net Assessment, July 1992, reprinted at Center for Strategic and Budgetary Assessments, http://www.csbaonline.org/wp-content/uploads/2011/03/2002.10.02-Military-Technical-Revolution.pdf.

Krepenevich posited a world on the brink of a revolution in military affairs, or RMA:

> Over the next several decades, the systems and operations characteristic of present-day military establishments will probably be superseded by new, far more capable means and methods of warfare...It is not clear how quickly this change will come about, particularly since the level of competition that spurred military innovation during the Cold War appears unlikely to recur, at least in the near future. Nonetheless, our initial examination of the MTR [military-technical revolution, the Soviet name for an RMA] strongly supports the hypothesis that, sooner or later, leading military powers will exploit available and emerging technologies, making major changes in the way they prepare and conduct operations in war, and realizing dramatic gains in military effectiveness.[53]

This RMA, Krepenevich argued, began with the invention of the transistor in the 1950s, continued with guided munitions development in the 1960s and 1970s, and was first demonstrated on an operational level during the Gulf War in 1991. But he believed that the revolution was still in its infancy, and had a ways to run before it was fully accomplished.[54]

The implications of Krepenevich's analysis were ground-breaking. If long-range sensors could see everything of military significance on the battlefield, then battle would devolve into a simple exercise of targeting and destroying enemy forces with precision-guided munitions. Information, rather than armor, would be the key to military superiority. Rather than closing with and destroying the enemy, military units would seek to gain information superiority over their opponents, which could then be eliminated with long-range, precision fires. Cruise missiles and unmanned aerial vehicles would assume more prominent roles than manned aircraft. At sea, smaller vessels that could fire guided missiles would supplant much more expensive carrier battle groups. Satellite guidance and other space systems would become increasingly important, especially for command and control and intelligence purposes.[55]

For a revolution in military affairs to be achieved, however, organizations and doctrine would also have to adjust to the new possibilities created by advanced technology. Operations would become increasingly multidimensional, putting

53 Andrew F. Krepinevich, Jr., "The Military-Technical Revolution: A Preliminary Assessment," 1.

54 Ibid., 8.

55 Ibid., 17–18.

a premium on the interoperability of joint forces. Long-range precision fires would enable the targeting of enemy centers of gravity from the outset of a campaign. Space would become an increasingly important medium for the control of military operations. Organizations that could take advantage of sensors and intelligence and engage the enemy with non-line of sight weapons would be optimized for this new type of warfare. Information dominance, rather than physical occupation of terrain, would be the key to victory.[56]

Despite his belief that a revolution in military affairs was underway, Krepenevich warned that non-peer adversaries would use unconventional warfare to counter the conventional superiority of those states with the means necessary to equip their forces with advanced information systems and guided weapons:

> Low-intensity warfare, comprising primarily but not exclusively insurgency, terrorism, and subversion, has been the most prevalent form of conflict in the post-World War II era. It seems likely that these conflicts, which are characterized by unconventional operational concepts, will continue as a dominant form of warfare in the post-Cold War era. It also is highly probable that non-peer competitors will engage in low-intensity warfare, employing the unconventional operational concepts characteristic of those kinds of conflicts, albeit modified somewhat by the infusion of more advanced military systems, as a means of frustrating peer competitor adversaries.[57]

Krepenevich was describing hybrid warfare, operations that melded conventional and unconventional forces and the fusion of unconventional forces with advanced military technology.[58] It was a perceptive warning, albeit one that was completely ignored by U.S. military leaders in the next decade.

Ironically, by the time the Office of Net Assessment drew wider attention in U.S. military circles to the revolutionary potential of advanced information systems and guided weapons, the information-precision RMA was already well underway. Although technological advances in guided weapons and information systems would follow, the new concepts that made victory over Iraqi forces so decisive, such as the U.S. Army's AirLand Battle doctrine and the concepts in the 1987 edition of U.S. Marine Corps Manual 1–1, *Warfighting*, were already

56 Andrew F. Krepinevich, Jr., "The Military-Technical Revolution: A Preliminary Assessment," 19–22.

57 Ibid., 31.

58 For an examination of hybrid warfare throughout history, see Williamson Murray and Peter Mansoor, eds., *Hybrid Warfare: Fighting Complex Opponents from the Ancient World to the Present* (Cambridge: Cambridge University Press, 2012).

in place. The services had grounded their new weapons and organizations in an examination of problems faced in the Vietnam and Yom Kippur Wars against real enemies. Krepenevich believed that the precision-information RMA was at the same stage as mobile, combined arms operations were in 1920, just after the introduction of the tank during World War I. But it is more likely that the RMA was closer to the point reached by the German military after the fall of France in 1940. Although armored warfare technologies, tactics, and techniques would continue to evolve after that point, the "blitzkrieg" RMA had essential been consummated; the Allies now understood the potential of armored operations, and they would in time catch up to the capabilities of the Wehrmacht in that regard. Likewise, the real question after the Gulf War was not how to perfect the information-precision RMA, but what would come next.[59]

Technological developments over the course of the 1990s, spurred on by massive increases in computing power, were truly impressive. The miniaturization of microprocessors and quantum leaps in their capacity enabled the development of a new generation of satellites, sensors, and guided weapons. The constellation of global positioning system (GPS) satellites, still in its infancy (albeit still critical to navigation in the trackless deserts of the Middle East) during the Gulf War, blossomed into a robust network that enabled the determination of precise locations and time signatures for both military and civilian applications.[60] The Department of Defense began research into navigation satellites in the 1960s, and launched two NAVSTAR satellites in the mid-1970s to test the concept. By 1985 there were eleven first generation GPS satellites in orbit; the first second-generation satellites were launched in 1989, with the constellation of twenty-four satellites completed in 1994.[61]

GPS was a godsend for guided weapons aimed at fixed facilities; instead of being directed to their targets by an operator, GPS-enabled weapons could be given geographic coordinates and set loose to guide themselves to their targets. The most significant beneficiary of GPS was the Joint Direct Attack Munition, or JDAM. The JDAM is merely a gravity bomb retrofitted with a GPS-enabled tail kit, not unlike the Fritz X, but without the need for an operator to steer the bomb to its target. The lack of a target designator also gave JDAMs all-weather capability. First used in combat against Serbian targets during Operation

59 For Krepenevich's assertion that the RMA was still in its infancy, see Krepinevich, "The Military-Technical Revolution," 51. For the argument that the precision-information RMA was essentially over, see Williamson Murray and MacGregor Knox, "The Future Behind Us," in Knox and Murray, *The Dynamics of Military Revolution*, 189–190.

60 GPS satellites are also equipped with sensors to detect nuclear detonations.

61 Scott Pace et al, *The Global Positioning System: Assessing National Policies* (Santa Monica, CA: RAND, 1995), 241–246.

Allied Force in 1999, JDAMs scored a remarkable circular error probable (CEP) of just 13 meters on 90 percent of its hits.[62] As impressive, this improvement was achieved at the cost of just $18,000 per guidance kit.[63]

Some senior American military leaders breathlessly announced that a new era of warfare had dawned. The precision-information RMA would enable U.S. forces to achieve "rapid dominance" in a future conflict waged with "shock and awe."[64] In a book entitled *Lifting the Fog of War*, retired Admiral Bill Owens argued that recent technological advances would make it possible for commanders to see both friendly and enemy forces on the battlefield from the vantage point of a computer screen in their command center. "I believe the technology that is available to the U.S. military today and now in development can revolutionize the way we conduct military operations," Owens wrote. "That technology can give us the ability to see a 'battlefield' as large as Iraq or Korea – an area 200 miles on a side – with unprecedented fidelity, comprehension, and timeliness; by night or day, in any kind of weather, all the time."[65] Such technology "challenges the hoary dictums about the fog and friction of war, and all the tactics, operational concepts, and doctrine pertaining to them."[66] In other words, the new revolution in military affairs would annul the military philosophy of the Prussian theorist Carl von Clausewitz along with nearly two centuries of writings based on his authoritative work, *On War*.[67] If only the cumbersome Pentagon bureaucracy would bend to the will of those with a like mindset, Owens argued, the United States military would be transformed into an information age superpower, with high-tech sensors and precision-guided munitions reducing the need for boots on the ground (See Illustration 16.5).

Technological determinism gripped the U.S. military up to (and indeed beyond) the invasion of Iraq in 2003. The concept of "rapid, decisive operations" and "effects-based operations" dominated the thinking of military planners in the run-up to the Iraq invasion.[68] In referring to these ideas, historian and

62 Watts, *Six Decades of Guided Munitions and Battle Networks*, 13.

63 Mets, "The *Force* in US Air Force," 70.

64 Harlan Ullman and James Wade, Jr., et al, *Shock & Awe: Achieving Rapid Dominance* (Washington: Institute for National Strategic Studies, National Defense University, 1996), http://www.dodccrp.org/files/Ullman_Shock.pdf.

65 Admiral Bill Owens with Ed Offley, *Lifting the Fog of War* (New York: Farrar, Straus,and Giroux, 2000), 14.

66 Ibid., 15. Owens was the vice chairman of the Joint Chiefs of Staff from 1994–1996.

67 For a stinging rebuttal to such notions, see Williamson Murray, "Clausewitz Out, Computer In: Military Culture and Technological Hubris," *The National Interest*, June 1, 1997 (48): 57–64. https://www.clausewitz.com/readings/Clause&Computers.htm.

68 For Iraq War planning, see Michael R. Gordon and Bernard E. Trainor, *Cobra II: The Inside Story of the Invasion and Occupation of Iraq* (New York: Pantheon, 2006). For a critique of

ILLUSTRATION 16.5 A Paveway IV laser-guided bomb is released from an F-35 Lightning II during trials in the United States. (By VX (https://www.dvidshub.net/image/2010920) [Public domain], via Wikimedia Commons - https://en.wikipedia.org/wiki/Paveway_IV#/media/File:F-35B_Aces_First_Release_of_a_UK_Paveway_IV_Bomb_150612-D-MJ303-310.jpg)

decorated military officer H.R. McMaster noted shortly after the invasion of Iraq, "The persistent belief in near-certainty in future war, however, elevated anticipated capabilities of information technologies to the level of strategy, encouraged linear thinking, and undermined the positive features of the new concepts."[69] The result of relying on information dominance was a significant reduction of the robust coalition force structure of Operation Desert Storm (956,000 U.S. and coalition troops), which was whittled down by nearly two-thirds (to 380,000 troops) for Operation Iraqi Freedom. Even with far fewer forces, U.S. commanders were given a much more expansive mission, to march to Baghdad (destroying the Iraqi Army along the way), depose Saddam Hussein and the Ba'ath Party from power, and install a government in Baghdad more amenable to U.S. interests. The available forces proved sufficient to achieve the first two objectives; they were wholly inadequate to achieve the final and most important goal. Cast adrift in Iraq with too few forces and doctrine unsuited to stabilizing a country and countering a growing insurgency, U.S. military leaders learned the limitations of the information-precision revolution in military affairs the hard way.[70]

rapid, decisive operations, see Antonio J. Echevarria II, "Rapid Decisive Operations: An Assumptions-Based Critique," U.S. Army War College Strategic Studies Institute, November 2001, http://www.strategicstudiesinstitute.army.mil/pdffiles/pub218.pdf.

69 H.R. McMaster, "Crack in the Foundation: Defense Transformation and the Underlying Assumption of Dominant Knowledge in Future War," Center for Strategic Leadership, U.S. Army War College, November 2003, 74, http://www.au.af.mil/au/awc/awcgate/army-usawc/mcmaster_foundation.pdf.

70 For a critique of U.S. military operations in the Iraq War, see Peter R. Mansoor, *Surge: My Journey with General David Petraeus and the Remaking of the Iraq War* (New Haven, CT: Yale University Press, 2013).

What went wrong? On the way to achieving the next great step forward in military innovation, U.S. military leaders overlooked the fundamental nature of war, as well as the fact that the enemy has some agency over the course of battle (i.e., the enemy always gets a vote). Six years before the invasion of Iraq, historian Williamson Murray lamented:

> The danger in the belief that technology will offer us total battlespace and foreign policy dominance in the next century does not lie in the technology itself. Technology can indeed offer us substantial leverage against future opponents. What is dangerous about the new technocratic view is the same thing that was dangerous about the older version: It is wholly disconnected from what others think, want, and can do. Precisely because we Americans have a long track record of overestimating our technological superiority and underestimating the ability of our opponents to short-circuit our advantages, this is a form of hubris we cannot afford to indulge again. This is also why many of the overtones we hear today about the coming "revolution in military affairs" are so disheartening.[71]

The historical record shows that truly effective military innovation is thoroughly grounded in the past, is aimed at solving concrete problems presented by actual enemies, and is infused with an appreciation of the art of the possible. Innovation that has not been tied to a real enemy and a discrete set of military challenges has all too often caused armed forces to waste time and resources on theoretical dead ends. By 2003, the U.S. military was well on its way to perfecting a method of warfighting ideally suited to combating itself.

The wars in Iraq and Afghanistan have demonstrated the limitations of the information-precision RMA. The "high-tech, low numbers" operational approach may suffice against an inept opponent in favorable geography, but it has proven insufficient to overcome irregular opponents fighting in urban areas or on more complex terrain. Trained to fight with overwhelming firepower delivered precisely to targets identified by advanced sensors and coordinated by sophisticated information systems, the U.S. military struggled to overcome old-fashioned, low-tech enemies that used guerrilla tactics and terrorism to prolong conflicts in space and time. After the Iraqi invasion, McMaster observed, "The belief in certainty, rooted as it is in technological hubris, masks the human dimension of war and creates the illusion that sensors and long-range fires can solve complex military and political problems."[72] In the Iraq

71 Murray, "Clausewitz Out, Computer In."
72 McMaster, "Crack in the Foundation," 94–95.

War, it took four years before U.S. commanders relearned how to wage an old-fashioned counterinsurgency struggle, but by then much of the strategic damage to American interests in the Middle East had already been done.

The U.S. military today still enjoys an enormous advantage in information systems, sensors, and guided munitions. A revolution in military affairs has undoubtedly occurred and may still be underway. But, this reality should not blind us from the fact that the precision-information RMA is tuned to wars against near-peer state competitors. Potential adversaries that lack the capability to compete with the United States on equal terms – that is, much of the rest of the world – will opt for different strategies and operational concepts, such as hybrid warfare or the development of nuclear weapons, to level the playing field. These may not be the type of wars the U.S. military would like to wage, but they will likely be the type of wars the United States is forced into if it chooses to use military force in the near future. The technologies of the information-precision RMA can be useful in these conflicts, as the use of armed unmanned aerial vehicles in Iraq and Afghanistan has shown, but they will certainly not be decisive. In the end, politics and strategy will still be determinative in sorting out winners from losers. Clausewitz, it appears, is not dead after all.

EPILOGUE

# Reflections On A Warrior and Scholar

*Alan Beyerchen*

My perspective among the speakers and participants in this memorial conference is rather unusual. I am not a military historian, although I sat on the generals and dissertation committees of many of Joe's military history students in this room. I was the one to whom he sent students to have their intellectual horizons "broadened" by exposure to topics, titles, and examination trials in modern European history.

I knew Joe more indirectly in the way two days of this conference have revealed him, namely through others. The conference presentations depict his engaging personality and multifaceted mind filtered through the work of his students—and, I must add in his good taste in selecting a lively spouse. This character is what I saw in interaction with his students in my classes, during doctoral examinations, when reading stages of dissertations, and when considering the program and prospects of a given student. He would plant an idea; I would press the students to think through the implications; they would explore with him something I presented to them; I would press them to elaborate on something he had laid out to them, and so on. Joe and I actually had scores of lively conversations. They were just not face-to-face. They were within the minds of his students.

From these "conversations," that is, from the way his students responded and grew under his tutelage, it was clear to me from early on that this man displayed the prime characteristic of a true leader. He made everyone around him better.

I find it interesting that technology is so very prominent in "Dr. G's" thinking, at the same time that Clausewitz figures so centrally there as well. Technology, after all, does not loom large in Clausewitz's writings. The human factor does. The common thread appears to be that, for Joe, technology, like war, is a fundamentally human endeavor. We change the world and war with technology, but we also change ourselves.

We need to pause for a moment and consider what we mean with the word and concept "technology." This is a word, like "strategy," which is used in myriad ways, often with overlapping but distinct meanings. The Greek root *teckhne* means simply "craft," "skill," or a way of doing things, and is the basis of our term "technique." Similarly, the Latin *ars*, is the root of "artifact" and "artificial."

Technology can be perhaps best understood as a system of systems, deeply contextually driven but always an extension of human mental and physical capabilities. The elaboration of the concept used by Joe and suffused throughout the work of his students has three elements that have appeared explicitly or implicitly at this conference. In Joe's 1994 article on "Technology and Strategy," the term means [my gloss in brackets]:

1. the application of knowledge to generate or achieve physical effects by means of an artifact or process to solve a problem [An artifact is a kind of precipitate from the flow we call "process."]
2. the artifact or process itself [which is one reason the "details" matter]
3. the context [social, cultural, political, economic] of the procedures needed to obtain, maintain and sustain the artifact or process. [This is often embodied in habits of mind or in institutions.]

I personally would add a fourth element that is implicit in his thinking: the feedback loops among the three elements above, which makes them operate in concert and produces evolution over time.

A straightforward example of a simple technology is the carpenter's hammer or a warrior's mace, which become extensions of the arm and fist of a human being. Most of us have used a hammer on a nail. But anyone who has ever seen a professional carpenter working can see how the swing of the hammer comes in a rhythm—first one blow to set the nail, then a second to pound it home. One blow to set, one to pound home, in a smooth motion that makes it clear the hammer is a part of the carpenter. It is not an add-on, but an extension of the body. Joe makes this point vividly in his autobiographical account of his helicopter crew's final runs into Saigon under fire. The technology available to him, including the voice of his crew chief in his earphone, became an extension of his senses as he flew at rooftop level around and over obstacles that he could not clearly see.

Let's consider for a moment a familiar technology and expose its less familiar dimensions. The technology of the automobile is for most people just the physical cars themselves, pieces of machinery that we use to get from point to point, sometimes carrying objects or people in addition to ourselves. But the elements of this technology mean we need to think through how an automobile solves the problem of transportation. We need to include in "automotive technology" such things as engineering and design, combustion and fuel consumption, steering and braking, appearance and part assemblage, consumer competition and market costs, and much more. We need also to consider how automobiles are actually generated. That is, we need to include the coal to feed the blast furnaces for making specialty steel, the chemical facilities for making specific polymers, the suppliers of all the parts and the machines and machine tools they use. These are all part of the "technology" of the automobile that we take for granted as "end users," but are daily requirements very much in the awareness of the producers of these artifacts.

What else does the context involve here? It means anything that would not exist without the automobile. This environment includes the assembly lines at the manufacturing plants, the truck carriers with eight or ten vehicles perched on their trailers on the way to dealers, the dealer showrooms, "zero" financing of purchases, and the maintenance and repair facilities scattered across the land. It includes the networks of roads and highways tailored to automotive traffic, the fuel stations, the signals and signs of the road, and the traffic laws in their myriad formulations. The context also involves the enforcement of those laws, the tickets and traffic courts and the insurance policies. It also includes the drivers' training courses, bureau of motor vehicles licensing procedures, skills behind the wheel, how we feel sitting up high in an SUV or down low in a race car, our social acceptance of tens of thousands of traffic deaths and hundreds of thousands of injuries each year—and much, much more.

Such thinking may sound like a bit of sophistry, but any historian of early and mid-twentieth century technology or industry would be able to document how much invention, innovation, and social adaptation by actual human beings went into the emergence and evolution of the technology of the automobile. All of these spreading, layered, convoluted, feedback-laden, evolving sets of mechanical and human interactions have been part and parcel of that technology.

Here is the point: Apply this kind of thinking to any artifact that ends up on a battlefield, and you have a sense of how Joe Guilmartin's mind worked. You also gain an insight into the minds of his students and into his legacy.

The papers presented at this conference, all of which reflect facets of Joe's mind, have no single theme. There is eclecticism about them that is faithful to his wide-ranging curiosity and knowledge. There is an enormous depth of chronology for the historical discipline. There is also an enormous geographical reach.

Without a single theme, one could nonetheless discern some broad lines running through the papers. In general terms, they demonstrate the limitations of technology, especially the cultural limitations that sometimes make technologies "prisoners of culture." In addition, there is clearly a set of contrasts in culture, because some cultures amplify the flexibility that technologies can exhibit. The papers also make clear the importance of niche technologies. And they almost all evince the human creativity, foibles, resilience, and reciprocity that emerge when two sides lock into combat. Combat itself is a central element, along with its concomitant violence.

In a narrower vein, one of the common aspects of these papers, and of Joe's students across the board, is that they emphasize details of artifacts and processes. It would be possible to imagine each of the papers here on the annual program of the Society for the History of Technology, although not on the annual program of the History of Science Society. The clear message is that details and stories about the details *matter*. For historians, truth is not as fungible as perhaps in some other fields, and for military historians truth often resides in the accuracy of detail. This focus allows for serious discussions of causation, correlation, and confusion.

Yet my experience with Joe's students and our discussions about them indicated that the artifacts and processes were never enough. In so far as we discussed students' work directly or during examination hours, Joe always cared about scaling to larger issues. Often, in fact, he encouraged a student to include me on a doctoral committee, because he knew my own interests emphasized the third, contextual element among his definitions. He wanted to stretch the minds of his students—not *quite* to the breaking point, mind you, but farther than many would have stretched without his guidance.

A second tangible aspect of the work of his students is that Joe seemed to instill in people the priority of getting to the point. This discipline is a rare trait among members of academic tribes, and is particularly difficult, but important, to achieve in dealing with political, social, and, especially, cultural constraints and conditions.

A third common technique underlined by Joe and evident among these papers is the use of comparison and contrast (often understood as a form of graphic perspective) to provide contours of relationships in technological innovation, adaptation, evolution, or revolution. This style means he trained people to think of intellectual mappings across time and geography.

Lastly, one of his students commented to me on a theme among the prefatory remarks by presenters that echoed his own experience: Dr. G as rescuer. Joe may not have flown Jolly Green Giant rescue helicopters all his life, but he seems to have retained his remarkable ability to step in at crucial moments and save people when things looked bleakest. I am reminded of an ancient judgment that I would like to adapt and paraphrase. It comes from the funeral oration by Pericles in the middle of the Peloponnesian War: What you leave behind is not what is engraved in stone monuments, but what is woven into the hearts and minds of others.

The legacy of Joe Guilmartin has been woven into the hearts and minds of family, friends, students, and colleagues—and in the blurring, and often removal, of the distinctions among these categories.

# Bibliography

## Primary Sources

### *Archival*

Austria

*Österreichisches Nazionalbiliothek, Wien*

Codex 3069. Fo. 2. (1411)

Manuscript CPV 3069, fo. 22v

Canada

*Library and Archives of Canada*

RG 24-D-1-c; RG 24-D-10

France

*Archives Municipales de Bordeaux*, IV:

Registres de la Jurade

Germany

*Bundesarchiv-Militärarchiv, Freiburg, Germany*

BA-MA N 246/7

BA-MA MSG 160/2

BA-MA MSG 160/8

BA-MA MSG 160/12

BA-MA MSG 160/13

*Heidelberg Universitätsbibliothek*

*Feuerwerkbuch.* Cod. Pal. germ. 502, fo. 4. Cf. the French translation of c. 1430

*Staatsbibliothek zu Berlin*

Ms. germ. qu. 1187. Fo. 15v

Portugal

*Arquivo da Defesa Nacional, Paço do Arcos*

F2/92/306/5

F3/10/23/38, F3/17/37/59

F3/17/37/64

F2/SR007/SSR1/61/35

report index Pº. 324.410-324.510-223.230 Nº. 4805

SGDN-1666

SGDN Cx 761. Annex A, Report index 1827/NI
*Arquivo Histórico da Força Aérea, Lisbon*
Região Aérea Manuais Collection
Relatórios Mensais do Comando, Anos 1968–1969 collection
Relatórios do Comando 1972–1974 collection
*Arquivo Histórico Militar, Lisbon*
AHM/Div/2/4/228/2
AHM/Div/2/4/314/2
*Arquivo Nacional Torre de Tombo, Lisbon*
Folder 5. PC 6992/6993. PIDE/DGS Collection
Folder 5. NT 6992–6993. PC 332-CR (2). PIDE/DGS Collection
*Fundação Mário Soares Arquivo e Biblioteca, Lisbon*
Amílcar Cabral Collection
Pedro Verona Pires Collection

### United Kingdom

*The National Archives of the United Kingdom* (*Kew*)
ADM 1/19308, ADM 1/23452, ADM 178/223, ADM 220/189
AIR2/2059, 2715, 2718, AIR 8, 257, AIR 10, AIR14/251, AIR20/9, 180, 425, AIR 22, Air 23, AIR 41, AIR 41/8
AVIA 10/1, AVIA 15/4
E 101/51/27, E101/393/11
FO 2/58, FO 403/224, FO 403/258, FO 403/277, FO 403/291, FO 403/307, FO 403/317, FO 403/329, and FO 403/339
MPK1/74
WO 201, WO 277
*Royal Armouries*
*Feuerwerkbuch*, Fo. 4v. MS I-34
*Royal Air Force Air Historical Branch*
Tedder Box 3
Tedder Box 4
Translations Box 7

### United States

*Air Force Historical Research Agency, Maxwell AFB, AL*
"Briefing slides, Headquarters Pacific Air Forces Hickam Air Force Base, HI" 26 May 1972
Record Group 519
Record Group 553

Record Group K113

*National Archives and Records Administration*

NARA I

Record Group 243

NARA II

MID report 2657-I-276

Record Group 165

Record Group 313

*United States Air Force Academy Library Special Collections Branch*

Section 5, Box 8, File 7

*United States Naval Academy Library, Annapolis, MD*

Fiche 3, 145–149. Microfiche, 1946

Reminiscences of Vice Admiral Lloyd M. Mustin, USN, interview by John T. Mason, Jr., 2003

*University of California-Riverside Library*

Ronald H. Chilcote Collection

### *Governmental Non-Archival*

#### Guinea

Cabral, Amílcar. *Aucune Manoeuvre ou Crime des Colonialists Portugais Ne Sera a Meme d'Eviter la Victoire Ineluctable de Notre People Africaine: Rapport Bref sur la Lutte in 1971* (Conakry, Guinea: Partido Africano da Independência da Guiné e Cabo Verde, January 1972.

Cabral, Amílcar. *Communiqué* (Conakry, Guinea: Partido Africano da Independência da Guiné e Cabo Verde, March 11, 1963a.

Cabral, Amílcar. *Communiqué* (Conakry, Guinea: Partido Africano da Independência da Guiné e Cabo Verde, June 1963b.

Direcção Geral de Segurança, Delegação da Guiné, "Actividades do 'PAIGC'", Lisbon.

*Qualques Crimes Perpetres par les Colonialistes Portugais contre Nosse Peuple en 1963* (Conakry, Guinea: Partido Africano da Independência da Guiné e Cabo Verde, 1964).

#### Mozambique

Direcção-Geral de Segurança, Delegação em Moçambique, "Relatório: Armas anti-aéreas inimigas.

#### Portugal

de Spínola, António, *Linha de Acção*. Lisbon: Agência-Geral do Ultramar, 1971) An English-language synopsis is available in American Embassy Lisbon, *Portuguese Guinea and Its Governor*, U.S. Department of State Airgram. 19 October 1971.

Commando Zona Aérea de Cabo Verde e Guiné weekly situation reports (SITREP), 1964–1973.

"Intensification of Armed Struggle," *Info Tchad* (Fort Lamy, Chad), 9 April 1973, translated and reprinted in Joint Publications Research Service, "Translations on Africa No. 1304," JPRS 58903 (May 2, 1974).

Machado, Pedro. "Liberation Movement Discussed by Guerrilla Leader." *Etumba*. Brazzaville, Congo. September 11, 1971. Translated and reprinted in Joint Publications Research Service. "Translations on Africa No. 1081." JPRS 54466 (November 11, 1971).

Morés" and "Aliu, o pescador." In *O Nosso Livro: 2ª Classe*. Regiões Libertades da Guiné: Serviços de Instrução do PAIGC, 1970.

"War Communiqué," *PAIGC Actualités Quarterly English Edition* nos. 45–48 (1972).

### United Kingdoms

Air Ministry, *Supply Organisation of the German Air Force, 1935–1945*. East Sussex, UK: The Naval & Military Press Ltd., 2009. Reprint of 1946 original. (original available at Air Historical Branch).

Dannatt, Dr. Cecil. "Report of Dr. C. Dannatt on American Fire Control Practice and Developments (Visit of U.S.A. and Canada - July, 1943)." 1943. B15 20. HMS Excellent.

*Proceedings and Ordinances of the Privy Council of England*. Nicolas, N. H., ed. 7 vols. London: Record Commission, 1834–7.

Ross, M.J. Commander RN. "American A.A. Fire Control Systems, Report of a Visit to U.S.A." July 1945, 31–35, B15 52, HMS Excellent.

### United States

Central Intelligence Agency. *The Guerrilla War in Portuguese Guinea*. Weekly Review and Special Report. 20 December 1968.

Fischer, Hannah. "A Guide to U.S. Military Casualty Statistics: Operation Inherent Resolve, Operation New Dawn, Operation Iraqi Freedom, and Operation Enduring Freedom." Congressional Research Service. Washington DC: US Government Printing Office, 2014.

National Security Council. *Implementation of Arms Embargo on South Africa and Portuguese African Territories*. National Security Council Memorandum 81. August 17, 1970.

Nicholson, Charles A. "Project CHECO Southeast Asia Report – The USAF Response to the Spring 1972 NVN Offensive: Situation and Deployment." Saigon: Headquarters Seventh Air Force, 1972.

Office of Technology Assessment. *Technologies for NATO's Follow-On Forces Attack Concept*. Washington: Government Printing Office, July 1986.

Rowland, Burford, and William Boyd. *U.S. Navy Bureau of Ordnance in World War* II. Washington, D.C.: Department of the Navy, 1953.

U.S. Congress House of Representatives. Subcommittee on Africa of the Committee on Foreign Affairs. *Report on Portuguese Guinea and the Liberation Movement.* (91st Congress, 2nd session, 26 February 1970).

U.S. Department of State. Teletype, 100930Z APR 74, American Embassy Dakar to Secretary of State, 10 April 1974.

U.S. Department of State. US *Arms in Portuguese African Territories*. U.S. Department of State telegram. December 3, 1963.

*War of the Rebellion: The Official Records of the Union and Confederate Armies.* Series I. Volume XXXVIII. Part III. Washington, D.C.: U.S. Government Printing Office, 1890–1901.

### *Manscript*

L'Assiedo di Pisa (1405–1406).

### *Online*

Commander, Cruiser-Destroyer Group 2. "Formal Investigation into the Circumstances Surrounding the Attack on the USS Stark (FFG 31) on 17 May 1987." June 12, 1987. Accessible at: http://web.archive.org/web/20120323222902/http://www.dod.gov/pubs/foi/operation_and_plans/USS_Liberty_Pueblo_Stark/65rev.pdf.

Comptroller General. "Report to the Congress: Decisions to Be Made in Charting the Future of DOD's Assault Breaker." February 28, 1981. Accessible at: http://www.gao.gov/assets/140/132235.pdf.

Krepinevich, Jr., Andrew F. "The Military-Technical Revolution: A Preliminary Assessment." Office of Net Assessment. July 1992. Reprinted at Center for Strategic and Budgetary Assessments. Accessible at: http://www.csbaonline.org/wp-content/uploads/2011/03/2002.10.02-Military-Technical-Revolution.pdf.

*Jagdgeschwader 27*, http://www.ww2.dk/air/jagd/jg27.htm.

*Jagdgeschwader* 53, http://www.ww2.dk/air/jagd/jg53.htm.

United States Merchant Marine Records, Royal Navy Ships Sunk, http://www.usmm.org/tankers.htmla, and https://www.battleships-cruisers.co.uk/merchant_navy_losses.htm.

*Petit Livre Noir* (Registre consulaire de Périgueux. 1360 - 1449. Archives départementales: BB. 13). Fo. 47v. Transcription by Jean Roux online at https://web.archive.org/web/20150405145020/http://www.perigordoccitan.fr/Paleographie/Livre_Noir/LNtexte.htm, accessed 24 Jan. 2017.

### *Newspapers and Media*

*ABC News.*

*CBS News.*

*The Columbus (Ohio) Dispatch.*

*Daily Monitor.*
*Financial Times.*
*Le Monde.*
*The New York Times.*
*Revue de Défense Nationale.*
*USA Today.*
*The Washington Post.*
*The Washington Post-Times Herald.*

## Secondary Sources

### *Foreign Language Sources*

Abecasis, Major General José Duarte Krus. *Bordo de ataque: memórias de uma caderneta de voo e um contributo para a história.* Coimbra, Portugal: Coimbra Editora, 1985.

Ágoston, Gábor. "Firearms and Military Adaptation: The Ottomans and the European Military Revolution, 1450–1800." *Journal of World History.* Vol. 25. 2014.

Afonso, Lieutenant Colonel Aniceto, and Colonel Carlos de Matos Gomes. Portuguese Army. "Guiné: Maio de 1973 – O Inferno." Working paper. Avenida da Liberdade/Associação 25 de Abril, 22 February 2008.

Angelucci, Angelo. *Delle artiglierie da fuoco Italiane. Memorie storiche con documenti inediti.* Turin: G. Cassone, 1862.

Angelucci, Angelo. *Documenti inediti per la storia delle armi da fuoco Italiane.* Turin: G. Cassone, 1869.

Anon. *Déliberations de 1414 à 1416 et de 1420 à 1422.* Bordeaux: Imprimerie G. Gounouilhou, 1883.

Anon. *Istoria dell'assedio e della ricupera di zara fata da'Veneziani nell'anno* MCCCXLVI*, scritta da autore contemporane,* in Morelli, Jacopo, ed. *Monumenta Veneziani di varia Letteratura.* Venice: Carlo Palese, 1796.

Aron, Raymond. *La Coexistence: Mai 1955 à Février 1965.* Edited Georges-Henri Soutou. 3 vols., Raymond Aron: Les Articles de politique Internationale dans Le Figaro de 1947 à 1977, vol. II. Paris: Editions de Fallois, 1993.

Associação de 25 de Abril (April 25th Association), "Operação Vulcano," http://www.guerracolonial.org/index.php?content=396.

Barthet, Laure. "La prise de la barbacane de Montségur (Ariège) en février 1244: une introduction à l'archéologie de la poliorcétique," in Nicolas Prouteau, Emmanuel de Crouy-Chanel and Nicolas Faucherre, eds., *Artillerie et Fortification, 1200–1600.* Rennes: Presses Universitaires de Rennes, 2011.

Basin, Thomas. *Histoire de Charles* VII *et Louis* XI. J. Quicherat ed. vol. 1 Paris: SHF, 1855.

Beffeyte, Renaude. *L'arte de la guerre au Moyen Âge.* Rennes: Éditions Ouest-France, 2005.

Bellaguet, Louis ed., *Chronique du religeux de St. Denys (1380–1422).* 6 vols. Paris: Crapelet, 1839–52.

Bertin, J. "Le siège du château de Vellexon en 1409." *Bulletin de la société d'agriculture. Ssciences et arts du département de la Haute-Saône,* 3d Ser., 29, 1898.

Bolton, Jim. "'The World Upside Down'. Plague as an Agent of Economic and Social Change." In *The Black Death in England.* Edited by E.M. Ormrod and P.G. Lindley. Stamford: Paul Watkins, 1996.

Brandão, João José. Lieutenant Colonel Portuguese Air Force. *Cronologia da Guerra Colonial: Angola-Guiné-Moçambique 1961–1974.* Lisbon: Prefacio, 2008.

Cabral, Luís. *Crónica da Libertação.* Lisbon: O Jornal, 1974.

Carpentier, Elisabeth. "The Plague as a Recurrent Phenomenon." In *The Black Death: A Turning Point in History?* Edited by William M. Bowsky. New York: Holt, Rinehart and Winston, 1971.

Catarino, Manuel. *As Grandes Operações da Guerra Colonial, 1961–1974,* XIV*: O Terror does Mísseis – Guiné, 1973.* Lisbon: Correio da Manhã, 2009.

Cavignac, Jean. *Jean Pellet, commercant de gros, 1694–1772.* Paris: S.E.V.P.E.N., 1967.

Cervelló, Josep Sanchez. "La inviabilidad de una victoria Portuguesa en la Guerra Colonial: el caso de Guinea-Bissau." *Hispania* (September–December 1989).

Coelho, Paulo Borges. "African Troops in the Portuguese Colonial Army, 1961–1974: Angola, Guinea-Bissau and Mozambique." *Portuguese Studies Review.* (September 2002).

Conde, José Antonio. *Historia de la dominacion de los Arabes en España, sacada de varios manuscritos y memorias Arabigas.* Paris: Baudry, 1840.

Contamine, Philippe. *Guerre, état et société à la fin du moyen âge. Etudes sur les armées des rois de France, 1337–1494.* Paris: Mouton, 1972.

Contamine, Philippe. "Les Industries de guerre dans la France de la Renaissance: l'exemple de l'artillerie." *Revue historique* 271 (1984).

Coreia, José Manuel. "Os Fiat G.91 ao Ataque na Guiné: Operação Barracuda." *Mais Alto* 45. No. 367 (May–June 2007).

Da Costa Cardoso, Edgar. Colonel Pereira Portuguese Air Force. *História da Força Aérea Portuguesa. 3º Volume.* Lisbon: Cromocolor, 1984.

De Abreu, António Graça. Lieutenant Portuguese Army. *Diário da Guiné: Lama, Sangue e Água Pura.* Lisbon: Guerra e Paz, 2007.

De Guzmán, Fernán Pérez. *Cronica del serenissimo rey don Juan el segundo deste nombre.* Valencia: Benito Monfort, 1779.

De Hocsem, Jean. *Chronique*. Godefroid Kurth, ed. Brussels: Commission royale d'histoire, 1927.

De la Fons-Mélicocq, Alixandre. *De l'artillerie de la ville de Lille aux* XIVe, XVe *et* XVIe *siècles: archers, arbalétriers, canonniers*. Lille: Lefebvre-Ducrocq, 1854.

de Fraga, Luís Alves. Colonel Portuguese Air Force. *A Força Aérea na Guerra em África*. Lisbon: Prefácio, 2004.

De Lettenhove, Kervyn ed. *Istore et croniqes de Flandres*. 2 vols. Brussels: F. Hayez, 1879–80.

De Melo, João. *Os Anos da Guerra, 1961–1975: Os Portugueses em África; Crónica, Ficção e História*. Lisbon: Publicações Dom Quixote, 1988.

De Miranda, Jose M. Lieutenant General Brochado Portuguese Air Force. "El empleo del arma aérea en la lucha antiguerrilla." *Revista de Aeronautica y Astronautica*. No. 715 (July–August 2002).

de Vasconcelos, Francisco. "O 'Inferno' de Guidage." *Público*. No. 5 (November 1995).

De Waurin, Jean. *Recueil des Croniques et anchiennes istories de la Grant Bretaigne, a present nomme Engleterre*. 5 vols. London: Rolls Series, 1864–1891.

Dufour, Guillaume Henri. *Mémoire sur l'artillerie des anciens et sur celle du moyen âge*. vol. 1 Paris: A. Cherbuliez, 1840.

Fabião, Carlos. General Portuguese Army. "A descolonização da Guiné-Bissau. Spínola: a figura marcante da guerra na Guiné." In *Seminário: 25 Abril 10 anos dopois*. Lisbon: Associação 25 de Abril, 1984.

Farmer, David. "Prices and Wages, 1041–1350." In *The Agrarian History of England and Wales* Vol. I,II. Edited by Joan Thirsk. Cambridge: Cambridge University Press, 1988.

Fereira, José Lemos. General Portuguese Air Force. "No ceú da Guiné." In José Freire Antunes. *A guerra de África (1961–1974)*. Vol. 2. Lisbon: Temas e Debates, 1996.

Ferreia, José Manuel Pinto. Lieutenant Colonel Portuguese Air Force. "A introdução do 'strela' na guiné." *Luís Graça e Camaradas da Guiné*. (blog). 9 February 2009. https://blogueforanadaevaotres.blogspot.com/2009/02/guine-6374-p3859-fap-6-introducao-do.html.

Ferreira, José Pinto, and Miguel Pessoa. "A Introdução do Míssil Russo SAM-7 Strela no CTIG." *Luis Graça & Camaradas da Guiné*. (blog). 9 February 2009. https://blogueforanadaevaotres.blogspot.com/2009/02/guine-6374-p3859-fap-6-introducao-do.html.

Fonseca, Ana Mónica. "Dez anos de relações luso-alemãs." *Relacões Internacionais* 11. (2006).

Garnier, Joseph. *L'Artillerie des Ducs de Bourgogne, d'après les documents conservés aux archives de la Cote-d'Or*. Paris: Honoré Champion, 1895.

Gerhardt, M.R.B. *Allgemeiner Contorist*. Pt. 1. Berlin: Arnold Wever, 1791.

Goetz, Dorothea. *Die Anfänge der Artillerie*. Berlin: Militärverlag der DDR, 1985.

Guerra, João Paulo. *Memória das Guerras Coloniais*. 2nd ed. Porto, Portugal: Edições Afrontamento, 1994.

Gundelach, Karl. *Die deutsche Luftwaffe im Mittelmeer 1940–1945*. Vol. I. Frankfurt am Mann: Peter D. Lang, 1981.

Hardy, Michel. ed. *Ville de Périgueux. Inventaire sommaire des archives communales antérieures à 1790*. Périgueux: R. Delage & D. Joucla, 1894.

Hassenstein, Wilhelm. *Das Feuerwerkbuch von 1420. 600 Jahre Deutsche Pulverwaffen und Büchsenmeisterei.* Munich: Verlag der Deutschen Technik, 1941.

Heisenberg, Werner. "Über den Anschaulichen Inhalt der Quantentheoretischen Kinematic und Mechanik." *Zeitschrift für Phisik*. Number 43.

Huguenin, Jean-François ed. *Les chroniques de la Ville de Metz, 900–1552*. Metz: S. Lamort, 1838.

Jacobs, Karl. *Das Aufkommen der Feuerwaffen am Niederrheine bis zum Jahre 1400.* Bonn: Peter Hanstein, 1910.

Jähns, Max. *Geschichte der Kriegswissenschaften, vornemlich in Deutschland.* Pt. 1. Munich: R. Oldenbourg, 1889.

Köhler, Gustav. *Die Entwickelung des Kriegswesens und der Kriegführung in der Ritterzeit*, 3 vols. in 5. Breslau: W. Koebner, 1886–1887.

Lacabane, L."De la poudre à canon et de son introduction en France." *Bibliothèque de l'école des chartes*, 2e ser., I (1844), 45–46n.

Lalande, Denis ed. *Le livre des fais du Mareschal Bouciquaut.* Geneva: Droz, 1985.

Lartéguy, Jean. *Les Prétoriens.* Paris: Presses de la Cité, 2013.

Lefevre-Pontalis, Germaine. "Petit chronique de Guyenne jusqu'à l'an 1442." *Bibliothèque de l'école des chartes. Vol.* 47. 1886.

Leng, Rainer ed. *Anleitung Schiesspulver zu bereiten, Büchsen zu laden und zu beschiessen: Eine kriegtechnische Bilderhandscrhift im cgm 600 der Bayrischen Staatsbibliothek München.* Wiesbaden, Reichert Verlag 2000.

Leuchs, Johann. *Der Contorwissenschaft*. Pt. 3, 4th ed. Nuremberg: C. Leuchs und Comp., 1834.

*Magyar történelmi tár: a történelmi kútfők ismeretének előmozdítására.* vol. 9. Pest, 1861.

Manucci, Niccolo. *Storia do Mogor, or Mogul India.* Translated by William Irvine. London: Asiatic Society of Bengal, 1907.

Martín, Francisco Javier López. "La evolución de la artillería entre los siglos XIV y XVI, con especial atención a los manuscritos de Walter de Milemete y los primeros usos de la artillería en Europa." *Fortificações e Território na Península Ibérica e no Magreb. séculos* VI *a* VXI, vol. 2, Isabel Cristina F. Fernandes, ed. Lisbon: Edições Colibri, 2013.

Mateus, Dalila Cabrita. *A PIDE/DGS na Guerra Colonial (1961–1974)*. Lisbon: Terramar, 2004.

Minden. *Scriptorum rerum Brunsvicensia.* Hanover: Nicolas Förster, 1710.

Mozhnik, Franz. *Lehrbuch des gesammten Rechnens für die vierte Classe der Hauptschulen in den k.k. Staaten.* Vienna: Verlage der k.k. Schulbücher Verschleiß-Administration, 1848.

Napoléon, III and I. Favé. *Études sur le passé et l'avenir de l'artillerie.* 6 vols. Paris: J. Dumaine, 1846–1871.

Neto, General Manuel Diogo. "Voo em Três Frentes." In *A Guerra de África (1961–1974)*. vol. 1. Edited by José Freire Antunes. Lisbon: Temas e Debates, 1996.

Ottsen, Johann. "Über den derzeitigen Stand unserer Kentnisse von den Anfängen der Pulverwaffen." *Beiträge zur Geschichte der Technik und Industrie.* Vol. 13. 1923.

Pépin, Guilhem. "The French Offensives of 1404–1407 against Anglo-Gascon Aquitaine." *Journal of Medieval Military History. Vol.* 9. 2011.

Pereira, Aristides. *O Meu Testemunho: Uma Luta, Um Partido, Dos Países.* Lisbon: Editorial Notícias, 2003.

Piper, Otto. *Burgenkunde.* Munich: R. Piper, 1905.

Pessoa, Miguel. "Um 'Fiat' abatido," in *A guerra de África (1961–1974)*. Vol. 1. Edited by José Freire Antunes. Lisbon: Temas e Debates, 1996.

Prouteau, Nicolas, Emmanuel de Crouy-Chanel, and Nicolas Faucherre, eds., *Artillerie et Fortification, 1200–1600.* Rennes: Presses Universitaires de Rennes, 2011.

Rathgen, Bernhard. *Das Geschütz im Mittelalter.* Berlin: V.D.I. Verlag, 1928.

Robin, Georges. *Commandant Rebelle: Algérie, 1958 - de l'obéissance à la Revolte.* Saint-Amand-Montrond: Editions JC Lattès, 1998.

*Rolandini Patavini Chronica,* ed. Philippo Jaffé in MGH, *Scriptores,* G.H. Pertz, ed. Vol 19. Hannover: MGH, 1866.

Sack, C.W., Emil Ferdinand Vogel, and Ferdinand Spehr, *Alterthümer der Stadt und des Landes Braunschwei.*, 2d. ed. Brunswick: Friedrich Otte, 1861.

Salamagne, Alain. "L'Artillerie de la Ville d'Arras en 1369." in Nicolas Prouteau, Emmanuel de Crouy-Chanel and Nicolas Faucherre, eds. *Artillerie et Fortification, 1200–1600.* Rennes: Presses Universitaires de Rennes, 2011.

Schmidtchen, Volker. *Bombarden, Befestigungen, Büchsenmeister.* Düsseldorf: Droste Verlag, 1977.

*Scritti e documenti inediti.* Corazzini, Giuseppe Odoardo ed. Florence: Ulisse Diligenti, 1885.

Stella, Giorgio. *Annales Genuenses,* in *Rerum italicarum scriptores,* Muratori, ed. vol. 17, Milan: n.p., 1730.

Stevenson, Joseph ed. *Chronicon de Lanercost, MCCI–MCCCXLVI.* Edinburgh: The Bannatyne Club, 1839.

Stora, Benjamin. *Appelés en guerre d'Algérie.* Evreux: Gallimard, 1997.

Viard, Jules ed. *Grandes chroniques de France.* 9 vols. Paris: SHF, 1920–53.

Von Hefner, J., and J.W. Wolf, *Die Burg Tannberg und ihre Ausgrabungen.* Frankfurt am Main: Verlag der S. Schmerber'schen Buchhandlung, 1850.

De Waale, M.J. *Arkelse oorlog, 1401–1412. Een politieke, krijgskundige en economische analyse.* Den Haag: Koninklikje Bibliotheek, 1990.

Walsingham, Thomas. Annales Henrici Quarti, in *Chronica et annales regnantibus Henrico Tertio, Edwardo Primo, Edwardo Segundo, Ricardo Segundo, et Henrico Quarto.* Henry Thomas Riley ed. London: Rolls Series, 1868.

Walter of Hemingburgh, *Chronicon Domini Walteri de Hemingburgh.* Hamilton, H.C., ed. London: English Historical Society, 1849.

Woollacott, John. "A luta pela libertação nacional na Guiné-Bissau e a revolução em Portugal." *Analise Social* XIX. Nos. 77–78–79 (1983).

### *English-Language Secondary*

Aberth, John. *The First Horseman: Disease in Human History.* Upper Saddle River, NJ: Pearson Education, 2007.

Adamsky, Dima. *The Culture of Military Innovation: The Impact of Cultural Factors on the Revolution in Military Affairs in Russia, the* US, *and Israel.* Palo Alto, CA: Stanford University Press, 2010.

Adas, Michael. *Machines as the Measure of Man: Science, Technology, and Ideologies of Western Dominance.* Ithaca: Cornell University Press, 1989.

Ágoston, Gábor. *Guns for the Sultan: Military Power and the Weapons Industry in the Ottoman Empire.* Cambridge: Cambridge U.P., 2005.

al-Fazl, Abu. *Ain-i-Akbari.* H. Blochmann. Translated. Online Edition. Los Altos, CA: Packard Humanities Institute, 2006.

Anon. 1430: *A Parisian Journal, 1405–1449.* Shirley, Janet tr. Oxford: Clarendon Press, 1968.

Anon. *Chroniques de la ville de Metz.* Edited by Hugenin, J.F. Metz: S. Lamort, 1838.

Anon. *Das Marienburger Tresslerbuch der Jahre 1399–1409.* Joachim, Erich, ed. Königsberg: Thomas & Opperman, 1896.

Ansel, Walter. *Hitler and the Middle Sea.* Durham, NC: Duke University Press, 1972.

Arnold, Thomas F. *The Renaissance at War.* London: Cassell, 2001.

Arnold, Thomas F. "War in Sixteenth Century Europe: Revolution and Renaissance." In *European Warfare 1453–1815.* Jeremy Black ed. New York: St. Martin's, 1999.

Audio Visual Instructional Technology Division. ".58 Rifled Musket" TVT 21–84. United States Military Academy.

Audio Visual Instructional Technology Division. "Flintlock Smoothbore Musket." TVT 21–82. United States Military Academy.

Audio Visual Instructional Technology Division. "The Kentucky Rifle in the American Revolution." TVT 21–83. United States Military Academy.

Bachrach, David S. "English Artillery 1189–1307 : The Implications of Terminology." *English Historical Review,* Vol. 121, 2006.

Barber, James. "The Moving Frontier of British Imperialism in the Lake Rudolf Region 1890–1919." *Ethnohistory* 53. No. 1. (Winter, 2006): 143–172.

Batchelor, John, and Ian Hogg. *Artillery*. New York: Scribner, 1972.

Beachey, R.W. "The East African Ivory Trade in the Nineteenth Century" *Journal of African History* 8. No 2 (1967).

Bear, Laura. *Lines of the National: Indian Railway Workers, Bureaucracy, and the Intimate Historical Self.* New York: Columbia University Press, 2007.

Beaver, Daniel R. "Deuce and a Half, Selecting United States Army Trucks, 1920–1945." In *Feeding Mars: Logistics in Western Warfare from the Middle Ages to the Present*. Edited by John A. Lynn. Boulder, CO: Westview Press, 1993.

Beffeyte, Reynaud. *War Machines in the Middle Ages*. Rennes: Éditions Ouest-France, 2004.

Bermeo, Nancy. "War and Democratization: Lessons from the Portuguese Experience." *Democratization* 14. No. 3 (June 2007).

Bhacker, M. Reda. *Trade and Empire in Muscat and Zanzibar*. London: Routledge, 1992.

Bichler, Andreas. "Testing Reconstructed Medieval Crossbows Featuring Composite Bows." http://www.historiavivens1300.at/biblio/beschuss/beschuss1-e.htm, accessed 2 September 2012.

Bidwell, Shelford, and Dominick Graham. *Fire-Power: British Army Weapons and Theories of War 1904–1945*. London: George Allen and Unwin, 1982.

Bijker, Wiebe E. *Of Bicycles, Bakelites, and Bulbs: Toward a Theory of Sociotechnical Change*. Cambridge, Massachusetts: MIT Press, 1995.

Bijker, Wiebe E., Thomas P. Hughes, and Trevor Pinch, editors. *The Social Construction of Technological Systems*. Cambridge, MA: The MIT Press, 1989.

Bishop, Edward. *The Wooden Wonder*. London, Max Parrish, 1959.

Blevin, C. and J. Bouczo. "Nairobi: A Century of History 1898–1997." In *Les Cahiers de l'IFRA*. Nairobi: Institut français de recherché en Afrique, 1997.

"Boeing/Lockheed Martin (Rockwell/Martin Marietta) AGM-114 Hellfire." Directory of U.S. Rockets and Missiles. Accessible at: http://www.designation-systems.net/dusrm/m-114.html.

Bormann, Charles. *The Shrapnel Shell in England and in Belgium*. 2nd ed. Brussels: Louis Truyts, 1862.

Brantley, Cynthia. *The Giriama and Colonial Resistance in Kenya, 1800–1920*. Berkeley: University of California Press, 1981.

Brass, L., and W. Page. "Stroke in former Prisoners of War (POW)." *Journal of Stroke and Cerebrovascular Disease*. No. 6 (1996).

Bremer, J. Douglas. *Does Stress Damage the Brain? Understanding Trauma-Related Disorders from a Mind-Body Perspective*. New York: W.W. Norton & Co., 2002.

Brewin, Bob. "Army Reserve and Guard Suicide Rates more than Doubled Last Year." *Government Executive.* 19 January 2011. http://www.govexec.com/dailyfed/0111/011911bb2.htm.

Bridger, Baldwin. "An Earlier Kind of Computer." *IEEE Industry Applications Magazine.* (October 2007) doi:10.1109/MIA.2007.902394.

Brown, David. *Nelson to Vanguard: Warship Development, 1923–1945.* Annapolis, MD: Naval Institute Press, 2000.

Burkholder, Peter J. "The Manufacture and Use of Cannons at the Siege of St-Sauveur-le-Vicomte, 1375." MA Thesis, U. of Toronto, 1992.

Byrne, Joseph P. *The Black Death.* Westport, CT: Greenwood Press, 2004.

Cabral, Amílcar. *A Acção Armada e os Métodos Militares.* Bissau: Departamenta de Informação, Propaganda e Cultura do C.C. do PAIGC, 1988.

Callwell, Colonel C.E. *Small Wars: Their Principles and Practice.* London: Harrison and Sons, 1906.

Calvert, Albert F. *German East Africa.* New York: Negro Universities Press, 1970.

Carr, Jeffrey. *Inside Cyber Warfare.* Sebastopol, CA: O'Reilly, 2012.

Cavaiola, Lawrence, David Gompert, and Martin Libicki. "Cyber House Rules: On War, Retaliation, and Escalation." *Survival*, Vol. 57. No. 1 (February–March, 2015)

Ceva, Lucio and Andra Curami. "Air Army and Aircraft Industry in Italy, 1936–1943." In *The Conduct of the Air War in the Second World War.* Horst Boog, Editor. New York, Berg, 1992.

*Calendar of Fine Rolls, Edward* III, *1337–1347.* London: His Majesty's Stationary Office, 1913–1923.

Chabal, Patrick. *Amilcar Cabral: Revolutionary Leadership and People's War.* Trenton, NJ: Africa World Press, 2003.

Chandler, David. *Atlas of Military Strategy.* New York: Free Press, 1980.

Chartier, Jean. *Chronique de Charles* VII. Vallet de Viriville ed. Paris: P. Jannet, 1858.

Chevedden, Paul. "The Hybrid Trebuchet: The Halfway Step to the Counterweight Trebuchet." in Donald J. Kagay and Theresa M. Vann, eds., *On the Social Origins of Medieval Institutions: Essays in Honor of Joseph O'Callaghan.* Leiden: Brill, 1998.

Chi, Hsi-Sheng. *Nationalist China at War, Military Defeats and Political Collapse, 1937–45.* Ann Arbor: University of Michigan Press, 1982.

Chief of Naval Operations, "How Radio-Controlled Bombs Were Jammed." *C.I.C. (Combat Information Center).* December 1945. Reprinted at Lone Sentry blog, September 19, 2010. Accessible at: http://www.lonesentry.com/blog/how-radio-controlled-bombs-were-jammed.html.

Cipolla, Carlo. *Guns Sails and Empires: Technological Innovation and the Early Phases of European Expansion, 1400–1700.* New York: Pantheon, 1966.

Citino, Robert Michael. *Death of the Wehrmacht: The German Campaigns of 1942.* Lawrence: University Press of Kansas, 2007.

Clark, David C. ed. *Arms for the Nation.* Greensburg, PA: South Greensburg Printing Co., Inc, 1994.

Clark, Kim. *The Redemptive Work: Railway and Nation in Ecuador, 1895–1930.* Wilmington, Delaware: SR Books, 1998.

Clymer, A. Ben. "The Mechanical Analog Computers of Hannibal Ford and William Newell." *IEEE Annals of the History of Computing* 15. No. 2 (1993).

Coelho, Paulo Borges. "African Troops in the Portuguese Colonial Army, 1961–1974: Angola, Guinea-Bissau and Mozambique." *Portuguese Studies Review.* (September 2002b).

Contamine, Philippe. *War in the Middle Ages.* tr. Michael C. Jones. Oxford: Basil Blackwood Ltd., 1987.

Cook, Weston. "Warfare and Firearms in Fifteenth Century Morocco, 1400–1492." *War & Society.* Vol. 11. 1993.

Cooper, Randolf G.S. *The Anglo-Maratha Campaigns and the Contest for India.* New York: Cambridge University, 2003.

Corum, James S., and Wray R. Johnson. *Airpower in Small Wars: Fighting Insurgents and Terrorists.* Lawrence, KS: University Press of Kansas, 2003.

Coss, Edward J. *All for the King's Shilling: The British Soldier under Wellington.* Norman: University of Oklahoma Press, 2010.

Craddock, John. "The Tank is Dead – Long Live the Tank." USAWC Military Studies Program Paper, April 15, 1993. Accessible at: http://www.dtic.mil/dtic/tr/fulltext/u2/a264578.pdf.

Crusis, Martin. *Annalium Sveveicorum.* Frankfurt: Nicolai Bassaei, 1596.

Currie, William. "New 'smart' weapons make 'Star Wars' appear tame." *Chicago Tribune.* January 21, 1979. Accessible at: http://archives.chicagotribune.com/1979/01/21/page/32/article/new-smart-weapons-make-star-wars-appear-tame.

Curtis, Duncan. *North American F-86 Sabre.* Marlborough, UK; The Crowood Press, 2000.

Dale, Stephen. The Garden of the Eight Paradises: Babur and the Culture of Empire in Central Asia, Afghanistan and India, 1483–1530. Boston: Brill, 2004.

Dalleo, Peter T. "The Somali Role in Organized Poaching in Northeastern Kenya, 1909–1939." *International Journal of African Historical Studies* 12, No 3. (1979)

Davison, Basil. "Guinea-Bissau and the Cape Verde Islands: The Transition from War to Independence." *Africa Today* 21. No. 4 (Autumn 1974).

De la Garza, Andrew. "Command of the Coast: The Mughal Navy and Regional Strategy." *World History Connected.* Vol. 12, No. 1 (February 2015a).

De la Garza, Andrew. *The Mughal Empire at War: Babur, Akbar and the Indian Military Revolution, 1500–1605.* London: Routledge, 2015b.

De Fenin, Pierre. *Mémoires de Pierre de Fenin.* Mlle. Dupont ed. Paris: SHF, 1837.

DeJean, Joan. *Ancients against Moderns: Culture Wars and the Making of a Fin de Siècle.* Chicago: University of Chicago Press, 1997.

De Jongh, Geleynssen. Journal of.Ttranslated and excerpted by Jos Gommans in *Mughal Warfare*. New York: Routledge, 2000.

DeLeon, Peter. *The Laser-Guided Bomb: Case History of a Development.* Santa Monica, CA: RAND, 1974.

De Pizan, Christine. *The Book of Deeds of Arms and of Chivalry*. Translated by Sumner and Charity C. Willard. University Park: Penn. State U.P., 1999.

De Saint Marc, Hélie. *L'aventure et L'espérance.* Paris: Editions des Arènes, 2010.

De Waurin, John *A Collection of the Chronicles and Ancient Histories of Great Britain, Now Called England (1422–31)*. William and Edward L.C.P. Hardy, trs. London, 1887.

Dhada, Mustafah. "The Liberation War in Guinea-Bissau Reconsidered." *Military History* 62. No. 3 (July 1998).

Dhada, Mustafah. *Warriors at Work: How Guinea Was Really Set Free.* Niwot, CO: University Press of Colorado, 1993.

DiGiulian, Tony. "The British High Angle Control System (HACS)." Naval Technical Board. Accessed September 18, 2016. http://www.navweaps.com/index_tech/tech-066.htm#7.

Douglas, Sarah K. "The Black Death and Nation-State Wars of the Fourteenth Century: Environment, Epigenetics, Excess and Expiation, 1346–1450." In *Epidemics and War: The Impact of Disease on Major Conflicts in History*. Prager: 2018.

Douglas, Sarah K. "The Price of Pestilence: England's response to the Black Death in the face of the Hundred Years War." Unpublished PhD Thesis. The Ohio State University, 2015.

Douglas-Hamilton, James. *The Air Battle for Malta: The Diaries of a Spitfire Pilot.* Barnsley, UK: Pen and Sword Aviation, 2006.

Douhet, Guilio. *The Command of the Air*. Tuscaloosa, AL: University of Alabama Press, 2010.

Drea, Edward. *Japan's Imperial Army: Its Rise and Fall, 1853–1945.* Lawrence: University Press of Kansas, 2009.

Dreyer, Edward L. *China at War, 1901–1945.* New York: Longman, 1995.

Duke, Neville. *The War Diaries of Neville Duke 1941–1944.* Edited by Norman Franks. London: Grub Street, 1995.

Ehlers, Robert S., Jr. *The Mediterranean Air War: Airpower and Allied Victory in World War* II. Lawrence, KS: University Press of Kansas, 2015.

Ehlers, Robert S., Jr. *Targeting the Third Reich: Air Intelligence and the Allied Bombing Campaigns.* Lawrence: University Press of Kansas, 2010.

Elliot, H.M. ed. *The History of India as Told by its Own Historians.* Volume 6, Chpt. 80 (*Akbarnama* of Shaikh Illahdád).

Ellis, L.F. *The War In France and Flanders 1939–1940.* London, HMSO, 1953.

Embry, Basil. *Mission Completed.* London: Quality Book Club, 1957.

"Epinephrine," Accessed 1 December 2011. http://eilat.sci.brooklyn.cuny.edu/newnyc/DRUGS/EPINEPHR.HTM.

Evans, David, and Mark Peattie. *Kaigun: Strategy, Tactics, and Technology in the Imperial Japanese Navy, 1887–1941*. Annapolis, MD: Naval Institute Press, 1997.

Faria, Dutra. "Governor Tells of Independence Party Activity." In *ABC-Diario de Angola* (Luanda). 7 December 7 1969. Translated and reprinted in Joint Publications Research Service. "Translations on Africa No. 867." JPRS 49903 (February 20, 1970).

Farquharson-Roberts, Mike. *A History of the Royal Navy: World War* I. New York: Palgrave Macmillan, 2014.

Flatman, Joe. *Ships and Shipping in Medieval Manuscripts*. London: The British Library, 2009.

Foran, Robert W. *A Cuckoo in Kenya: The Reminiscences of a Pioneer Police Officer in British East Africa*. London: Hutchinson & Company, 1936.

"Ford Instrument Company," *Journal of Jet Propulsion* 26, no. 5 (May 1956).

France, John. *Western Warfare in the Age of the Crusades*, 1000–1300. Ithaca: Cornell U.P., 1999.

Fraser, Benjamin, and Steven D. Spalding. *Trains, Culture, and Mobility: Riding the Rails*. Lanham, Maryland: Lexington Books, 2012.

Freeman, Michael J. *Railways and the Victorian Imagination*. New Haven: Yale University Press, 1999.

Friedman, Norman. *Naval Anti-Aircraft Guns and Gunnery*. (Annapolis, MD: Naval Institute Press, 2013.

Friel, Ian. "Oars, Sails and Guns: The English and War at Sea, c.1200–c.1500." In *War at Sea in the Middle Ages and the Renaissance* Edited by John B. Hattendorf and Richard W. Unger. Rochester, NY: The Boydell Press, 2003.

Fritzsche, Peter. *A Nation of Fliers*. Cambridge, MA: Harvard University Press, 1992.

Froissart, Jean. *Chronicles*. Translated by Thomas Johnes. Vol. 2 London: Routledge, 1874.

Fudge, Thomas, ed. and tr. *The Crusade against Heretics in Bohemia, 1418–1437: Sources and Documents for the Hussite Crusades*. Aldershot: Ashgate, 2002.

Furse, Anthony. *Wilfrid Freeman:the genius behind Allied survival and air supremacy 1939–1945*. Staplehurst, Spellmount, 2000.

Galdorisi, George. "U.S. Navy Missile Defense: Getting Surface-to-Air Missile Development Started." Defense Media Network. October 18, 2012. Accessible at: http://www.defensemedianetwork.com/stories/u-s-navy-missile-defense-getting-surface-to-air-missile-development-started/.

Gallois, Pierre. *The Balance of Terror: Strategy for the Nuclear Age*. Trans., Richard Howard. Boston: Houghton Mifflin Company, 1961.

Gawassa, G.C.K. "The German intervention and African resistance in Tanzania." In *A History of Tanzania*. Edited by I.N. Kimambo and A.J. Temu. Nairobi: East African Publishing House, 1969.

Gillingham, John. "Richard I and the Science of War in the Middle Ages." In John Gillingham & J.C. Holt, eds. *War and Government in the Middle Ages.* Boydell Press, Woodbridge 1984.

Gladman, Brad William. *Intelligence and Anglo-American Air Support in World War Two: The Western Desert and Tunisia, 1940–43.* New York: Palgrave MacMillan, 2009.

Gleick, James. *Chaos: Making a New Science*. New York, NY: Viking Press, 1987.

Gleijeses, Piero. "The First Ambassadors: Cuba's Contribution to Guinea-Bissau's War of Independence." *Journal of Latin American Studies* 29. No. 1 (February 1997).

Glete, Jan. *War and the State in Early Modern Europe: Spain, the Dutch Republic and Sweden as fiscal-military states, 1500–1660.* New York: Routledge Publishing, 2002.

Goda, Norman J.W. *Tomorrow the World: Hitler, Northwest Africa, and the Path toward North America.* College Station, TX: Texas A&M University Press, 1998.

Goldrick, James. "For Want of a Nail … Royal Naval Anti-Aircraft Gunnery 1919–1939." *Journal of the Royal United Services Institution of Australia* 6 (April 1983).

Gongora, Thierry, and Harald von Riekhoff, eds. *Toward a Revolution in Military Affairs? Defense and Security at the Dawn of the 21st Century.* Westport: Greenwood Press, 2000.

Gordon, Michael. *Five Days in August: How World War* II *Became a Nuclear War*. Princeton, NJ: Princeton University Press, 2015.

Grancsay, Stephen. "Just How Good Was Armor?" *True.* (April 1954), p. 90.

Grandmaison, Ch. L. ed. *Documents inédits pour servir à l'histoire des arts en Touraine.* Tours: Soc. Arch. de Touraine, 1870.

Grant, U.S. *Personal Memoirs of U.S. Grant.* Volume I. Charles L. Webster & Company, 1885.

Grayling, A.C. *Among the Dead Cities: the History and Moral Legacy of the WW*II *Bombing of Civilians in Germany and Japan*. New York, NY: Walker and Company, 2006.

Grossman, David. *On Killing.* New York: Little, Brown and Company, 1996.

Grossman, David. Cited in Adam Linehan, "This is your Brain on War." *Task and Purpose.* Accessed on 21 June 2016. http://taskandpurpose.com/this-is-your-brain-on-war/.

"Guided German air to ground weapons in WW2." http://www.1jma.dk/articles/1jmaluftwaffegroundweapons.htm.

Guilmartin, John F., Jr. "The Aircraft that Decided World War II: Aeronautical Engineering and Grand Strategy, 1933–1945, The American Dimension." The Harmon Memorial Lecture in Military History. No. 44. United States Air Force Academy, 2001.

Guilmartin, John F., Jr. "The military revolution in warfare at sea during the early modern era: technological origins, operational outcomes and strategic consequences." *Journal for Maritime Research* 13. No. 2. (November 2011a).

Guilmartin, John F., Jr. "The Military Revolution in Warfare at Sea During the Early Modern Era: Technological Origins, Operational Outcomes and Strategic Consequences." *Revue d'Histoire Meritime: Marine, État et Politique.* No. 14 (Paris: Presses de l'université, Paris Sorbonne, 2011b).

Guilmartin, John F., Jr. *Gunpowder and Galleys: Changing Technology and Mediterranean Warfare at Sea in the 16th Century.* Revised Edition. Annapolis, MD: Naval Institute Press, 2003.

Guilmartin, John F., Jr. "Technology and Asymmetrics in Modern Warfare." In *Challenging the United States Symmetrically and Asymmetrically: Can America Be Defeated?* Edited by Lloyd J. Matthews. Carlisle Barracks, PA: US Army War College – Strategic Studies Institute, July 1998.

Guilmartin, John F. Jr., "Technology and Strategy: What Are the Limits?" In *Two Historians in Technology and War.* Strategic Studies Institute, 1994. http://www.au.af.mil/au/awc/awcgate/ssi/2hist.pdf.

Gray, Thomas. "Scalacronica 1271–1363." Edited and translated by Andy King. *The Publications of the Surtees Society.* Vol. 209. Rochester, NY: The Boydell Press, 2005.

Gwynn, Charles W. Major "The Frontiers of Abyssinia: A Retrospect." *Journal of the Royal African Society.* 38. (1937).

Gwynn, Charles W. Major. "A Journey in Southern Abyssinia." *Geographical Journal.* 38 (1911).

Hagecii, Wenceslai. *Böhmische Chronica.* Vol. 2 Cadan: J.S. Zluticensem, 1596.

Hall, Bert S. *Weapons and Warfare in Renaissance Europe.* Baltimore: Johns Hopkins U.P., 1997.

Hall, Bert S., and Kelly DeVries. "Essay Review – the 'Military Revolution' Revisited." *Technology and Culture.* No. 31 (1990).

Hall, R. Cargill. "To Acquire Strategic Bombers: The Case of the B-58 Hustler." *Air University Review* XXXI. No. 6 (September–October 1980): 1–20.

Hamner, Christopher. *Enduring Battle: American Soldiers in Three Wars, 1776–1945.* Lawrence: University of Kansas Press, 2011.

Hansen, Peter Vemming. "Experimental Reconstruction of a Medieval Trébuchet." *Acta Archaeologica. Vol.* 63, 1992. Accessed at http://web.archive.org/web/20070105190530/http://www.middelaldercentret.dk/acta.html.

Hanson, Victor Davis. *Carnage and Culture: Landmark Battles in the Rise of Western Culture.* New York: Anchor Books, 2001.

Harmsen, Peter. *Shanghai 1937: Stalingrad on the Yangtze.* Havertown, PA: Casemate Publishers, 2013.

Haydon, F.S., ed. *Eulogium historiarum.* 3 vols. London: Rolls Series, 1858–63.

Hayward, Joel, Editor. *Air Power, Insurgency, and the "War on Terror".* Cranwell, UK: Royal Air Force Center for Air Power Studies, 2009.

Headrick, Donald. *Tools of Empire.* New York: Oxford University Press, 1981.

Hebron, Malcolm. *The Medieval Siege: Theme and Image in Middle English Romance.* Oxford: Clarendon Press, 1997.

Held, Werner, and Ernst Obermaier. *The Luftwaffe in the North African Campaign, 1941–1943.* West Chester, PA: Schiffer, 1992.

Henderson, W. *Studies in German Colonial History.* London: F. Cass, 1962.

Henry VI. *Letters and Papers Illustrative of the Wars of the English in France during the Reign of Henry the Sixth, King of England.* Joseph Stevenson ed. 2 vols. London: RS, 1861–4.

Herlihy, David. *The Black Death and the Transformation of the West.* Cambridge: Harvard University Press, 1997.

Hewitt, J. "Mons Meg, the Ancient Bombard Preserved at Edinburgh Castle." *Archaeological Journal.* Vol. 10. 1853.

Higham, Robin. *Bases of Air Strategy: Building Airfields for the RAF, 1914–1945.* London: Airlife, 1998.

Hill, M.F. *Permanent Way: The Story of the Kenya and Uganda Railway.* London: Hazell Watson and Viney Ltd, 1949.

Hoagland, Jim. "Portugal's 'Unwinnable' War Heats Up." *The Washington Post-Times Herald.* 21 February 1971.

Hoffman, Michael. "Guard, Reserve Suicide Rate sees Big Spike." 19 January 2011. http://www.armytimes.com/news/2011/01/army-guard-reserve-suicide-rate-sees-big-spike-011911w/.

Hollingsworth, L.W. *Zanzibar Under the Foreign Office 1890–1913.* London: MacMillan & Company, 1953.

Houser, George M. *No One Can Stop the Rain: Glimpses of Africa's Liberation Struggle.* New York: The Pilgrim Press, 1989.

Hugill, Peter J. *World Trade Since 1431: Geography, Technology and Capitalism.* Baltimore, Maryland: Johns Hopkins University Press, 1993.

Humbaraci, Arslan, and Nicole Muchnik. *Portugal's African Wars: Angola, Guinea Bissao, Mozambique.* New York: The Third Press, 1974.

Hunyadi. In *Monumenta Hungariae historica.* Budapest: Kiadja a Magyrar Tudományos Akadémia, 1907.

Hussey, Brian. "The U.S. Navy, the Neutrality Patrol, and Atlantic Fleet Escort Operations, 1939–1941." Trident Scholar Report, Annapolis, MD, 1991.

Huzel, Dieter K. and David H. Huang. *Design of Liquid Propellant Rocket Engines.* 2nd edition. NASA Special Pamphlet SP-125, Washington, DC, 1971.

Hyde, H. Montgomery. *British Air Policy Between The Wars 1918–1939.* London, Heinemman, 1976.

*Ingulph's Chronicle of the Abbey of Croyland: With the Continuations by Peter of Blois and Anonymous Writers.* Translated by Henry Thomas Riley. London: Bohn's, 1854.

*The Interview.* http://www.imdb.com/title/tt2788710/?ref_=fn_al_tt_1 (accessed 22 July 2015).

Iriye, Akira. *The Origins of the Second World War in Asia and the Pacific.* New York: Longman, 1987.

Jackson, A.J. *de Havilland Aircraft Since 1909.* London, Putnam, 1962.

Jordan, Donald A. *China's Trial by Fire: The Shanghai War of 1932.* Ann Arbor: University of Michigan, 2001.

Junger, Sebastian. *War.* New York: Hatchette Book Group, 2010.

Keegan, John. *The Face of Battle.* New York: Pimlico, 1976.

Kelly, P. *The Universal Cambist.* 2nd ed. Vol. 2. London: Printed for the Author, 1821.

Kerr, Ian J. *27 Down: New Departures in Indian Railway Studies.* New Delhi: Orient Longman, 2007.

Khan, Iqtidar Alam. *Gunpowder and Firearms: Warfare in Medieval India.* New Delhi: Oxford, 2004.

Kirby, William C. *Germany and Republican China.* Stanford: Stanford University Press, 1984.

Knox, MacGregor. *Hitler's Italian Allies: Royal Armed Forces, Fascist Regime, and the War of 1940–1943.* New York: Cambrige University Press, 2000.

Kopp, Carlo. "The Sidewinder Story: The Evolution of the AIM-9 Missile." *Australian Aviation.* April 1994. Accessible at: http://www.ausairpower.net/TE-Sidewinder-94.html.

Kramer, Gerhard W. ed. *The Fireworkbook: Gunpowder in Medieval Germany.* Klaus Leibnitz, tr. London: Arms and Armour Society, 2001.

Kuehn, John. "The Ostfriesland, the General Board of the Navy, and the Washington Naval Treaty: A Relook at a Historic Sinking." In *New Interpretations in Naval History: Selected Papers from the Sixteenth Naval History Symposium Held at the United States Naval Academy, 10–11 September 2009*, edited by Craig Felker and Jones Marcus, U.S. Naval War College Historical Monograph Series. Newport, RI: Naval War College Press, 2012.

Kuhn, Thomas. *The Structure of Scientific Revolutions.* Chicago, IL: University of Chicago Press, 1970.

Lains, Pedro. "An Account of the Portuguese African Empire, 1885–1975." *Revista de Historia Económica* 15. No. 1 (January 1998).

Lambert, Craig L. *Shipping the Medieval Military: English Maritime Logistics in the Fourteenth Century.* (ochester, NY: The Boydell Press, 2011.

Lartéguy, Jean. *The Centurions.* Trans., Xan Fielding. New York: Penguin Books, 2015.

Lavalle, A.J.C. Major USAF, ed. *Airpower and the 1972 Spring Invasion.* Washington, DC: Office of Air Force History, 1985.

Lawrence, T.E. *The Seven Pillars of Wisdom: A Triumph.* New York: Anchor Books, 1991.

Levin, Ed. "The Highland Fling." *Timber Framing: Journal of the Timber Framers Guild.* 50, 1998.

Lewis, Anthea M. "A Tribute to Air-Vice Marshal GG Dawson, 1895–1944, CMSO ME & N Africa 1941–1944." Air Historical Branch.

Libicki, Martin. *Cyberdeterrence and Cyberwar*. Santa Monica, CA: RAND, 2009.

Libicki, Martin, and David Gompert. "Waging Cyberwar the American Way." *Global Politics and Strategy* 57. No. 4 (July 2015): 7–28.

Liddell-Hart, Basil. *Strategy.* New York, NY: Praeger, 1967.

Little, Roger. "Buddy Relations and Combat Performance." Edited by Morris Janowitz. *The New Military.* New York: Russell Sage Foundation, 1964.

Lloyd-Jones, William. *KAR: Being an Unofficial Account of the Origin and Activities of the King's African Rifles*. London: Arrowsmith, 1926.

Lobato, António. *Liberdade o Evasão: O Mais Longo Captiveira da Guerra.* Amadora, Portugal: Erasmos, 1995.

Lopes, Mario Canongia. "Portugal's Ginas." *Air Enthusiast* 36. (May–August 1988).

Lopes, Mario Canongia. "High-Winged Workhorses: Broussards and Dorniers in Portuguese Service." *Air Enthusiast* 73. (May–June 1998).

Lorge, Peter A. *The Asian Military Revolution: From Gunpowder to the Bomb.* Cambridge: Cambridge University Press, 2008.

Lynn, John. *Battle: A History of Combat and Culture.* Boulder: Westview Press, 2003.

MacQueen, Norrie. "Portugal's First Domino: 'Pluricontinentalism' and Colonial War in Guinea-Bissau, 1963–1974." *Contemporary European History* 8. No. 2 (1999).

Mallett, Michael. *Mercenaries and their Masters.* Totowa: Towman and Littlefield, 1974.

Mallett, Michael E., and J.R. Hale. *The Military Organisation of a Renaissance State: Venice c. 1400 to 1617.* Cambridge: Cambridge U.P., 1984.

Mansoor, Peter R. *Surge: My Journey with General David Petraeus and the Remaking of the Iraq War.* New Haven, CT: Yale University Press, 2013.

Marshall, Ian. *Armored Ships,* Charlottesville, Virginia: Howell Press, 1993.

Martello, David. *1974: Cessar-Fogo em África.* Mem Martins, Portugal: Publicações Europa-América, 2001.

Matthews, Michael. "Cognitive and Non-Cognitive Factors in Soldier Performance." In *The Oxford Handbook of Military Psychology*. Editors Janice H. Laurence and Michael D. Matthews. Oxford: Oxford University Press, 2012.

Maxwell, Kenneth. "The United States and the Portuguese Decolonization (1974–1976)." Working paper. Instituto Portugûes de Relaçãos Internacionais, Universidade Nova de Lisboa, 2002.

McAulay, John D. *Rifles of the U.S. Army.* Lincoln, RI: Andrew Mowbray Publishers, 2003.

McCaffrey, Patrick. "Cerebral Lobes, Cerebral Cortex, and Brodmann's Areas." Accessed 11 January 2016. http://www.csuchico.edu/~pmccaffrey/syllabi/CMSD%20320/362unit4.html.

McDermott, P.I. *British East Africa or IBEA: A History of the Formation and Work of the Imperial British East Africa Company*. London: Chapman and Hall, 1893.

McFarland, Stephen L., and Wesley Phillips Newton. *To Command the Sky: The Battle for Air Supremacy over Germany, 1942–1944*. Washington, D.C.: Smithsonian Institution Press, 1991.

McGranahan, William J. "The Fall and Rise of Marshal Tukhachevsky." *Parameters*. Vol. VIII, No. 4 December, 1978.

McMahon, Martin T. "The Death of General John Sedgwick." In *Battles and Leaders of the Civil War: Retreat with Honor*. Volume IV. Secausus, NJ: Castle, 1983.

McMaster, H.R. "Crack in the Foundation: Defense Transformation and the Underlying Assumption of Dominant Knowledge in Future War." Center for Strategic Leadership. U.S. Army War College. November 2003. Accessible at: http://www.au.af.mil/au/awc/awcgate/army-usawc/mcmaster_foundation.pdf.

McNally, Brendan. "The Sinking of the Battleship Roma and the Dawn of the Age of Precision Guided Munitions." Defense Media Network. June 1, 2014. Accessible at: http://www.defensemedianetwork.com/stories/the-sinking-of-the-battleship-roma-and-the-dawn-of-the-age-of-precision-guided-munitions/.

McNeill, William Hardy. *The Pursuit of Power: Technology, Armed Force and Society since A.D. 1000*. Chicago: University of Chicago Press, 1982.

Medieval Gunpowder Research Group. "The Ho Experiments. Report Number 1: The Firing Trials." Nykøbring, F., Denmark: Middelaldercentret, 2002.

Medieval Gunpowder Research Group, "The Ho Experiments. Report Number 2." Nykøbring, F., Denmark: Middelaldercentret, 2003.

Mehta, Aalok. "What makes a 'supersoldier'?" Dana Foundation. 17 February 2009. https://danablog.org/2009/02/17/aaas-what-makes-a-supersoldier/.

Mendell, David. *Between Human and Machine: Feedback, Control, and Computing before Cybernetics*. Baltimore, MD: Johns Hopkins University Press, 2004.

Metcalf, Thomas R. *Imperial Connections: India and the Indian Ocean Arena, 1860–1920*. Berkeley: University of California Press, 2007.

Mets, David R. "The *Force* in US Air Force." *Aerospace Power Journal*. Fall 2000. Accessible at: http://www.au.af.mil/au/afri/aspj/airchronicles/apj/apjoo/faloo/mets.pdf.

Meyer, Retsina, et al, "A Ghrelin-Growth Hormone Axis Drives Stress-Induced Vulnerability to enhanced Fear." *Mol Psychiatry*. No, 19 (2012).

Mierzejewski, Alfred C. *The Collapse of the German War Economy, 1944–1945: Allied Air Power and the German National Railway*. Chapel Hill and London: University of North Carolina Press, 1988.

Molesworth, Sir Guilford. "The Uganda Railway." International Engineering Congress. Glasgow, 1901.

Morgan, Charles A., Kevin Colwell, and Gary A. Hazlett. "Efficacy of Forensic Statement Analysis in Distinguishing Truthful from Deceptive Eyewitness Accounts of Highly Stressful Events." *Journal of Forensic Sciences* 56. No. 4 (2011).

Morgan, Philip. *Fascism in Europe, 1919–1945*. New York: Routledge, 2003.

Moretz, Joseph. *The Royal Navy and the Capital Ship in the Interwar Period: An Operational Perspective*. London: Routledge, 2002.

Moritz, Bob, and David Burg. "How Corporate America can Fight Cybersecurity Threats." In *Fortune* (online, 17 February, 2015). http://fortune.com/2015/02/17/how-corporate-america-can-fight-cybersecurity-threats/.

*Moseby's Dictionary of Medicine, Nursing, & Health Professionals*. China, 2009.

Moskos, Charles. *The American Enlisted Man: the Rank and File in Today's Military*. New York: Russell Sage Foundation, 1970.

Mrozek, Donald J. *Air Power and the Ground War in Vietnam*. McLean, VA: Pergamom-Brassey's, 1989.

Mungeam, G.H. "Masai and Kikuyu Responses to the Establishment of British Administration in the East Africa Protectorate." In *Journal of African History* 11. No. 1 (1970): 127–143.

Murray, Williamson. "Clausewitz Out, Computer In: Military Culture and Technological Hubris." *The National Interest*. June 1, 1997 (No. 48): 57–64. Accessible at: https://www.clausewitz.com/readings/Clause&Computers.htm.

Murray, Williamson. *Strategy for Defeat: The Luftwaffe, 1933–1945*. Maxwell AFB, Alabama: Air University Press, 1983.

Murray, Williamson, and MacGregor Knox, eds. *The Dynamics of Military Revolution, 1300–2050*. Cambridge: Cambridge U.P., 2001. Partington, J.R. *A History of Greek Fire and Gunpowder*. Cambridge: W. Heffer & Sons, 1960.

Murray, Williamson and Peter Mansoor, eds. *Hybrid Warfare: Fighting Complex Opponents from the Ancient World to the Present*. Cambridge: Cambridge University Press, 2012.

Nanda, Admiral S.M. *The Man Who Bombed Karachi*. Noida, India: HarperCollins India, 2015.

n.a. "Researcher Shows Success with PTSD Treatment for Female Veterans." Wayne State University, 10 February, 2015. https://wayne.edu/newsroom/release/2015/02/10/researcher-shows-success-with-ptsd-treatment-for-female-veterans-5530.

National Public Radio. *When did Companies Become People? Excavating the Legal Evolution*, National Public Radio (NPR). 28 July 2014. http://www.npr.org/templates/transcript/transcript.php?storyId=335288388.

Neulen, Hans Werner. *In the Skies of Europe: Air Forces Allied to the Luftwaffe 1939–1945*. Wiltshire, UK: Crowood Press, 2000.

Neve, Kim A ed. *Dopamine Receptors*. 2nd edition. New York: Humana Press, 2009.

New Medical. "What is Dopamine," *New Medical*. Accessed 23 September 2011. http://www.news-medical.net/health/What-is-Dopamine.aspx.

Niccoli, Riccardo. "Portuguese Numerology: Serial Systems Used by the Aeronautica Militar and the Força Aérea Portuguesa." *Air Enthusiast* 75 (May–June 1998).

Nossov, Konstantin. *Indian Castles 1206–1526: The Rise and Fall of the Delhi Sultanate.* New York: Osprey, 2006.

Office of the Historian. U.S. Department of State. "U-2 Overflights and the Capture of Francis Gary Powers, 1960." Accessible at: https://history.state.gov/milestones/1953-1960/u2-incident.

Oman, C.W.C. *The Art of War in the Middle Ages.* Ithaca, NY: Cornell University Press, 1953.

Orange, Vincent. "Getting Together: Tedder, Coningham, and Americans in the Desert and Tunisia, 1940–43." In *Airpower and Ground Armies: Essays on the Evolution of Anglo-American Air Doctrine 1940–1943.* Edited by Daniel R. Mortensen. Maxwell AFB, AL: Air University Press, 1998.

Orr, W.G. "Armed Religious Ascetics in Northern India." In *Warfare and Weaponry in South Asia.* Jos Gommans and Dirk Kolff, ed. New Delhi: Oxford, 2001.

Overy, Richard. *Russia's War: A History of the Soviet War Effort, 1941–1945.* New York: Penguin, 1998.

Owen, Roderic. *The Desert Air Force; an Authoritative History Published in aid of the Royal Air Force Benevolent Fund.* London: Hutchinson, 1948.

Owens, Admiral Bill with Ed Offley. *Lifting the Fog of War.* New York: Farrar, Straus, and Giroux, 2000.

Pace, Scott et al. *The Global Positioning System: Assessing National Policies.* Santa Monica, CA: RAND, 1995.

Pagliery, Jose. "OPM Hack's Unprecedented Haul: 1.1 Million Fingerprints." *CNN Money.* 10 July 2015. http://money.cnn.com/2015/07/10/technology/opm-hack-fingerprints/.

Pape, Robert. *Bombing to Win: Airpower and Coercion in War.* Ithaca, NY: Cornell University Press, 2014.

Paret, Peter. *French Revolutionary Warfare from Indochina to Algeria: The Analysis of a Political and Military Doctrine.* New York: Frederick A. Praeger, Publishers, 1964.

Parker, Geoffrey. "The Military Revolution, 1560–1660—A Myth?," *Journal of Modern History*, 48 (1976).

Parker, Geoffrey. *The Military Revolution: Military Innovation and the Rise of the West.* Cambridge: Cambridge University Press, 1996.

Parker, Geoffrey. "The Western Way of War." In Geoffrey Parker, ed. *The Cambridge History of Warfare.* New York: Cambridge University Press, 2005.

Paris, Michael. "Air Power and Imperial Defence 1880–1919." *Journal of Contemporary History* 24. No. 2 (April 1989).

Patterson, Bruce D. *The Lions of Tsavo: Exploring the Legacy of Africa's Notorious Man-Eaters.* New York: McGraw-Hill, 2003.

Perras, Anne. *Carl Peters and German Imperialism.* Oxford: Clarendon Press, 2004.

Peter, C. "Okinawa Campaign, 1 April 1945–21 June 1945." World War II Database. Accessible at: http://ww2db.com/battle_spec.php?battle_id=15.

Picatinny Arsenal. "Smart Munitions." Accessible at: http://www.pica.army.mil/Picatinny/products_services/products01.aspx.

Pilcher, June, and Allen Huffcutt. "Effects of Sleep Deprivation on Performance: a Meta-analysis." *Journal of Sleep Research & Sleep Medicine* 19. No. 4 (1996).

Platt, Colin. *King Death: the Black Death and its aftermath in late-medieval England.* London: University College London Press, 1996.

Playfair, Major-General I. S. O. *The Mediterranean and Middle East. Vol.* I. *The Early Successes against Italy* [*to May 1941*]. Uckfield, UK: Naval and Military Press, 2004.

Porch, Douglas. *The Path to Victory: The Mediterranean Theater in World War* II. New York: Farrar, Straus, and Giroux, 2004.

Portal, Sir Gerald. *The British Mission to Uganda in 1893.* Westport, Connecticut: Negro Universities Press, 1970.

Pout, H.W. "Weapon Control in the Royal Navy, 1935–45." In *The Applications of Radar and Other Electronic Systems in the Royal Navy in World War 2.* Edited by F.A. Kingsley. Baskingstoke, UK: Palgrave Macmillan, 1995.

Postan, M.M., D Hay and J D Scott. *Design and Development of Weapons.* London, HMSO, 1964.

Pries, Sascha of the Kölnisches Stadtmuseum. Personal Communication with Clifford Rogers.

Prestwich, Michael. *Armies and Warfare in the Middle Ages.* New Haven: Yale U.P., 1999.

Prestwich, Michael. "The Trebuchets of the Tower," in Gregory I. Halfond, ed., *The Medieval Way of War: Studies in Medieval Military History in Honor of Bernard S. Bachrach.* Farnham: Ashgate, 2015.

Purton, Peter. *A History of the Late Medieval Siege.* Woodbridge: Boydell, 2010.

Ralston, David B. *Importing the European Army: The Introduction of Military Techniques and Institutions into the Extra-European World, 1600–1914.* Chicago: University of Chicago Press, 1990.

Reals, Tucker. "Stuxnet Worm a U.S. Cyber-Attack on Iran Nukes?" *CBS News.* 24 Sept., 2010. accessed 22 July 2015. http://www.cbsnews.com/news/stuxnet-worm-a-us-cyber-attack-on-iran-nukes/.

Reardon, Jim. *Upside: The New Sciences of Post-Traumatic Growth.* New York: Simon & Schuster, 2015.

Reveron, Derek ed. *Cyberspace and National Security.* Washington, DC: Georgetown University Press, 2012.

Rhodes, Richard. *Nuclear Renewal: Common Sense About Energy.* New York, NY: Viking, 1993.

Rich, Norman. *Hitler's War Aims: Ideology, the Nazi State, and the Course of Expansion.* New York: W.W. Norton and Company, 1992.

Richards, Denis and G. Saunders St. Hilary. *RAF 1939–1945. Vol. 2. The Fight Avails.* London: Her Majesty's Stationery Office, 1954.

Roberts, Joseph. *The Hand-Book of Artillery*. New York, NY: D. Van Nostrand, 1863, reprinted by Invictus, Decatur, MI, 1999.

Roberts, Michael. "The Military Revolution, 1560–1660," *Essays in Swedish History*. Minneapolis: University of Minnesota Press, 1967.

Robinson Jr., Clarence A. "Gulf War 20th: Apache Raid." Defense Media Network. February 2, 2011. Accessible at: http://www.defensemedianetwork.com/stories/gulf-war-20th-apache-raid/.

Roff, Peter. "Hacker's Paradise." *U.S. News and World Report*. 1 July 2015, http://www.usnews.com/opinion/blogs/peter-roff/2015/07/01/chinas-opm-hack-just-the-beginning-of-us-cybersecurity-problems.

Rodgers, Nicholas A.M. *The Safeguard of the Sea: A Naval History of Britain, 660–1649*. New York: W.W. Norton and Co., 1997.

Rogers, Clifford J. "The Artillery and Artillery Fortress Revolutions Revisited." In Nicolas Prouteau, Emmanuel de Crouy-Chanel and Nicolas Faucherre, eds., *Artillerie et Fortification, 1200–1600*. Rennes: Presses Universitaires de Rennes, 2011a.

Rogers, Clifford J. "The Longbow, the Infantry Revolution, and Technological Determinism," *The Journal of Medieval History*, Vol. 37, 2011b.

Rogers, Clifford J. "The Battle of Agincourt," *The Hundred Years War (Part II): Different Vistas*. L.J. Andrew Villalon and Donald J. Kagay, eds. Leiden: Brill, 2008.

Rogers, Clifford J. "By Fire and Sword: *Bellum Hostile* and 'Civilians' in the Hundred Years War." In *Civilians in the Path of War*. Mark Grimsley and Clifford J. Rogers eds. Lincoln: University of Nebraska Press, 2002.

Rogers, Clifford J. "The Idea of Military Revolutions in Eighteenth and Nineteenth Century Texts." *Revista de História das Ideias* 30, 2009.

Rogers, Clifford J, ed. *The Military Revolution Debate*. Boulder: Westview, 1995a.

Rogers, Clifford J. *The Military Revolution Debate*. Boulder, CO: Westview Press, 1995b.

Rogers, Clifford J. "The Military Revolutions of the Hundred Years' War," in Clifford J. Rogers, ed. *The Military Revolution Debate*. Boulder, CO: Westview, 1995c.

Rogers, Clifford J. "The Vegetian 'Science of Warfare' in the Middle Ages." *Journal of Medieval Military History. Vol.* 1 (2003): 1–20.

Rogers, Peter. "The British and the Kikuyu 1890–1905: A Reassessment." In *The Journal of African History* 20. No. 2 (1979): 255–269.

Romjue, John L. "The Evolution of the Airland Battle Concept." *Air University Review*. Vol. 35, No.4. May–June, 1984. Accessible at: http://www.airpower.maxwell.af.mil/airchronicles/aureview/1984/may-jun/romjue.html.

Rommel, Erwin. *The Rommel Papers*. Edited by B.H. Liddell Hart. New York: Harcourt, Brace and Company, 1953.

Rose, Susan. *Medieval Ships and Warfare*. Burlington, VT: Ashgate Publishing, Routledge Press, 2008.

Rosen, Stephen Peter. *Societies and Military Power: India and its Armies*. Ithaca: Cornell University, 1996.

Rosen, Stephen Peter. *War and Human Nature*. Princeton: Princeton University Press, 2005.

Roskill, Stephen. *Naval Policy Between the Wars. Volume* II*: The Period Reluctant Rearmament, 1930–1939*. South Yorkshire, UK: Seaforth Publishing, 2016.

Sadler, E.H. "Notes on the Geography of British East Africa." In *Journal of the Royal African Society* 1. No. 2 (January, 1912): 173–186.

Saimire, Tanel. "Trebuchet—A Gravity-Operated Siege Engine. A Study in Experimental Archaeology." *Estonian Journal of Archaeology. Vol.* 10, 2006.

Salerno, Reynolds M. *Vital Crossroads: Mediterranean Origins of the Second World War, 1935–1940*. Ithaca, NY: Cornell University Press, 2002.

Sanders, Damien. "Knowing the Ropes: The Need to Record Ropes and Rigging on Wreck-Sites and Some Techniques for Doing So." *The International Journal of Nautical Archaeology*. (2010).

Schantzberg, Eric. "Ideology and Technical Choice: The Decline of the Wooden Airplane in the United States, 1920–1945." *Technology and Culture* 35. No. 1 (January 1994).

Schivelbusch, Wolfgang. *The Railway Journey: The Industrialization of Time and Space in the 19th Century*. Berkley: University of California Press, 1986.

Schlenger, William E. et al. "The Prevalence of Post-Traumatic Stress Disorder in the Vietnam Generation: A Multimethod, Multisource Assessment of Psychiatric Disorder." *Trauma Stress*. No. 5 (1992).

Schmider, Klaus. "The Mediterranean in 1940–1941: Crossroads of Lost Opportunities?" *War and Society* 15, no. 2 (October 1997).

Schreiber, Gerhard, Bernd Stegemann, and Detlef Vogel. *Germany and the Second World War Volume* III*: The Mediterranean, South-east Europe, and North Africa, 1939–1941*. Translated by Dean S. McMurry, Ewald Ossers, and Louise Willmot. Oxford: Clarendon Press, 1995.

Science Daily. "Sympathetic Nervous System." *Science Daily*. Accessed 12 January 2016, http://www.sciencedaily.com/articles/s/sympathetic_nervous_system.htm.

*Science News Digest for Physicians and Scientists*. "Neuropetide Y and Body Weight." Accessed 9 January 2012. http://www.bioscience.org/news/scientis/obesity2.htm.

Setton, Kenneth M. *The Papacy and the Levant, 1204–1571*. Vol. 2 Philadelphia: American Philosophical Society, 1978.

Sharp, C. Martin, and Martin Bowyer. *Mosquito*. London, Faber and Faber, 1967.

Showell, Jak P. Mallmann. Foreword. *Fuehrer Conferences on Naval Affairs 1939–1945*. London: Chatham Publishing, 2005.

Shores, Christopher, and Hans Ring. *Fighters over the Desert: The Air Battles in the Western Desert, June 1940–December 1942*. New York: Arco, 1969.

Shulte, John C. "An analysis of the historical effectiveness of anti-ship cruise missiles in littoral warfare." Masters Thesis. Naval Postgraduate School. 1994. Accessible at: http://calhoun.nps.edu/bitstream/handle/10945/27962/analysisofhistorooschu.pdf?sequence=1.

Siddle, Bruce. "The Stress Paradox: Safe and Rational Protocols for Pre-hospital Tourniquet Use." *War on Trauma: Lessons Learned from a Decade of Conflict.* (October, 2008) http://www.lawofficer.com/article/training/stress-paradox.

Silva, Tomás George Conceição. General Portuguese Air Force. "The Portuguese Air Force in the African Wars of 1961–1974." In *Memories of Portugal's African Wars, 1961–1974: Proceedings of a Conference, King's College London, 10 June 1997*. ed. John P. Cann. Quantico, VA: Marine Corps University Foundation, 1998.

Simpson, George L. "Frontier Banditry and the Colonial Making Process: The East Africa Protectorate's Northern Borderland prior to the First World War." *International Journal of African Historical Studies* 29, No 2 (1996).

Simpson, Michael. "Superhighway to the World Wide Web: The Mediterranean in British Imperial Strategy, 1900–1945." In *Naval Policy and Strategy in the Mediterranean: Past, Present, and Future*. Edited by John B. Hattendorf. (London: Cass, 2000).

Singer, Peter. *Ghost Fleet*. Boston, MA: Houghton Mifflin Harcourt, 2015.

Sinnott, Colin. *The Royal Air Force and Aircraft Design.* London, Frank Cass, 2001a.

Sinnott, Colin. *The RAF and Aircraft Design, 1923–1939: Air Staff Operational Requirements.* London: Frank Cass, 2001b.

Smith, Malcolm. *British Air Strategy between the Wars.* Oxford, Clarendon, 1984.

Smith, Merritt Roe. "Technological Determinism in American Culture." In Merritt Smith and Leo Marx, editors, *Does Technology Drive History? The Dilemma of Technological Determinism.* Cambridge, MA: The MIT Press, 1994.

Smith, Robert Douglas. *Rewriting the History of Gunpowder.* Nykøbing Falster: Middelaldercentret, 2010.

Smith, Robert D. and Ruth Rhynas Brown. *Bombards: Mons Meg and her Sisters.* London: Royal Armouries, 1989.

Solon, Paul. "Charles VII and the *Compagnies d'Ordonnance*, 1445–61: A Study in Medieval reform." Ph.D. Dissertation, Brown University, 1970.

Solon, Paul. "Valois Military Administration on the Norman Frontier, 1445–61: A Study in Medieval Reform." *Speculum*. Vol. 51 (1976): 91–111.

*Southern African*. "Guinea-Bissau: Advances on the Military Front," *Southern Africa: A Monthly Survey of News and Opinion* 6, No. 9 (November 1973).

Spence, Jonathon D. *The Search for Modern China.* New York: W.W. Norton, 1991.

Spiers, Edward M. "Introduction: The Reconquest Reappraised." In *Sudan: The Reconquest Reappraised.* London: Frank Cass, 1998.

Springer, Paul. *Cyber Warfare.* Santa Barbara, CA: ABC-Clio, 2015.

Stanley, Elizabeth. "Neuroplasticity, Mind Fitness, and Military Effectiveness." *Bio-Inspired Innovation and National Security*. Washington DC, 2010.

Stanley, Elizabeth, and Amishi Jha. "Mind Fitness: Improving Operational Effectiveness and Building Warrior Resilience." *Joint Forces Quarterly*. No. 55, 4th quarter (2009).

Stanley, Henry Morton. "The Uganda Railway." *Saturday Review*. 1 June 1895.

Stanley, Roy M. II. *World War II Photo Intelligence*. New York: Scribner, 1981.

Steinhart, Edward I. "Elephant Hunting in 19th Century Kenya: Kamba Society and Ecology in Transformation." In *Journal of African Historical Studies* 33. No. 2 (2000).

Sterling, Michal J. "Soviet Reactions to NATO's Emerging Technologies for Deep Attack." RAND, August 1985. Accessible at: http://www.rand.org/content/dam/rand/pubs/notes/2009/N2294.pdf.

Stewart, Richard W. ed., *American Military History*. Vol. II. Washington: Center of Military History, 2005. Accessible at: http://www.history.army.mil/books/amh-v2/amh%20v2/chapter12.htm.

Stone, Michael L. "Organized Poaching in Itui District: A Failure of District Authority, 1900 to 1960." *International Journal of African Historical Studies* 5, No 3 (1972).

Stouffer, Samuel, et al. *The American Soldier: Combat and Its Aftermath*. Vol. 2. *Studies in Social Psychology in World War* II. 3 vols. Princeton: Princeton University Press, 1949–50.

Strawson, John. "The Shape and Course of the Mediterranean War, 1940–43." In Royal Air Force Historical Society. *The End of the Beginning: Bracknell Paper No. 3, A Symposium on the Land/Air Cooperation in the War, 1940–43*. Edited by Derek Wood. Bracknell, UK: Royal Air Force Historical Society, 20 March 1992.

Strzyz, Piotr. "Archaeological Traces of Use of Firearms During Sieges. Selected Examples from the Territory of Poland." *Fasisculi Archaeologiae Historicae*. Vol. 28, 2015.

"Stress: Constant Stress puts your Health at risk." http://www.mayoclinic.com/health/stress/SR00001.

Strzyz, Piotr, Piotr Czubla, and Adam Mackiewicz, "Cannonballs from the Olsztyn Turret," *Fasisculi Archaeologiae Historicae*. Vol. 28. 2015.

"Submarine Armament: Cruise Missiles." National Museum of American History. Accessible at: http://www.americanhistory.si.edu/subs/weapons/armament/missiles/index.html.

Sullivan, Brian R. "Downfall of the Regia Aeronautica, 1933–1943." In Robin Higham and Stephen J. Harris, Editors. *Why Air Forces Fail: The Anatomy of Defeat*. University Press of Kentucky, 2006.

Sumida, Jon Tetsuro. "British Naval Procurement and Technological Change, 1919–1939." In *Technology and Naval Combat in the Twentieth Century and Beyond*, edited by Phillips Payson O'Brien. London: Frank Cass, 2001.

Sumida, Jon Tetsuro. "Forging the Trident: British Naval Industrial Logistics, 1914–1918." In John A. Lynn, ed. *Feeding Mars: Logistics in Western Warfare from the Middle Ages to the Present*. Boulder, Colorado: Westview Press, 1993.

Sumption, Jonathan. *The Hundred Years War I: Trial by Battle.* Philadelphia: University of Pennsylvania Press, 1990.

Sun, Tzu. *The Art of War*. London, UK: Oxford University Press, 1963.

Sweetman, Bill. *Stealth Aircraft: Invisible Warplane, Black Budget.* Osceola, Wisconsin: Motorbooks International, 1989.

Taleb, Nassim Nicholas. *The Bed of Procrustes: Philosophical and Practical Aphorisms.* New York: Random House, 2010.

Tavernier, Jean-Baptiste. *Travels in India.* Translated by V. Ball. Macmillan: London, 1889.

Taylor, Jay. *The Generalissimo: Chiang Kai-shek and the Struggle for Modern China.* Cambridge MA: Belknap Press of Harvard University Press, 2009.

Tedder, Arthur. *With Prejudice: The War Memoirs of Marshal of the Royal Air Force Lord Tedder G.C.B.* Boston : Brown and Company, 1966.

Thompson, Mark. "Army Suicide Rate Hits New High." *Time Magazine, Battleland* (online). 12 August 2011. http://battleland.blogs.time.com/2011/08/12/army-suicide-rate-hits-new-high/.

Thucydides. *History of the Peloponnesian War.* New York, NY: Penguin, 1974.

Tilford, Earl H. Jr, *Crosswinds: The Air Force's Setup in Vietnam*, College Station, Texas: Texas A&M University Press, 1993.

Till, Geoffrey. *Air Power and the Royal Navy, 1914–1945 : A Historical Survey.* London: Macdonald and Jane's, 1979.

Till, Geoffrey. "Retrenchment, Rethinking, Revival, 1919–1939." In *The Oxford Illustrated History of the Royal Navy*, edited by J.R. Hill. Oxford, UK: Oxford University Press, 1995.

Tomes, Robert. "The Cold War Offset Strategy: Assault Breaker and the Beginning of the RSTA Revolution." War on the Rocks. November 20, 2014. Accessible at: http://warontherocks.com/2014/11/the-cold-war-offset-strategy-assault-breaker-and-the-beginning-of-the-rsta-revolution/.

Tooze, Adam. *The Wages of Destruction: The Making and Breaking of the Nazi Economy.* New York: Viking, 2007.

Topsfield, Andrew. *Paintings from Mughal India*. London: Oxford, 2008.

Tout, T.F. "Firearms in England in the Fourteenth Century." *English Historical Review,* Vol. 26 (1911), 690.

Towle, Philip Anthony. *Pilots and Rebels: The Use of Aircraft in Unconventional Warfare, 1918–1988.* London: Brassey's UK, 1989.

Toy, Sidney. *The Fortified Cities of India.* London: Heinemann, 1965.

Trenchard, Hugh. *Air Power*. London, UK: Directorate of Staff Studies, Air Ministry, 1946.

Ullman, Harlan and James Wade, Jr., et al. *Shock & Awe: Achieving Rapid Dominance.* Washington: Institute for National Strategic Studies, National Defense University, 1996. Accessible at: http://www.dodccrp.org/files/Ullman_Shock.pdf.

"USS Savannah (CL-42)." Naval Warfare. March 10, 2009. http://navalwarfare.blogspot.com/2009/03/uss-savannah-cl-42.html.

"USS Savannah (CL-42), 1938–1960 – Damaged by German Bomb, 11 September 1943." Naval Historical Center. Accessible at: https://www.ibiblio.org/hyperwar/OnlineLibrary/photos/sh-usn/usnsh-s/cl42-l.htm.

Van de Ven, Hans J. *War and Nationalism in China.* New York: Routledge, 2012.

Van der Waals, W.S. *Portugal's War in Angola, 1961–1974.* Rivonia, South Africa: Ashanti, 1993.

Vasterling, Jennifer, et al. "Neuropsychological Outcomes of Army Personnel following Deployments to the Iraq War." *Journal of the American Medical Association* 296. No. 5 (2006).

Venter, Al J. *The Chopper Boys: Helicopter Warfare in Africa.* London: Greenhill Books, 1994.

Venter, Al J. "Portugal's Forgotten War." *Air Enthusiast* 2. No. 2 (February 1972).

Von Clausewitz, Carl. *On War.* Editors and translators Michael Howard and Peter Paret. Princeton, NJ: Princeton University Press, 1976.

Wagner, Amy, and Matthew Jakupcak. "Combat-Related Stress Reactions among U.S. Veterans of Wartime Service." In *The Oxford Handbook of Military Psychology*. Edited by Janice Laurence and Michael Matthews. Oxford: Oxford University Press, 2012).

Wakeman, Fredrick. "A Revistion view of the Nanjing Decade: Confucian Fascism." *The China Quarterly.* No. 150 (June 1997).

Waldrop, M. Mitchell. *Complexity: The Emerging Science at the Edge of Order and Chaos.* New York, NY: Simon and Schuster, 1992.

Waller, Richard. "The Maasai and the British, 1895–1905 and the Origins of an Alliance." In *Journal of African History* 17. No. 4 (1976): 529–553.

Watts, Barry D. "The Evolution of Precision Strike." Center for Strategic and Budgetary Assessments, 2013. Accessible at: http://www.voltairenet.org/IMG/pdf/The_Evolution_of_Precision_Strike.pdf

Watts, Barry D. *The Foundations of U.S. Air Doctrine: The Problem of Friction in War.* Maxwell AFB, Alabama: Air University Press, 1984.

Watts, Barry D. *Six Decades of Guided Munitions and Battle Networks: Progress and Prospects.* Washington: Center for Strategic and Budgetary Assessments, 2007.

Webster, Sir Charles, and Noble Frankland. *The Strategic Bombing Offensive Against Germany.* Vol 4. London, HMSO, 1961.

Weinberg, Gerhard L. *Visions of Victory: The Hopes of Eight World War* II *Leaders.* Cambridge, UK: Cambridge University Press, 2005.

Weinberg, Gerhard L. *A World at Arms: A Global History of World War* II. Cambridge, UK: Cambridge University Press, 1994.

Welch, Stuart. *India: Art and Culture, 1300–1900.* New York: Metropolitan Museum, 1985.

White, Jr., Lynn. "The Crusades and the Technological Thrust of the West," *Medieval Religion and Technology: Collected Essays* (Los Angeles: University of California Press, 1978.

Wildenberg, Thomas. *Destined for Glory: Dive Bombing, Midway, and the Evolution of Carrier Air Power.* Annapolis, MD: Naval Institute Press, 1998.

Wilkins, Tony. "Is the Main Battle Tank Obsolete?" Defence of the Realm. March 18, 2016. Accessible at: https://defenceoftherealm.wordpress.com/2016/03/18/is-the-main-battle-tank-obsolete/.

Williams, Alan R. *The Knight and the Blast Furnace: A History of the Metallurgy of Armour in the Middle Ages & the Early Modern Period.* Leiden: Brill, 2003.

Williamson, A., and Anne-Marie Feyer. "Moderate Sleep Deprivation Produces Impairments in Cognitive and Motor Performance Equivalent to Legally Prescribed Levels of Alcohol Intoxication." *Occup Environ Med 57.* (2000).

Wong, Leonard, et al. *Why They Fight: Combat Motivation in the Iraq War.* Carlisle: Strategic Studies Institute, 2003.

Wynn, Humphrey. "The RAF in the Mediterranean Theatre." In Royal Air Force Historical Society. *The End of the Beginning: Bracknell Paper No. 3. A Symposium on the Land/Air Cooperation in the War, 1940–43.* Edited by Derek Wood. Bracknell, UK: Royal Air Force Historical Society. 20 March 1992.

Yadin, Yigael. *The Art of War in Biblical Lands in the Light of Archaeological Study.* 2 vol. London: McGraw-Hill, 1963.

"Yom Kippur War: The Battle of Latakia." Jewish Virtual Library. http://www.jewishvirtuallibrary.org/jsource/History/latakia.html.

Yudofsky, Stuart, and Robert E. Hales. (2007). *The American Psychiatric Publishing Textbook of Neuropsychiatry and Behavioral Neurosciences.* 5th edition. New York: American Psychiatric Publishing, Inc., 2007.

Żabiński, Grzegorz. "Technology of Manufacture of Firearms in the Teutonic Order's State in Prussia—Gun Barrels and Metal Projectiles." *Fasisculi Archaeologiae Historicae.* Vol. 28 2015.

Zetter, Kim. *Countdown to Zero Day: Stuxnet and the Launch of the World's First Digital Weapon.* New York, NY: Crown Publishers, 2014.

# Index